CYBERSECURITY FOR EXECUTIVES

CYBERSECURITY FOR EXECUTIVES

A Practical Guide

Gregory J. Touhill

C. Joseph Touhill

AIChE

WILEY

For general information on our other products and services or for technical support, please contact our
Customer Care Department within the United States at (800) 762-2974, outside the United States at (317)
572-3993 or fax (317) 572-4002.

Wiley also publishes its books in a variety of electronic formats. Some content that appears in print may
not be available in electronic formats. For more information about Wiley products, visit our web site at
www.wiley.com.

Library of Congress Cataloging-in-Publication Data

Touhill, Gregory J.
Cybersecurity for executives : a practical guide / by Gregory J. Touhill and C. Joseph Touhill.
   pages   cm
  Includes bibliographical references and index.
  ISBN 978-1-118-88814-8 (cloth)
1. Computer networks–Security measures.   I. Touhill, C. J., 1938–   II. Title.
  TK5105.59.T67 2014
  658.4'78–dc23

                                                          2014002691

To our wives and children

CONTENTS

Foreword   xiii

Preface   xvii

Acknowledgments   xxiii

1.0 INTRODUCTION   1

1.1 Defining Cybersecurity   1

1.2 Cybersecurity is a Business Imperative   2

1.3 Cybersecurity is an Executive-Level Concern   4

1.4 Questions to Ask   4

1.5 Views of Others   7

1.6 Cybersecurity is a Full-Time Activity   7

2.0 WHY BE CONCERNED?   9

2.1 A Classic Hack   9

2.2 Who Wants Your Fortune?   12

2.3 Nation-State Threats   13

   2.3.1 China   13

   2.3.2 Don't Think that China is the Only One 17

2.4 Cybercrime is Big Business   20

   2.4.1 Mercenary Hackers   20

   2.4.2 Hacktivists   25

   2.4.3 The Insider Threat   26

   2.4.4 Substandard Products and Services   29

2.5 Summary   36

3.0 MANAGING RISK   37

3.1 Who Owns Risk in Your Business?   37

3.2 What Are Your Risks?   38

3.2.1 Threats to Your Intellectual Property and Trade Secrets 38
3.2.2 Technical Risks 42
3.2.3 Human Risks 47
3.3 Calculating Your Risk 54
3.3.1 Quantitative Risk Assessment 55
3.3.2 Qualitative Risk Assessment 63
3.3.3 Risk Decisions 71
3.4 Communicating Risk 77
3.4.1 Communicating Risk Internally 78
3.4.2 Regulatory Communications 79
3.4.3 Communicating with Shareholders 86
3.5 Organizing for Success 89
3.5.1 Risk Management Committee 89
3.5.2 Chief Risk Officers 90
3.6 Summary 91

4.0 BUILD YOUR STRATEGY 95
4.1 How Much "Cybersecurity" Do I Need? 95
4.2 The Mechanics of Building Your Strategy 97
4.2.1 Where are We Now? 99
4.2.2 What Do We Have to Work With? 103
4.2.3 Where Do We Want to Be? 104
4.2.4 How Do We Get There? 107
4.2.5 Goals and Objectives 108
4.3 Avoiding Strategy Failure 111
4.3.1 Poor Plans, Poor Execution 111
4.3.2 Lack of Communication 113
4.3.3 Resistance to Change 114
4.3.4 Lack of Leadership and Oversight 117
4.4 Ways to Incorporate Cybersecurity into Your Strategy 118
4.4.1 Identify the Information Critical to Your Business 119
4.4.2 Make Cybersecurity Part of Your Culture 119
4.4.3 Consider Cybersecurity Impacts in Your Decisions 119
4.4.4 Measure Your Progress 120
4.5 Plan For Success 121
4.6 Summary 123

5.0 PLAN FOR SUCCESS — 125

5.1 Turning Vision into Reality — 125
    5.1.1 Planning for Excellence — 127
    5.1.2 A Plan of Action — 128
    5.1.3 Doing Things — 131
5.2 Policies Complement Plans — 140
    5.2.1 Great Cybersecurity Policies for Everyone — 140
    5.2.2 Be Clear about Your Policies and Who Owns Them — 188
5.3 Procedures Implement Plans — 190
5.4 Exercise Your Plans — 191
5.5 Legal Compliance Concerns — 193
5.6 Auditing — 195
5.7 Summary — 196

6.0 CHANGE MANAGEMENT — 199

6.1 Why Managing Change is Important — 199
6.2 When to Change? — 201
6.3 What is Impacted by Change? — 205
6.4 Change Management and Internal Controls — 209
6.5 Change Management as a Process — 214
    6.5.1 The Touhill Change Management Process — 215
    6.5.2 Following the Process — 216
    6.5.3 Have a Plan B, Plan C, and maybe a Plan D — 220
6.6 Best Practices in Change Management — 220
6.7 Summary — 224

7.0 PERSONNEL MANAGEMENT — 227

7.1 Finding the Right Fit — 227
7.2 Creating the Team — 229
    7.2.1 Picking the Right Leaders — 230
    7.2.2 Your Cybersecurity Leaders — 233
7.3 Establishing Performance Standards — 237
7.4 Organizational Considerations — 240
7.5 Training for Success — 242
    7.5.1 Information Every Employee Ought to Know — 242
    7.5.2 Special Training for Executives — 246

7.6 Special Considerations for Critical Infrastructure Protection   249

7.7 Summary   258

8.0 PERFORMANCE MEASURES   261

8.1 Why Measure?   261

8.2 What to Measure?   267

   8.2.1 Business Drivers   267

   8.2.2 Types of Metrics   271

8.3 Metrics and the C-Suite   272

   8.3.1 Considerations for the C-Suite   273

   8.3.2 Questions about Cybersecurity Executives Should Ask   275

8.4 The Executive Cybersecurity Dashboard   277

   8.4.1 How Vulnerable Are We?   277

   8.4.2 How Effective Are Our Systems and Processes?   282

   8.4.3 Do We Have the Right People, Are They Properly Trained,
            and Are They Following Proper Procedures?   286

   8.4.4 Am I Spending the Right Amount on Security?   287

   8.4.5 How Do We Compare to Others?   288

   8.4.6 Creating Your Executive Cybersecurity Dashboard   289

8.5 Summary   291

9.0 WHAT TO DO WHEN YOU GET HACKED   293

9.1 Hackers Already Have You Under Surveillance   293

9.2 Things to do Before it's Too Late: Preparing for the Hack   295

   9.2.1 Back Up Your Information   296

   9.2.2 Baseline and Define What is Normal   296

   9.2.3 Protect Yourself with Insurance   297

   9.2.4 Create Your Disaster Recovery and Business
            Continuity Plan   298

9.3 What to do When Bad Things Happen: Implementing Your Plan   299

   9.3.1 Item 1: Don't Panic   300

   9.3.2 Item 2: Make Sure You've Been Hacked   301

   9.3.3 Item 3: Gain Control   302

   9.3.4 Item 4: Reset All Passwords   303

   9.3.5 Item 5: Verify and Lock Down All Your External Links   304

   9.3.6 Item 6: Update and Scan   305

   9.3.7 Item 7: Assess the Damage   305

   9.3.8 Item 8: Make Appropriate Notifications   307

9.3.9 Item 9: Find Out Why It Happened and Who Did It 309

9.3.10 Item 10: Adjust Your Defenses 310

9.4 Foot Stompers 310

9.4.1 The Importance of Public Relations 310

9.4.2 Working with Law Enforcement 315

9.4.3 Addressing Liability 317

9.4.4 Legal Issues to Keep an Eye On 318

9.5 Fool Me Once… 319

9.6 Summary 320

10.0 BOARDROOM INTERACTIONS **323**

Appendix A: Policies **347**

Appendix B: General Rules for Email Etiquette: Sample Training Handout **357**

Glossary **361**

Select Bibliography **371**

Index **373**

FOREWORD

I always have thought of myself as a savvy businessman. I worked for or served on the boards of some of the best known and most successful companies in the world. Additionally, I started several companies, beginning small and growing them into successful enterprises. I've been CEO of a New York Stock Exchange member firm and a board member of 20 publicly owned companies. I believed that I had a pretty good handle on how technology benefits the management of businesses of all sizes. In fact, I prided myself at being an early adopter of computers, incorporating them into my businesses where they quickly became indispensable to our operations. Computers enabled us to be more productive and efficient, improving the value proposition of our businesses.

I'll be the first to admit that I am not an expert on computers. Like other senior executives, I recognize their great value and look for opportunities to improve my businesses through automation. As computers became more integral to our businesses, I developed a healthy respect for those who understood the mysteries that lurked within that box. While I liked to fiddle with my computer from time to time, I never deluded myself into believing I was a "computer expert." When I needed help, I went to the professionals.

I saw the explosion of innovative technology in the 1990s, and took the advances pretty much in stride, but then something alarming happened. At first it was an annoyance, but as time went on, a scary scenario. Reports of "hackers" penetrating businesses to steal consumer's personal and financial information started appearing in the newspapers. First it was just isolated attacks, often accomplished by insiders with an axe to grind. But the reports kept coming from all business sectors with increasingly negative effects, including interruptions to operations, expensive lawsuits, and regulatory fines. The computer systems that my colleagues and I had installed to improve our productivity and efficiency were now under threat, or siege by hackers.

Obviously all businesses today heavily rely on computers. Most cannot operate in today's highly competitive markets without trusted, timely, and accurate information. That's why it is very important for executives to have a solid understanding of the emerging role of "cybersecurity" as a prime mechanism to control and manage risk. Like many executives, I needed to become smarter and better versed on cybersecurity quickly.

I was fortunate to be introduced to Greg Touhill by my friend John Maluda, a Telos Corporation director. John told me that Greg was retiring from the U.S. Air Force where he was the general in charge of cybersecurity and information technology for one of the nation's ten combat commands. John told me that Greg was an expert on cybersecurity and led his team to the Rowlett Award, which is given to the organization that has the best cyber defense in the Department of Defense. After a long and distinguished military career, Greg was taking his experience as a CIO and cybersecurity professional to the business world.

Greg joined the Corporate Director's Group, an organization I founded, on John's recommendation and quickly earned his Professional Director certification. His transition from the battlefield to the boardroom has been smooth and transparent as he possesses not only the leadership skills typical of generals, but he also has mastered business principles and maintains his technical certifications in cybersecurity. He now is a highly successful consultant and an adjunct professor at Washington University in St. Louis teaching (of course) cybersecurity.

Cybersecurity is a hot topic among the Corporate Director's Group members and executives in general. While at a recent CDG seminar on cybersecurity, Greg told me that he was in the process of writing a book for people like me. He said it would be entitled, *Cybersecurity for Executives: A Practical Guide*. I told Greg that I couldn't wait to get my hands on it as most writing on cybersecurity is focused on simply scaring people or is written in technical jargon that is nearly undecipherable to the common person. Greg assured me that not only would I understand the message of his book but also I would be able to put it to very practical use immediately.

I am delighted that Greg and his father Joe, a long-time executive, CEO, and board member, wrote this book. It was an easy read; in fact at times I found it hard to put down. They present the material with clarity, humor, and flair. When I finished the draft he gave me, I said, "Greg, I believe you have a real winner here! Executives and directors ought to read this book!"

I meant what I said. I believe this book ought to be the Cybersecurity Bible on every executive's desk. It lays out what the threat to business is from unscrupulous intruders; it frames the problem in terms of risk management; it tells you how to build an appropriate corporate strategy to deal with attempts to steal or alter data and information; it sets out in detail the policies and procedures you need to protect your organizations; and it tells you what changes you need to make with software, hardware, and personnel to make your plan work. It also tells you how to measure the success of your defenses. Additionally, it addresses unique threats to critical infrastructure. Until I read the book, I didn't realize that there are many legal requirements and responsibilities that must be complied with if you are hacked.

There are two chapters that really resonated for me. The first is Chapter 9. In fact, I anticipate some readers may skip to that chapter first for obvious reasons. But if you do, please go back and read from the beginning. You will be glad you did. The formulation of the Disaster Recovery and Business Continuity Plan and the steps called for in implementation are worth the price of the book. When you are hacked (and most experts, including the Touhills, believe that everyone will be sometime), read the chapter carefully and go down the list of recommendations carefully.

The last Chapter 10 had special meaning to me, on several levels. First, I have served on numerous boards of all types over the years—public and private companies, and hospitals and charitable organizations. To me successful programs happen only when you have a fully informed and fully engaged board. Second, I believe the creative setting of the chapter captures the essence of most board meetings I attend. I was fascinated, for instance, by the story of the Kilcawley Chemical Corporation. An eye opener.

In summary, this is a terrific, important and well-written book by experts. I believe it will be your standard reference, as well, when you encounter tricky cybersecurity issues. Read it carefully and use it well.

Clint Allen
A.C. Allen & Company
Needham, MA

PREFACE

Cybersecurity is the deliberate synergy of technologies, processes, and practices to protect information and the networks, computer systems and appliances, and programs used to collect, process, store, and transport that information from attack, damage, and unauthorized access.

Brigadier General Gregory J. Touhill,
United States Air Force (retired)

As my retirement from the United States Air Force was nearing, I contemplated what I would do in my next career. The Air Force had prepared me well to serve in a number of senior executive roles in the private sector. As I went through the excellent transition class the Air Force offers its departing Airmen, I looked at my resume and saw a lot of opportunity. I have extensive leadership and management experience in electronics, telecommunications, software development, finance, program management, information technology, and cybersecurity. I commanded at the squadron, group, and wing levels. I managed the Air Force's US $22B information technology budget at the Pentagon. I served as a diplomat when I was the defense attaché to the State of Kuwait during our nation's crucial transition from Iraq. I was the base commander (equivalent to a chief executive officer) of Keesler Air Force Base, with an annual budget of US $1.3B and 12,000 personnel under my command. I have been a chief information officer (CIO) several times and maintain my technical certifications as a certified information systems security professional (CISSP). In my last assignment as the United States Transportation Command CIO, my team and I were recognized by the National Security Agency with the 2013 Rowlett Award for the best Information Assurance Program in the United States Department of Defense. As my military career came to its conclusion, I was well prepared to do many different things yet I was confronted by a new problem: choosing what to do next.

I did not have to wait too long to find my answer. While I was still in uniform, I had countless discussions with the CEOs and CIOs that my units did business with over how they protected and secured *my* information. I was keenly aware that my information needed to be protected from inadvertent disclosure to those who didn't have "a need to know." One of my duties was to make sure our partners properly protected our military information. I found that more often than not I ended up educating many of our business partners on how to protect *our* vital information, how to implement best practices in cybersecurity, how to educate their work force, how to audit for cybersecurity compliance, and how to create a culture with cybersecurity in mind. As I entered the business world, I found that nearly every business executive I talked with was eager to discuss with me their information technology and concerns over how to secure their information. They found my information technology and cybersecurity experience was valuable to helping them protect their information and competitive advantage. Several of them even suggested that I write a book about cybersecurity. My second career as an information technology and cybersecurity consultant was born.

While it surprises many executives in the private/commercial sector, the Air Force already had given me an excellent foundation in business because in many respects it is managed much like a major corporation and, as a general officer, I rose to one of its senior executive positions. Nonetheless, before I left military service, I recognized that I needed to expand my knowledge of contemporary business practices, terminology, and procedures. I joined the Corporate Directors Group, a public company director education organization, where I earned my Professional Director certification and was introduced to Clint Allen, the group's president. Clint is a highly experienced senior executive and board member who has been a great source of knowledge and advice as I made the transition from the military. When I told him that I was thinking of writing this book, not only was he encouraging, but also he wanted to know when I'd get it done as he said he was so eager to read it. His enthusiasm and interest in educating executives on cybersecurity issues inspired me to invest in this writing effort. He was equally generous in volunteering to write the foreword.

While I had great credentials and experience in information technology and cyber-security, I knew that I needed an expert in business to complement my technical skills. I did not have to look far and turned to my father, Dr. Joe Touhill, who, in addition to being a renowned technologist, is a highly successful CEO, board member, and senior executive. His experience in creating and managing companies, both large and small, was invaluable in filling any gaps in my resume. He has been a corporate officer for 41 years, 29 years of which he has been a CEO. Additionally, he has had extensive board and high-level committee experience. For example, he has been on the board of directors or a trustee of a hospital, a regional MRI facility, a publicly traded bank where he was chairman for several years, a municipal authority, and a major engineering certification organization. He also served on advisory committees for a leading technological university. In other words, he has been around and has an in-depth knowledge and understanding of what executives need and want.

As we did while writing our first book, *Commercialization of Innovative Technologies: Bringing Good Ideas to the Marketplace*, in this book we collaborated over long distances through the countless emails and phone calls, synchronizing our research, outlines, text, and edits. During the first writing effort, I was deployed to Iraq, Afghanistan, and throughout the Middle East for a year, so my father had to patiently wait for my contributions, which I worked on during my increasingly rare off-duty time. This time was different as we were able to talk several times during the day and exchange manuscripts in near real time, permitting us to partner extremely well. My dad's extensive business experience was critical as we focused on the business aspects of cybersecurity rather than jumping into the trap taken by others of being too enamored by the information technology itself. *Our position is that information technology is a tool used by businesses to create value and that cybersecurity is about risk management.*

As we prepared to write this book, we recognized that the Internet has become a powerful ecosystem teaming with data and a myriad of practical and very helpful applications. In fact, it is said that every day in today's so-called "Cyber Age" the human race generates more data than all recorded history before 2003. All that data is being mined to uncover your shopping habits, what web sites you tend to visit, with whom you associate, your recreational and political interests, where you like to travel, and how you spend your money. The Cyber Age has spawned a whole new industry just to collect, collate, and analyze "Big Data." However, "Big Data" equates to "Big Money" which in turn attracts people who seek to gain access to your information or may act in a manner that adversely affects your business.

Some people use terms like hackers and cyber terrorists to describe these people. We prefer to use a term commonly used by most cybersecurity professionals: *bad actors*. A wide variety of individuals can adversely affect your business using information technologies. Some act in a malevolent manner while others cause damage inadvertently. Successful cybersecurity programs protect against all threats, providing businesses and individuals the resiliency to maintain the confidentiality, integrity, and availability of their information. Therefore, when we refer to bad actors, we refer to anyone who is acting in a manner adverse to your business and its information.

Our analysis led us to conclude that your information, especially that which is proprietary, is the lifeblood of your business and your information technology systems have become the circulatory system that keeps your business healthy and vibrant. Regrettably, the same information technology that delivers competitive advantage to businesses also presents serious threats if not managed closely and well. Your information needs to be secure for your business to survive. Unfortunately, we find many executives view cybersecurity as an unnecessary cost and a topic solely for their information technology staffs. We believe this is a mistake.

Think about your business.

- As an executive, how long can you do your job without access to information technology?[1] How long can your business survive without a trusted, stable, and reliable information technology system?[2]
- Can your financial team operate without access to sensitive information? What if their information is tainted?
- How about your production facilities? What do they do if the electricity shuts off? Can they continue to operate? What if their supervisory control and data systems, which are the control units that operate machinery, are corrupted and the machines don't work properly?

---

[1] For purposes of this book, we define "executive" thusly: an executive is someone who has administrative and managerial responsibility for a shareholder-owned business, or a publicly-owned organization committed to the protection and promotion of the health, welfare, and safety of its constituents.

[2] Also for purposes of this book, we will use the word "business" in its broadest sense. "Business" can mean any operation or organization falling under the purview of an executive as previously defined above.

- What if your competitors have access to all of your company research? In addition to the risk that others may use your plans to field your desired product and bring it to market before you, others may decide maliciously to poison your information and sabotage your plans so you deliver a flawed product to market instead.
- What if your shareholders and potential investors lose confidence in your business because of your company's inability to safeguard its information and systems?

For many years, information technology was the always valued, often derided, and frequently misunderstood part of a company's business. Corporate executives appreciated the ability of information technology to enhance business operations through its capacity to manipulate information with ever-increasing speed and precision. Yet, for many corporate executives, information technology was the realm of the "geeks"; those technical wizards who appear more enamored with the technology than the bottom-line that drives the business. Indeed, for most businesses, information technology has been considered an important supporting arm of the business, but rarely a key component.

Perhaps because information technology has been considered a supporting role to the business, many corporate executives delegated much of the oversight and many of the decisions regarding information technology to their information technology department heads. However, as businesses became more and more dependent on information technology, the stakes of a failure involving information and the information system magnified in importance to where a single failure could be an existential event that could doom a company. The stakes are so high that the security of the company can no longer be entrusted to the "geeks" in the server room. That is why we believe our book will fill the need of executives to understand fully the cybersecurity risks they will encounter within the context of management and mitigation.

We contend that cybersecurity is about risk management. It is about protecting shareholders and their business, maintaining competitive advantage, and protecting assets. It is not just about computer technology. Rather, it is a multidisciplinary approach to managing risk; a principal concern of executives.

If you are looking for a book that will make you a technical expert, we certainly can help there, yet most corporate executives don't need to be a technical expert to make good decisions. They need to understand their business and its needs. They especially need to understand the risks their business faces and determine best courses of action to take to mitigate risks. They need to understand the value of their information as much as the value of their inventory and manage both with effectiveness, efficiency, and security. They need to know how to build and sustain great teams composed of the right talent and motivation to best posture the company for success. They need to lead people to perform at peak levels and continuously seek to innovate and improve the business. They need to be able to create an environment where they can gather the right information from the right sources at the right time to make the right decisions.

This book will help and guide you to do that.

Remember, this book will not make you a cybersecurity expert. Instead, we seek to make you *Cyber-Aware* and prepared to make the key decisions to make your business better, and effectively manage the risks inherent in the Cyber Age.

As you read this book, please carefully consider the following:

Cybersecurity is not just a technical issue, it is a business imperative.

While we have done our best to eliminate redundancies, you may see some information that may appear more than once in the book. There are several reasons that this information may be presented in multiple areas. Firstly, because some concepts, such as cybersecurity insurance, transcend many of the key themes of the book, they are discussed within the context of the appropriate sections. Secondly, we recognize that many readers will use the content of the book as reference material as they adjust their cybersecurity programs. We anticipate they may use individual chapters to address pressing concerns. As such, we have included all relevant subject matter that enables the chapters to serve as stand-alone references. Thirdly, there are some items of interest that are so important that they bear repeating for emphasis.

We hope that you find our work informative and valuable. We recognize that cybersecurity is a multi-disciplinary subject, replete with its own idioms, acronyms, and phrases. As such, we include a glossary of terms and an appendix to help add clarity to some of the concepts discussed in the text.[3] Unless attributed to other sources, the work, and any errors we and our editors did not catch in the many reviews leading to publication, are ours and ours alone.

We look forward to hearing your feedback. Please feel free to write us at cyber@touhill.com.

Gregory J. Touhill
Jamison, Pennsylvania
December 2013

---

[3] These will help you to communicate with the "computer aficionados" without being snowed.

ACKNOWLEDGMENTS

We would like to thank our reviewers, including Aaron Call, Chief Information Security Officer for IO.com; Sean Kern of the National Defense University's Cybersecurity and Information Technology cadre; Sekhar Prabhakar, founder and Chief Executive Officer of CEdge Software Consultants; and Jack Zaloudek, Program Director for Information Management and Cybersecurity at Washington University in St. Louis. Their insightful comments, suggestions, and encouragement were invaluable and made this a better book. Many thanks for your sage advice and investment in your precious time.

We also would like to thank Clint Allen, who contributed the foreword for this book. His enthusiasm and interest in educating executives on cybersecurity issues inspired us to invest in this writing effort. Thank you for your encouragement and the foreword's kind words.

Kate McKay of STM Publishing Services was instrumental in launching this book. She provided great advice throughout the publication process. Thank you for your assistance and counsel.

Finally, we also would like to thank Andrew M. Touhill for his contributions to this book. He was instrumental in assisting us with research, editing, and contributed the glossary of terms. His crisp edits and critical analysis helped us reduce and (hopefully) avoid ramblings, worthless opinions, and errors. Thank you Drew for your great work. With your help, this truly is a father and son and grandson effort.

1.0

INTRODUCTION

There are two kinds of companies. Those that have been hacked,
and those that have been hacked but don't know it yet.[1]
House Intelligence Committee Chairman Mike Rogers

1.1 DEFINING CYBERSECURITY

When Congressman Mike Rogers included the words above in a press release to announce new legislation designed to help better defend American business against cyber threats, many executives were alarmed over the prospect that their businesses likely were already victims of hackers. They were shocked.

We weren't.

For over 30 years, we have been deeply involved in not only building, integrating, and defending complex information technology (IT) systems but also in running and managing businesses that have come to rely on IT to create value and deliver profits.

[1] Mike Rogers and Dutch Ruppersberger, "Rogers & Ruppersberger Introduce Cybersecurity Bill to Protect American Businesses from "Economic Predators," November 30, 2011, http://intelligence.house.gov/sites/ intelligence.house.gov/files/documents/113011CyberSecurityLegislation.pdf. Accessed on December 13, 2013.

Cybersecurity for Executives: A Practical Guide, First Edition. Gregory J. Touhill and C. Joseph Touhill.
© 2014 The American Institute of Chemical Engineers, Inc. Published 2014 by John Wiley & Sons, Inc.

During our professional careers, we have seen IT systems grow from stand-alone computers to today's globally connected information ecosystem that permits users to access information anytime, anywhere. We also have seen the increase in the numbers of hackers and others who attempt to gain access to information for reasons that include curiosity, personal profit, or competitive advantage. Threats to your vital information are real and intensifying.

While the term "cybersecurity" is creeping into discussions in boardrooms around the world, we find that most executives, while certainly cognizant of the importance of IT to their businesses, need help to understand what cybersecurity is, how to integrate it into their businesses to provide best value, and how to invest wisely to protect their vital information.

Cybersecurity is a relatively new discipline. It is so new that there is no agreed-upon spelling of the term nor is there a broadly accepted definition. Many people believe cybersecurity is something you can buy in increments, much like a commodity. Others believe cybersecurity just refers to technical measures, such as using password protection or installing a firewall to protect a network. Still, others believe it is an administrative and technical program solely in the realm of IT professionals. Some think it refers only to protection against hackers. We view it differently and define it as follows:

> Cybersecurity is the deliberate synergy of technologies, processes, and practices to protect information and the networks, computer systems and appliances, and programs used to collect, process, store, and transport that information from attack, damage, and unauthorized access.

We view cybersecurity as a holistic set of activities that are focused on protecting an organization's vital information. Cybersecurity includes the technologies employed to protect information. It includes the processes used to create, manage, share, and store information. It includes the practices such as workforce training and testing to ensure information is properly protected and managed. Effective cybersecurity preserves the confidentiality, integrity, and availability of information, protecting it from attack by bad actors, damage of any kind, and unauthorized access by those who do not have a "need to know." *In today's business environment, cybersecurity is not just a technical issue, it is a business imperative.*

1.2 CYBERSECURITY IS A BUSINESS IMPERATIVE

Executives across every business sector are increasingly concerned about cybersecurity. After all, reports indicate hacking incidents are on the rise with an estimated nearly one billion hacking attempts in the final quarter of 2012 alone.[2] New governmental laws and regulations place a premium on cybersecurity controls. Lawsuits lodged in the wake of cybersecurity breaches continue to mount in volume and damages. Customers,

---

[2] Nick Summers, "Hacking Attempts will Pass One Billion in Q4 2012, Claims Information Assurance Firm," November 12, 2012, http://thenextweb.com/insider/2012/11/12/hacking-attempts-to-pass-one-billion-in-final-quarter-of-2012-claims-information-assurance-firm/#!pQLJh. Accessed on December 13, 2013

shareholders, and potential investors increasingly are demanding that effective controls are put in place to protect sensitive information and avoid liabilities. Clients expect that their personal and financial information will be protected from unauthorized disclosure and possible exploitation. Executives recognize that their vital corporate information, such as their intellectual property and trade secrets, provides a powerful competitive advantage for their businesses and needs to be protected. They want to invest wisely in cybersecurity, but don't want to break the bank. Many don't know how their investments in cybersecurity draw positive returns. Additionally, because many cybersecurity measures rely on complex technical controls, many feel uncomfortable with the terminology of the information technologists, many of whom often focus more on the technology than the business it supports. The resulting language gaps create barriers that sometimes produce organizational friction, lack of communication, and poor decision-making. Discussions with our clients convince us there is an acute and growing need to help executives understand and cope with the problems posed by cybersecurity issues.

George Polya, a famous twentieth-century mathematician, said the first step in solving a problem is to understand it.[3] We agree and wrote this book in the hope that it would help executives from all business sectors better understand the nature and extent of cybersecurity and learn how to train personnel to combat cyber attacks, how to recover from such attacks, how to prevent infections, and how to best manage their business to incorporate best practices in cybersecurity.

We propose that the best way to address cybersecurity is to do so from the perspective of a manager rather than a technologist. Cybersecurity is not solely a technical issue. It affects every business function. Every activity in virtually every business relies on information to maintain a competitive advantage. Managers at every level need to understand how investing in cybersecurity produces effective, efficient, and secure results. That, in turn, produces value.

As senior executives ourselves, we recognize that a discussion of cybersecurity with fellow executives should not be too "technical," because such discussions could diminish this book's utility.[4] Executives run the entire organization, and they don't need to be focused on the coding techniques of their computer programmers. Rather, their job is to optimize the human and physical resources and assets of the organization in order to fulfill its mission safely, profitably, and beneficially. We understand that a prime focus of executives is risk management, and that is where discussions of cybersecurity should begin.

Cybersecurity is about risk management. It is about protecting your business, your shareholders' investments, and yourself while maintaining competitive advantage and protecting assets. It is not just about IT. Rather, it is a multidisciplinary approach to managing risk, a principal concern of every executive. Note that in addition to Chapter 3.0's emphasis on risk management, discussions of risk and risk management are prominently interspersed throughout this book.

---

[3] George Polya, *How to Solve it, A New Aspect of Mathematical Method*, Princeton University Press. Princeton, NJ, 1945, p. 6.

[4] Some people may refer to that meaning "don't make it too geeky," while others may say that the focus should be on the ends (i.e., the effects) rather than the means (i.e., the technology). In this case, we believe both are apropos.

1.3 CYBERSECURITY IS AN EXECUTIVE-LEVEL CONCERN

In our professional dealings, we have had interactions regarding cybersecurity with numerous senior executives and board members. All are highly intelligent and exceptionally talented individuals who understand their businesses inside and out. Nonetheless, many express great frustration in understanding cybersecurity and integrating it into their management processes. Here are some noteworthy concerns from some of our clients:

> I have several people who do cybersecurity for us, but we don't speak the same language. I don't understand what they say and I'm not sure they understand me either. I guess we just have to trust each other.

> I know that cybersecurity is important, but I don't know how well we are doing. How do I measure it?

> Sure, we have a cybersecurity program. How good is it? Okay, I think.

> I am concerned because I don't know whether I am spending too much, too little, or just the right amount on cybersecurity. I don't like playing Goldilocks.

> I am not sure what questions about cybersecurity I should be asking.

Some of these concerns might sound familiar to you. Perhaps you share some of these same concerns. If you do, you are not alone. According to IBM, who manages IT services for customers around the world, their clients average 1400 cyber-based attacks per week.[5] Malicious activity continues to increase from what are commonly called "bad actors," those who attempt to collect, disrupt, deny, degrade, or destroy information or the systems that collect, process, store, transport, and secure that information. Executives at all levels and in all business sectors need to be on heightened alert to threats, understand their vulnerabilities, and take appropriate action to protect their information. Cybersecurity has emerged as a leading concern of executives around the world, and it appears it will remain so for the foreseeable future.

1.4 QUESTIONS TO ASK

Have you ever noticed that great executives often seem to know the right questions to ask? Peter Drucker said, "The leader of the future will be a person who knows how to ask."[6] We submit that such a future is upon us right now. In today's business environment, asking the right questions is indispensible for executives at all levels. According to Gary Cohen, leaders can't know everything, especially today. With information accumulating at such a rapid pace and with so many ways to access information, our coworkers routinely know

---

[5] IBM Global Technology Services, *IBM Security Services Cyber Security Intelligence Index*, July 2013, p. 3.
[6] Peter Drucker, The Drucker Institute at Claremont Graduate University, April 22, 2011, http://thedx.drucker-institute.com/2011/04/the-fab-five/. Accessed on December 13, 2013.

more about their work than their executives do.[7] Therefore, wise executives routinely ask a lot of questions about cybersecurity as part of their management rhythm.

Asking the right question, often to several different people, helps identify vulnerabilities, exposes defects, discovers potential areas of improvement, and illuminates budding talent that should be nurtured. Not only will asking the right questions make you appear smarter, but also listening closely to the answers will make you smarter and better prepare you to ask even smarter questions the next time.

Like our client who said, "I am not sure what questions about cybersecurity I should be asking," many of our clients ask us to help them identify the questions that they should be asking their technical staff, fellow executives, finance staff, legal counsel, and other advisors. Every business is different, and in our consulting efforts, we typically prepare specific sets of targeted questions for each client. Nevertheless, you may find the following sample generic questions helpful as you keep cybersecurity on your agenda:

- Is my computer system infected?
- How did you find out it was infected?
- If it is infected, what do I do about it?
- How did the infection happen?
- If it is not infected, what did I do right, and how can I keep it up?
- If it is infected and we have an IT system administrator, isn't he supposed to keep that from happening?
- If the system administrator isn't enough to keep me safe, who else and what else do I need?
- Do I need outside help?
- How much is this going to cost to permit us to stay safe?
- What is the extent of possible damage in dollars and cents and to our reputation?
- If I am shut down, what happens?
- What do I tell the board of directors?
- Are regular audits by insiders and outsiders a good idea?
- How does one go about these audits?
- How much will the audits cost?
- What is the cost–benefit ratio for audits?
- How do I keep my people from making dumb decisions and doing stupid things that allow "bad guys" into our systems?
- What is the best way to train my people to be safe?
- When I started in business, I remember "safety first" signs and the positive impact they had. Can I do the same thing with computers?
- If I train my employees how to protect their home computers, will it raise their awareness at work?

[7] Gary B. Cohen, *Just Ask Leadership: Why Great Managers Always Ask the Right Questions*, McGraw-Hill. New York, NY, 2009, p. 1.

- Is there any especially high-rated protection software that I should be using?
- What kind of vulnerabilities do we have?
- How often do you check for vulnerabilities?
- Is all our software up to date with the latest version and patches?
- How do I develop an overall cybersecurity strategy?
- Is there any way to be 100% safe?
- If I can't be 100% safe, how do I mitigate risk?
- To what extent is redundancy a help?
- Is our information backed up? How often? Where are back-ups stored?
- To what extent do multiple locations help?
- In the event of a major disaster to my system, how do I recover?
- Do we have a disaster recovery and business continuity plan? When was the last time it was tested? What was the result?
- Do I have internal spies and saboteurs?
- If there are internal spies and saboteurs, how do I know who they are and how do I catch them?
- What are my liabilities if my hidden viruses infect somebody else?
- Are there any public relations (PR) firms who can help me handle cybersecurity problems?
- How about having my own hackers?
- How much should I pay my computer "geeks"?[8]
- How should I hire, train, and vet my cybersecurity staff and the people who use my computer systems?
- How do I keep track of what my cybersecurity crew and the people who use my computer systems are doing?
- Are my other machines at risk? How about my pumps, chemical dosing equipment, product quality control instrumentation, and all processing systems? What about my distribution and logistics plans? Are they at risk?
- How do I protect the organization across multiple devices, that is, mobile phones and tablets?
- Who do I call for help?

We recognize that we just presented you with a lot questions. As you read through subsequent chapters of this book, you'll see that we address the issues behind these questions in greater detail and present information that likely will inspire you to ask other questions as you make cybersecurity part of your corporate culture. One of the key objectives in our book is to attempt to answer as many of the questions listed earlier as we possibly can. Upon reviewing our manuscript, we believe that we did a pretty good job in doing so. We hope you agree.

---

[8] Some may view use of the term "geek" to be a pejorative term; however, many IT professionals freely use the term to denote that they have achieved a high level of technical competence. In this question, our intent is to convey the latter meaning.

1.5 VIEWS OF OTHERS

We are not the only ones who have thought about key questions regarding cybersecurity. The U.S. Department of Homeland Security (DHS) is doing some exceptional work in the cybersecurity realm and created a list of cybersecurity questions for CEOs that we find to be very helpful. To avoid duplication, we deliberately didn't include the DHS questions in our aforementioned list; however, we believe that they are highly pertinent and are listed below. Here are the U.S. DHS's cybersecurity questions for CEOs[9].

Five questions CEOs should ask about cyber risks

1. How is our executive leadership informed about the current level and business impact of cyber risks to our company?
2. What is the current level and business impact of cyber risks to our company? What is our plan to address identified risks?
3. How does our cybersecurity program apply industry standards and best practices?
4. How many and what types of cyber incidents do we detect in a normal week? What is the threshold for notifying our executive leadership?
5. How comprehensive is our cyber incident response plan? How often is it tested?

1.6 CYBERSECURITY IS A FULL-TIME ACTIVITY

You can't let your cybersecurity guard down when you leave the office. Today's executives seemingly are always connected to the Internet in one way or another. When they aren't connected, many of them don't view it as a respite; they view it as a calamity. While in their office, they rely upon a host of IT to conduct their daily business. Mobile devices such as smart telephones, tablet computers, and other such devices have untethered executives, permitting them to access information while commuting, traveling on business, or even while they are on vacation.[10] While many executives complain about being slaves to their emails and other electronic exchanges, nonetheless, they insist upon having the capability to be continually accessible.

Such accessibility requirements create interesting cybersecurity challenges. Many executives and those who work for them frequently perform work on their personal computing devices. The resulting exchange of information between home and work IT devices exposes both to potential cybersecurity threats and creates its own class of vulnerabilities. As such, we propose that *executives should treat home computing systems with the same due care and due diligence as they would their computing systems at the office*. During the course of this book, we will share tactics, techniques, and procedures that will be helpful both at home and in the office as you protect your vital information.

---

[9] U.S. Department of Homeland Security Publication, https://www.us-cert.gov/sites/default/files/publications/DHS-Cybersecurity-Questions-for-CEOs.pdf.

[10] We recommend you avoid doing work on vacation. Nevertheless, the reality of today is that in your executive role that you need to maintain connectivity to perform your duties effectively.

Throughout this book, we emphasize that cybersecurity is about risk management and executives are in the risk management business. By making cybersecurity part of your corporate strategy and culture; by implementing comprehensive plans, policies, and procedures; and by instituting the positive management practices outlined in this book, we believe you will be best postured to manage risk, protect yourself and your business, and deliver to your customers, clients, and constituents results that are effective, efficient, and secure.

2.0

WHY BE CONCERNED?

The loss of intellectual property due to cyber attacks amounts to the greatest transfer of wealth in human history.[1]

General Keith Alexander
Commander, United States Cyber Command

2.1 A CLASSIC HACK

Do you know who Robert Fortune was?

If you are an American, chances are fairly good that you don't, but you most likely recognize the product of his work.

Robert Fortune was the perpetrator of what author Sarah Rose calls, "the greatest single act of corporate espionage in history."[2]

We submit that Robert Fortune was a nineteenth-century hacker, arguably one of the most notorious, whether he knew it or not.

[1] http://www.infosecisland.com/blogview/21876-Cyber-Espionage-and-the-Greatest-Transfer-of-Wealth-in-History.html. Accessed on August 29, 2013.

[2] Sarah Rose, *For All the Tea in China: How England Stole the World's Favorite Drink and Changed History*, Penguin Group, New York, NY, 2010.

Cybersecurity for Executives: A Practical Guide, First Edition. Gregory J. Touhill and C. Joseph Touhill.
© 2014 The American Institute of Chemical Engineers, Inc. Published 2014 by John Wiley & Sons, Inc.

Robert Fortune was a Scottish botanist and adventurer who is credited with the introduction of tea to India, whose subsequent mass production of tea produced great wealth for the British Empire. Truly, Fortune changed the fortunes of the British Empire.

According to Rose's research and surviving documents authored by Fortune and his contemporaries, in 1843, Fortune journeyed to China under the employ of the Royal Horticultural Society. While on this trip, he discovered that green and black teas were derived from the same plant, which caused quite a stir in botanical circles. Upon his return from Far East Asia, he was recognized by many as an expert on teas.

At that time, the United Kingdom had enjoyed a love affair with tea that extended back over 200 years. China was the sole source of tea and costs were high. After all, despite the relatively cheap purchase price from the producers, distributors had to contend with transportation costs, insurance, handling fees, tariffs, and other costs to bring the tea to the demanding European market. If there was a better way of doing business, the British were interested in finding it.

Fortune's reports from Asia spurred interest by the British East India Company. With the British firmly ensconced in India, executives of the company saw an opportunity to supplant the Chinese as the world's primary producer of tea. Blessed with an amiable climate, plenty of land, and an inexpensive labor pool, all the company executives needed were tea plants and the appropriate procedures needed to convert the leaves into the valuable product.

Naturally, the Chinese were loathe to share their monopoly. Moreover, the tea region and how they processed the tea were considered state secrets. There was no legal mechanism to export tea plants for cultivation in potential competing markets and no plans to change. Faced with such a dilemma yet encouraged by prospects of success, the company decided the risks were worth contracting an agent to do their bidding.

Robert Fortune filled the bill. His experience in the region, his botanical skills, and his familiarity with the tea plants made him an ideal candidate. According to Rose,[3] his tasking was simple: return to India sufficient quantity and quality of tea plants to serve as nursery stocks that would launch the Indian-based tea plantations and learn the process for manufacturing tea.

Fortune made his way to China, where he enlisted the help of a local man named Wang, who served as his guide and translator. Wang helped Fortune, who allegedly wore a disguise so that he would not be recognized as a European, gain admission to tea manufacturing facilities where he discovered and carefully documented the process for converting freshly picked leaves into the cured product coveted by millions.

Interestingly, while he was observing the cultivation and processing of tea leaves, Fortune noted that some of the workers handling the leaves had blue marks stained into their hands. In another area, he observed gypsum being added to the tea leaves. With the help of Wang, he soon deduced that the Chinese, believing that their British clients expected their green tea to actually look green, were adding iron ferrocyanide (which has a color sometimes known as Prussian blue) and gypsum (a yellowish color) to produce a green tint to the tea. Normally, their tea was produced naturally and safely yet they were responding to what they thought the demand was from the Europeans. Unfortunately, such a tonic ingested in sufficient amounts is toxic.

---

[3] Ibid.

Soon, laden with the requisitioned tea plants, the manufacturing techniques, and information that implicated the Chinese as suppliers of a tainted and unhealthy product, Fortune returned to India. The British East India Company soon used its vast resources to spawn the creation of tea plantations in India as well as its holdings in Ceylon, Southeast Asia, and elsewhere.

Plaintive cries for justice from China fell on deaf ears. After all, there was evidence that the Chinese were marketing a product (the tinted) green tea that contained toxins. The British East India Company, on the other hand, was able to produce the tea without toxins, and the price was very competitive. The Chinese monopoly soon was broken and the British East India Company, already fabulously rich and powerful, was even richer and (believe it or not) more powerful.

So what? What do you, a modern business executive, care about all the tea in China?

We submit it is because there is a Robert Fortune out there right now, watching your business, potentially looking to find a way to outmaneuver your business and achieve a competitive advantage. Corporate espionage has been around for a long, long time, and it does not look like it will fade away anytime soon. In fact, as we are writing this, Hollywood movies and television shows abound featuring corporate espionage as a theme.

Corporate espionage is a fact of life and is a noteworthy part of the corporate landscape, no matter what the size of your company. Unlike what you see in the movies, corporate espionage is increasingly easy to do in the Cyber Age. With so many companies and their resources connected to the Internet, today's Robert Fortune no longer has to stage a multiyear expedition to strange and exotic lands to conduct his reconnaissance and execute his mission. He can do it from the comfort of his local coffee shop over a Wi-Fi connection. He can enter your corporate network; rifle through all your files; copy, alter, or destroy all your important data; and even steal your secret family recipes. He can destroy your business right under your nose if you don't protect against him.

We don't think that Robert Fortune's actions of corporate espionage were the first the Chinese encountered. After all, we've read Sun Tzu's *Art of War*, which states:

> If you know the enemy and know yourself, you need not fear the result of a hundred battles. If you know yourself but not the enemy, for every victory gained you will also suffer a defeat. If you know neither the enemy nor yourself, you will succumb in every battle.
>
> Sun Tzu, *The Art of War*

This chapter is about the cybersecurity risks you and your company face—the enemies to you and your business. We submit that cybersecurity is an essential part of your risk management profile, both at work as well as at home, and it is essential that you have a thorough understanding of the risks you face. In this chapter, we will help you to "know your enemy," which is the first step in your risk management process.

You may not have known who Robert Fortune was before you read this section, but we bet there are many Chinese citizens who do.

2.2 WHO WANTS YOUR FORTUNE?

The top five cyber threat sources to your survival (in no particular order) are:

1. *Nation-states*
2. *Organized crime and hackers*
3. *Hacktivists*
4. *Insider threats*
5. *Substandard products and services*

In the Cyber Age, we all live in a potentially bad neighborhood. While at face value it is a wonderful place, with information from around the world available at your fingertips, evil lurks in the dark corners of the net. Those with whom you communicate on the net for your correspondence, commerce, and other intercourse may not be who they appear. Criminals abound, seeking to prey on the unsuspecting. Hackers navigate through the net, looking to exploit weaknesses in someone else's code and then brag about their victories in conferences hosted in Las Vegas. Foreign intelligence services play "spy versus spy," using cyberspace as their operating theater. With all the bad guys out there, is now the time to disconnect and take a few steps back to the good old days before the Internet? We think not.

Prior to the advent of the Internet, you faced many risks in your personal and business lives and seemed to do just fine. The advent of the Cyber Age just presents many of the same type of risks in a different format or means of delivery. It also presents amazing opportunities to communicate faster, analyze better, and act with velocity and precision. You need the Internet. You want the Internet. You can't live without the Internet.

You and your company didn't become successful by avoiding all risk. Rather, you learned how to manage risk and maintain a competitive advantage. You learned how to avoid risks that were too profound, how to mitigate risks that you could, and how to manage risks when you could. In the Cyber Age, cybersecurity is about risk management. You and your business can and will survive on the "mean streets" of the Internet provided you understand the risk environment, know your weaknesses, understand the risks, and make intelligent decisions to carefully avoid, mitigate, and accept risk.

As Sun Tzu advised, you must know yourself and your enemy. In this chapter, we identify your potential foes and share with you information about the risk environment. The Internet has a plethora of bad actors, including nation-states, organized crime, hackers, and hacktivists. While you may be aware of many of these threats, we'll also discuss threats posed by insiders, substandard products and services, and a poorly trained workforce, all of which pose significant risk to your organization.

Do you know "the Robert Fortune" looking at your company? What are his objectives? What methods will he use to achieve them? How can you thwart him? How can you mitigate your risk? To know him is the first step.

2.3 NATION-STATE THREATS

2.3.1 China

In 1998, Senior Colonels Qiao Liang and Wang Xiangsui of the Chinese People's Liberation Army wrote a brilliant study entitled *Unrestricted Warfare*[4] as part of their War College studies. In the aftermath of the stunning coalition victory in Operation DESERT STORM, the Chinese colonels looked at how technology changed the landscape of warfare and how potential foes could counter the technological juggernaut created by the United States and its allies.

Among the many conclusions in the study was to leverage the technical reliance of the Western militaries against them. Like judo and eastern martial arts, which seek to leverage the adversary's momentum and mass against them, the authors posited that reliance of the Western powers on technology could serve as an Achilles heel as well as an advantage. They concluded that a smaller, disadvantaged power could achieve a decisive advantage by exploiting weaknesses in the technological framework of the Western power, principally through such means as mastery of computer systems and computer-based attack. In their *Unrestricted Warfare* analysis, all the pillars of society, including the economy and critical infrastructure, supporting the adversary were fair game for attack.

Unrestricted Warfare caused quite a sensation in military and academic circles. While the colonels' work was ostensibly done without government sponsorship in the spirit of academic freedom, many folks took notice, including the Chinese government. In fact, in the aftermath of its publication, the Chinese government stepped up their creation of a "Cyber Corps" of IT professionals and incorporated many of the tenets spelled out in the work into their doctrine and strategy documents. Countless schools dedicated to creating workers highly skilled in computer programming, network computing, and other IT disciplines sprouted around China. Reports of a Chinese "Cyber Militia" sent western Sino-scholars into a tizzy.[5] Many western China watchers, including intelligence and military personnel, questioned the motives of the Chinese. Yet, we ask, why wouldn't they take this action? After all, from a military perspective, the colonels provided a brilliant analysis and presented options that were feasible, acceptable, suitable, and (importantly) affordable. Moreover, from an economic standpoint, it was readily apparent that the world was making the jump to the Cyber Age and China had to jump with it.

In ensuing years, reports of cyber attacks attributed as originating from China emerged with regular and ever-increasing frequency. Qiao and Wang, perhaps inspired to capitalize on the success of their groundbreaking monograph, authored another study that purported that preeminent control of the world's economy had become a goal for

[4] Qiao Liang and Wang Xiangsui, *Unrestricted Warfare*, PLA Literature and Arts Publishing House, Beijing, 1999.

[5] Shane Harris, China's cyber-militia, *National Journal*, 31 May 2008, http://www.nationaljournal.com/nationalsecurity/china-s-cyber-militia-has-penetrated-u-s-computer-networks-20080531?mrefid=site_search&page=1. Accessed on August 29, 2013.

the Chinese.[6] Soon after, businesses and governments around the world were pointing to China as the source of an endless onslaught of attacks against Western commerce, military and government secrets, and precious research and development.

Yet, what constitutes an attack? Is a port scan, which some consider the equivalent of checking doorknobs to see which ones are locked and which ones are not, an attack? Your organization most likely experienced the rapid increase in scans and attempted access to your computer systems that flooded the Internet starting at the turn of the century. Many people contend that these scans are not attacks because frequently they are implemented as part of wide sweeps of Internet address ranges, looking for unprotected addresses; they are not targeted. We disagree. We submit they ought to be considered as an attack because of several factors:

1. Intent: Whoever is scanning you is deliberately looking for weaknesses in your security posture. While they may be conducting only wide sweeps not specifically targeting you, they are done with a purpose and malicious intent.

2. Motive: Scanners are not banging on your digital front door out of altruism. Whoever is scanning you is doing so to achieve an advantage over you whether it means they will take the next step to attempt to control your network, access your information, blackmail you for not properly securing your information (imagine your clients' response to hearing from a third party that you don't lock your digital doors!), or sabotage you.

3. Means: With the proliferation of easy-to-use software such as **Netsploit**, people and organizations who run scans against your system can quickly find your weaknesses and run automated programs to exploit them. This is the equivalent of the adversary finding you have an unlocked door, opening the door, and using their tools to do what they want once inside your network!

4. Results: Some will argue a scan does not necessarily draw a response; therefore, it does not constitute an attack. We disagree. Whether you decide to take a deliberate reaction to a port scan, your organization expends resources to prevent adversaries from identifying and exploiting your security defenses. Expending resources to defend against port scans and other malicious activities is derivative of the attack threat, the potential consequences, and your risk appetite. Moreover, in the event the port scan reveals a vulnerability that is subsequently exploited, you will pay the consequences of having your network "owned" by the unauthorized entities.

Given the intent, motive, means, and results methodology of determining whether or not one is under attack, one can make the case that "bad actors" operating from locations in China have actively engaged in deliberate and widespread activities to attack Western business interests, including yours, as well as government, academic, research and development, and other organizations for the purpose of gaining a competitive advantage for Chinese interests.

[6]Qiao Liang and Wang Xiangsui, Fully calculating the costs and profits of war, in *On the Chinese Revolution in Military Affairs*, ed. Shen Weiguang, New China Press, Beijing, 2004.

Not everyone will agree with this conclusion, yet there are plenty of executives who will. For example, in early January 2010, Google executives announced that the company had been subjected to a "sophisticated cyber attack originating in China."[7] Lockheed Martin, Adobe, Dow Chemical, Coca-Cola, and the New York *Times* all have been the subject of highly publicized cyber attacks that were attributed to Chinese sources. In one CNN report, it was asserted one out of every three observed cyber attacks originated from China.[8]

Business executives soon recognized the threats. According to Timothy Thomas in his monograph, "Google Confronts China's 'Three Warfares'":

> People engaged in the world of business activities agree on one thing, that the Chinese are excellent at espionage. Most businesspersons readily understand that their Blackberrys, laptops, and cell phone are all compromised once they enter the mainland of China. They also come to expect the bugging of their cars, hotel rooms, and casual conversations. Businessmen feel neutered entering negotiations with the Chinese. Many have noted that it seems as if the Chinese knew every proposal they were going to make and had responses in hand.[9]

We submit that the Chinese knew every proposal because they made a point of doing their research. They used every means available, including but not limited to cyber-based reconnaissance, to gain the advantage. Given the widespread reporting of the tsunami of cyber attacks attributed to Chinese sources, it appears evident that the Chinese not only have demonstrated the intent, motivation, and means of launching cyber attacks against the business community, but also they have done so with great effect over many years. No business or organization, including yours, is immune!

Perhaps the most damning evidence of Chinese cyber threats to your business came with the publication of Mandiant's report on *Advance Persistent Threat 1*. Mandiant is a cybersecurity firm headquartered in the United States. Founded in 2004, Mandiant, like other companies in the growing cybersecurity market, offers its customers advanced cybersecurity intelligence and defense capabilities.

In February 2013, as a follow-up to previous reports it had done on cyber espionage, Mandiant released a report documenting evidence it had collected that demonstrated the Chinese People's Liberation Army (specifically Unit 61398 in Shanghai) had engaged in deliberate cyber attacks against at least 141 Western organizations extending as far back as 2006.[10] For many executives, this was the smoking gun evidence that fully implicated Chinese sources as a cybersecurity threat not only to government information but also to the commercial sector.

Many executives who heard about the report through various media outlets concluded the threat was real yet only applies to the very large corporations, especially those who

---

[7] *Cyber Attacks Blamed on China*, http://www.bbc.co.uk/news/world-asia-china-21272613. Accessed on August 29, 2013.

[8] Kevin Voigt, *Chinese cyber attacks on West are widespread, experts say*, February 1, 2013, http://www.cnn.com/2013/02/01/tech/china-cyber-attacks. Accessed on August 29, 2013.

[9] Timothy L. Thomas, *Google Confronts China's "Three Warfares"*, Summer 2010, 40 (2): 111, http://strategic-studiesinstitute.army.mil/pubs/parameters/Articles/2010summer/Thomas.pdf. Accessed on August 29, 2013.

[10] *APT 1: Exposing One of China's Cyber Espionage Units*, Mandiant Inc., Washington, DC, February 2013, http://intelreport.mandiant.com/. Accessed on August 29, 2013.

deal within the so-called military–industrial complex. Those who draw these conclusions are asking for trouble.

Regrettably, the scope and scale of the activities conducted by Chinese-based actors extends throughout the commercial domain to all sizes of business and all types of sectors. In fact, in April 2013, Symantec Corporation, a respected cybersecurity firm, issued its annual *Internet Security Report for 2013*, which stated that small businesses were increasingly the subject of cyber attacks. According to the report, targeted attacks saw a 42% increase in 2012 and businesses with fewer than 250 employees accounted for 31% of those attacks, up from 18% a year earlier.[11] Clearly, the cybersecurity threat to small business is growing.

People are paying increased attention to the threats to small business and you should too. Matthew Wocks, a technical journalist for the Canadian publication, *The Financial Post*, followed up on the Symantec report. Wocks interviewed Vikram Thakur, a researcher at Symantec, who states small businesses may be more vulnerable to attacks because malware creators understand that such firms have weaker security. Thakur says data indicates bad actors look to small business "as places they can acquire intellectual property because the smaller companies are dealing with the large scale organizations." When looking at cyber attacks by sector, manufacturing was targeted the most at 24%, said Mr. Thakur. For the first time, the government sector was not at the top of the list, ranking fourth at 12% of total attacks by industry.[12]

So why do Chinese sources engage in such controversial activities in cyberspace? Is it deliberate payback for the Fortune incident of the mid-nineteenth century? Probably not. We submit they do it because it is extremely profitable. Why bother investing in costly and potentially unproductive research and development when you can acquire someone's information at a fraction of the cost? If you are one of those who think that the theft involved is small, think again. According to Interpol, cyber espionage is responsible for the theft of intellectual property from businesses worldwide worth up to US $1 trillion.[13]

Why do nation-states like China risk condemnation in diplomatic circles and the court of public opinion? We submit they do because it is difficult to prove they do it and hold them accountable. Because it is extremely difficult to prove conclusively the source of cyber attacks, international organizations, such as the United Nations and trade organizations, are hesitant to openly accuse or condemn nations for engaging in malicious activities on the Internet. Even when they are confronted with evidence of complicity, the Chinese vehemently deny involvement.[14] With analysts saying that 90% of the probes

---

[11] *Internet Security Trends Report: 2013*, Symantec Corp., Mountain View, CA, April 2013, http://www.symantec.com/content/en/us/enterprise/other_resources/b-istr_main_report_v18_2012_21291018.en-us.pdf. Accessed on August 29, 2013.

[12] Matthew Wocks, *Cyberattacks increasingly targeting small businesses, report says*, http://business.financialpost.com/2013/04/16/cyberattacks-symantec-report/?__lsa=9c41-0ad6. Accessed on August 29, 2013.

[13] Interpol, *Cybercrime*, http://www.interpol.int/Crime-areas/Cybercrime/Cybercrime. Accessed on August 29, 2013.

[14] Thomas, op cit., pp. 105–108.

and scans of American defense and commercial computer networks originate in China, the resulting bounty of intellectual property has proved to be staggering.[15]

How much is this loss of intellectual property and what are the economic impacts in the United States? According to a 2013 study released by the Center for Strategic and International Studies, *Estimating the Cost of Cybercrime and Cyber Espionage*, researchers posit a US $100 billion annual loss to the U.S. economy and as many as 508,000 U.S. jobs lost as a result of malicious cyber activity.[16]

In summary, Chinese bad actors continue to pose a cyber threat to you and your organization because it pays for them to do so. To quote a colleague of mine, "The juice is worth the squeeze."[17]

Some executives may look at this information and say, "So what?! How does this apply to me and my business?" The lesson to be learned is that you and your business are at risk due to the widespread and aggressive efforts of Chinese-based bad actors to acquire information and resources that give them a competitive advantage. The evidence is compelling and overwhelming that these actors are targeting businesses of all sizes and sectors, including yours.

There is a modern day Robert Fortune out there, coveting your information and resources. As the Boy Scouts preach, "Be prepared!"

2.3.2 Don't Think that China is the Only One

While we use the Chinese-based bad actors as exemplars, the Chinese are not alone in flexing their muscles in cyberspace. According to Mike McConnell, who served as the U.S. director of National Intelligence from 2007 to 2009 and director of the National Security Agency (NSA) from 1992 to 1996, there are many countries that are capable of conducting cyber attacks. "Probably the best in the world in the cyber realm are the United States, then the Russians, the British, the Israelis and the French. The next tier is the Chinese."[18]

While attribution of cyber attacks frequently remains in question, mounting evidence from Iran's Stuxnet experience indicates that nation-states have employed targeted attacks against other nation-states.[19] But, again, for the executive, you may be wondering, "So what?" We know that governments often use skullduggery, espionage, and even outright

---

[15] Nathan Gardels, *Cyberwar with China: Former Intelligence Chief Says It Is Aiming at America's "Soft Underbelly,"* www.huffingtonpost.com/nathan-gardels/cyberwar-with-china-forme_b_452639.html?view=print. Accessed on August 29, 2013.

[16] *Estimating the Cost of Cybercrime and Cyber Espionage*, The Center for Strategic and International Studies, Washington, DC, July 2013, http://csis.org/publication/economic-impact-cybercrime-and-cyber-espionage. Accessed on August 29, 2013.

[17] Major Christopher I. Loftis, U.S. Army. Major Loftis is a highly decorated infantry officer who uses this phrase to describe situations where the difficulties are outweighed by the results.

[18] Gardels, op cit.

[19] Iain Thomson, *"Snowden: US and Israel Did Create Stuxnet Attack Code"*, July 8, 2013, http://www.the-register.co.uk/2013/07/08/snowden_us_israel_stuxnet/. Accessed on August 29, 2013.

attack against other governments. The "So what?" answer is that, like in kinetic armed conflict, cyberspace operations frequently inflict casualties upon innocent bystanders. You and your business may be susceptible to being an innocent bystander victim of one of these attacks.

An example of this innocent bystander concept is illustrated in the 2007 cyber attack on Estonia.

On April 27, 2007, the government of Estonia relocated a statue placed in downtown Tallinn, the capital of Estonia, to a location in the suburbs. The statue commemorated the Soviet soldiers who defeated German aggression during World War II and was installed during the time the country was under the control of the Soviet Union. For the quarter of the population that are of Russian lineage, the statue is a treasured symbol of the sacrifices of the Soviet soldiers, some of whom were Estonian conscripts. For much of the remaining population, the statue is a bitter, visible reminder of the oppression they felt during the control of the Soviet Union. When the government acted to relocate the statue, the ethnic Russian population and diaspora protested vehemently, yet what happened next is what made history and leads to our lessons learned.

Over the course of the three weeks following the relocation, the entire country of Estonia endured relentless cyber attacks. Using a series of "botnets," bad actors attacked the Estonian government, businesses, financial institutions, and critical infrastructure through successive waves of distributed denial-of-service (DDoS) attacks. This means that the bad actors hijacked countless computers from around the world, most likely those they previously had compromised through hacking or other means, and launched them in a coordinated attack on Estonian targets. The DDoS attack is a technique where multiple computers send an unending stream of traffic to a single computer address or range, flooding the receiver with traffic and overwhelming their ability to process the traffic. As a result, the receiving computer is unable to function.

The attacks were severe and affected all parts of Estonian society. Banks were unable to do business so the population didn't have access to their money. Newspapers and media outlets were besieged and were unable to report the news. Private companies, principally those associated with the communications industry, were attacked relentlessly. During the crisis, former White House Cybersecurity Advisor Howard Schmidt even went so far as to say, "Estonia has built their future on having a high-tech government and economy, and they've basically been brought to their knees because of these attacks."[20]

Beleaguered by the onslaught, as the attacks continued to mount, the Estonians reluctantly decided to block all traffic with the outside world, effectively disconnecting their entire country from the rest of the world and asked for help from NATO and cyber experts from around the world. Within three weeks, the attacks ceased, as if they were under command of a single entity.

After the initial finger-pointing and denials that frequently are associated with many of these types of attacks, the international community began analysis in an attempt to attribute the source of the attack and identify ways to prevent it from occurring again.

---

[20] Stephen Herzog, "Revisiting the Estonian Cyber Attacks: Digital Threats and Multinational Responses," *Journal of Strategic Security*, 2011, 4 (2): 49–60, http://scholarcommons.usf.edu/cgi/viewcontent.cgi?article=1105&context=jss. Accessed on August 29, 2013.

There is plenty of evidence that would lead you to believe Russian complicity in the attacks. For example, Joshua Davis' brilliant analysis of the attack as reported in *Wired* magazine observes that in the aftermath of the relocation of the statue in Tallinn, Russian chat rooms and online forums were flushed with rhetoric that was seen as having "stoked into fervor" the Russians who subscribed to the forums.[21] Soon, instructions were posted on the sites on how, when, and where to launch attacks against Estonian targets. These chat rooms served as the command and control (C2) for the attacks. It was a brilliant strategy and masked any attribution of the attack to a single source.

Meanwhile, the Russian government, which is not known for its light hand, wasn't shy in its rhetoric directed toward the Estonians. Russian President Vladimir Putin, in his May 9, 2007, speech at Russian ceremonies commemorating victory over the Germans in World War II stated, "Those who are trying today to… desecrate memorials to war heroes are insulting their own people, sowing discord and new distrust between states and people."[22] It should be noted the 9th of May saw the highest surge in attacks against Estonia.

Determining exactly who is responsible for cyber attacks is extremely difficult. The attacks on Estonia are widely believed to have originated in Russia, although the botnets used to deliver the denial-of-service attacks were located all over the planet. This evidence of botnet "sleepers" indicates that those who orchestrated the Estonian attack had hijacked numerous computers throughout the world prior to the event. By inference, one can assume that they stand at the ready for future attacks. To nobody's surprise, the Russian government has denied any responsibility, claiming that private citizens and groups were involved.[23] To date, no official attribution or sanction has been levied regarding this attack.

But does that matter to the innocent bystanders, the businesses, and the public caught in the cyber crossfire? If your business is unable to operate on the Internet, you may find yourself out of business for good. Consider what you would do if your company's computers were hijacked and used as part of a botnet attack on another company like what happened in Estonia. How would you know? Could your business withstand the pain of being an innocent bystander or, perhaps worse still, the potential liability of providing the means of attack for someone else?

What if an incident like that endured by Estonia occurs in a country where you have one of your critical suppliers? In this age of international business with just-in-time logistics networks spanning across multiple continents, it behooves the smart executive to closely monitor world events, pick your partners wisely, and have a plan to minimize your risk.

In the coming chapters, we will give you strategies, techniques, and tactics you should employ to avoid being crippled by an attack. Before we do, let's visit some of the other threats that you and your business face in the Cyber Age.

---

[21] Joshua Davis, *Hackers Take Down the Most Wired Country in Europe*, 21 August 2007, http://www.wired.com/politics/security/magazine/15-09/ff_estonia?currentPage=all. Accessed on August 29, 2013.

[22] Ibid.

[23] Henry S. Kenyon, *Cyber Attack Yields Lessons*, July 2009, http://www.afcea.org/content/?q=node/1990. Accessed on August 29, 2013.

2.4 CYBERCRIME IS BIG BUSINESS

2.4.1 Mercenary Hackers

Nobody knows for sure how much cybercrime is affecting business; however, estimates are staggering and continue to grow. According to Interpol, in 2007 and 2008, the cost of cybercrime worldwide was estimated at approximately US $8 billion.[24] In 2013, that cost is estimated to exceed US $100 billion per year and is increasing exponentially.[25] While the estimates may not be precise, the bottom line is clear: cybercrime is adversely affecting businesses and individuals around the world.

Perhaps, cybercrime already has touched you, your family, or your business. If it has, you likely know how difficult it is to recover compromised identities, purloined assets, or tarnished reputations. While law enforcement agencies have increased their abilities to respond to cybercrime cases, all too frequently, they find themselves outgunned and overmatched. Regrettably, by the time most folks recognize they are victims, perpetrators are long gone and have taken steps to erase their tracks.

Many Internet observers and theorists consider nation-state-sponsored activities that cause theft, disruption, or alteration of information to be criminal activity. If this is indeed the case, you may argue that there is a compelling argument that nation-states sponsor the largest amount of cybercrime today. While we acknowledge the gravity of these activities earlier in this chapter and do consider many of those activities fall into the cybercrime arena, it is our opinion most activity conducted by nation-state bad actors is more appropriately considered as cyber espionage.

As such, we submit most cybercrime is the product of the work done by mercenary hackers, many of whom work as "independent contractors." Such hackers are persons who deliberately attempt to find and exploit weaknesses in computer systems, networks, and the code that supports them. Unlike the stereotype that portrays hackers as Mountain Dew guzzling, Twinkie eating, antisocial, and unkempt overweight post-adolescent males who have never kissed a girl, today's hackers are more likely to be either male or female, highly intelligent, often charismatic, and highly sociable within their cyber ecosystem.

Hacking, that is, unauthorized entry into a computer system, is now big business for organized crime. According to a report in *The Economist*, organized crime now has access to the same sophisticated exploit kits and cloud-based software services (and perhaps even better than) used by Fortune 500 companies.[26] Hackers acting as "independent consultants" provide organized crime their services to hijack credit cards and banking information, enter the databases of businesses, provide surveillance on competitors and law enforcement, and tamper with records. Awarded hefty commissions, hackers are increasingly able to foil law enforcement by covering their tracks as they enjoy their "piece of the action" in a multibillion dollar criminal environment.

[24] Ibid.

[25] http://www.go-gulf.com/blog/cyber-crime/. Accessed on August 29, 2013.

[26] The Economist, *Organized Crime Hackers are the True Threat to American Infrastructure*, March 11, 2013, http://www.businessinsider.com/organized-crime-hackers-are-the-true-threat-to-american-infrastructure-2013-3. Accessed on August 29, 2013.

What type of threats do hackers pose to your business? Let's take a look at some famous hackers and what they did and see how your business may be at risk.

Kevin Poulsen is now a news editor at *Wired* magazine, an author, and noted cyber-security consultant. He also is a famous hacker who served over five years in prison. Poulsen, also known as "Dark Dante," used his knowledge of telephone systems and computers for a series of criminal activities. For example, he hacked into the Los Angeles KIIS 102FM radio station's telephone system to ensure he was the winning "102nd caller." Between him and his accomplices, they collected two Porsche 944 cars, US $20,000, and two Hawaiian vacations.

Unfortunately for him, before he executed that hack, he had hacked into the U.S. Defense Threat Reduction Agency and the Federal Bureau of Investigation (FBI). The feds were actively hunting him, and he was continually needling them by posting online information revealing their wiretaps of the American Civil Liberties Union (ACLU), foreign consulates, and suspected mobsters. To the senior federal law enforcement personnel, he was referred to as the "Hannibal Lecter of computer crime."[27] Poulsen had embarrassed the FBI and they were out to get him.

After evading the FBI for 17 months, Poulsen's story was shown on the American television show *Unsolved Mysteries*. As was the template for the show, at the conclusion of the report, Poulsen's image was shown on the screen along with a toll-free number for viewers to call if they had information about Poulsen or his whereabouts. Recall Poulsen's specialty was telephone systems. Sure enough, as Poulsen's image and the phone number were displayed, the show's toll-free telephone number went dead.

Poulsen's luck ran out days later as he went shopping for groceries. Fellow shoppers, who had been watching the show, took notice of him in the store, tackled him, and held him until police arrived. He was later convicted of wire and mail fraud. After serving his time in prison, he reportedly has stayed on the right side of the law and is now a respected journalist.

But how does the Poulsen case matter to you or your business? Poulsen hacked into telephone systems over 20 years ago and technology has leapt forward; surely, they are much more secure, right? Maybe. In the intervening years, telephony has evolved to include voice-over-Internet-protocol (VOIP) capabilities, mobile telephony, and integrated services that bundle voice, video, and data. Your telephone system is a computer system and one that needs to be secured.

You may be thinking, "Of course it needs to be secured! Otherwise, thieves could access your system to get unlimited long distance, right?" Well, yes, but they poten-tially also can use your telephone system as an attack vector to dive deeper into your company's data systems and access your financial systems, your correspondence, and your trade secrets. Many executives do not know who has access to their telephone systems and how their telephone system is linked to their corporate networks. Many also don't know whether their telephone system is protected by a properly configured and up-to-date firewall. Every telephone system should be password protected with regularly changed and complex passwords that are tightly controlled and only shared with those who have a need to know them. Ask the question, "Is my business protected against the Kevin Poulsens of the world?"

---

[27] http://www.nndb.com/people/453/000022387/. Accessed on August 29, 2013.

What about hackers who specialize in e-commerce? Chances are good that you or your business engages in e-commerce, either through sales, procurement, or both. You may not remember the name Albert Gonzalez, yet you likely remember his hacking handicraft.

Gonzalez, the son of parents who escaped Castro-controlled Cuba, was the mastermind of an elaborate scheme that resulted in the theft of over 170 million credit card and automated teller machine (ATM) numbers during 2005–2007. To put this in perspective, 170 million represents about half the number of people in the United States today. According to U.S. Attorney General Eric Holder, Gonzalez was responsible for over US \$400 million in financial damages to businesses before he was apprehended, tried, and incarcerated.

Gonzalez was a talented computer aficionado whose real talents lay in his leadership skills; he was a "hacker executive" who was able to forecast opportunities and recruit talented individuals to do his bidding.

According to court documents and details provided in an excellent New York *Times Sunday Magazine* article by James Verini,[28] Gonzalez's computer crime career began in his early youth. When he was 14, he hacked into a NASA computer system, drawing a visit to his high school by the FBI. As he entered adulthood, he already was immersed in the culture of hacking and cybercrime.

In 2003, Gonzalez was an emerging leader in the hacker group known as ShadowCrew when he was nabbed by police in New York for using stolen credit card and ATM numbers to harvest money from ATM machines. Caught red-handed, Gonzalez eagerly agreed to serve as an informant to federal law enforcement officials in exchange for immunity from prosecution.

ShadowCrew was an amorphous group that would serve as a brokerage house for stolen credit card and ATM numbers. Members would hack into computer systems around the world, steal credit card and ATM numbers, and then sell them to the highest bidder using ShadowCrew's venues. Several members of the ShadowCrew were among the greatest perpetrators of the theft, and U.S. law enforcement officials were eager to apprehend them. They needed insider information to gain the evidence they needed to gain convictions, and Albert Gonzalez was a willing assistant to identify the conspirators and evidence they needed.

Over the course of the next year, Gonzalez worked with the Secret Service to create what law enforcement officials called "Operation Firewall." On October 26, 2004, Gonzalez was brought to the Washington, D.C., command center of the Secret Service where he initiated actions to bring ShadowCrew members into a chat session, which would lead to their subsequent apprehension and arrest. By 9 p.m. that evening, 28 ShadowCrew members in eight U.S. states and six countries had been arrested and charged with the theft and distribution of over a million credit card and ATM numbers. At that time, it was the most successful cybercrime in U.S. history.

What officials did not know was that Gonzalez was a double agent. While he was setting up many of his peers to take the fall in "Operation Firewall," he and his band of

---

[28] James Verini, *The Great Cyberheist*, November 14, 2010, http://www.nytimes.com/2010/11/14/magazine/14Hacker-t.html?pagewanted=all&_r=0. Accessed on August 29, 2013.

carefully chosen acolytes had been exploring vulnerabilities in corporate wireless networks. Armed with powerful laptops and antennas, they would use a practice known as "war driving" to gain access to corporate networks through wireless signals. War driving got its name from hackers who would drive through neighborhoods looking for unprotected Wi-Fi signals to gain free Internet access. Gonzalez and his cohorts would use the same techniques to gain deeper access.

Gonzalez was clever. He would identify targets and send his teams out to collect the information he sought. He recruited well as those affiliated with him were generally much more technically talented than he yet lacked his leadership skills, imagination, and initiative; they were good fits for his business model.

Soon, his teams had successfully accessed the Wi-Fi signal from a local BJ's Wholesale Club and accessed corporate information to collect over 400,000 credit card transactions, including credit card numbers and other sensitive information. Emboldened, they used the same techniques at a DSW store to collect over a million numbers. Then, they moved on to the local Marshall's store, where they accessed the store's system; tunneled through the corporate network to the main office's of Marshall's parent company, TJX; and installed custom software they crafted to provide an enduring presence on the TJX network. Using a sniffer program to collect and copy transactions as they were reported from the stores, the custom software they planted compressed and encrypted the data and sent it back to the conspirators via a virtual private network (VPN). According to court documents, they harvested over 40 million transactions from TJX alone.

However, TJX was not alone. According to public reports, Gonzalez's team was able to access corporate networks at Office Max, Barnes & Noble, Target, Sports Authority, and Boston Market. Gonzalez and his team were collecting millions of credit card numbers, but they had to sell them to make a profit. This is where Gonzalez's prowess paid dividends for his technically talented team.

Gonzalez established an international syndicate to sell the card numbers in the United States, the Americas, Europe, and Asia. Using a variety of sources, he diffused the distribution of the card numbers, making it harder to track and attribute to the source. Additionally, he contracted for overseas storage of the stolen numbers, making it increasingly difficult for U.S. law enforcement authorities to track and recover the stolen information. As a paid informant of the U.S. Secret Service, Gonzalez knew the capabilities of law enforcement and organized his operation to protect himself.

He and his team also invested in new technologies. With the rapid expansion of Internet shopping, Gonzalez and his cohorts looked for opportunities to exploit flaws in the web pages of businesses. Using a technique known as SQL (often pronounced "sequel") injection, his technical teams were able to exploit coding weaknesses in the websites of several businesses and gain domain administration access to the business' computer systems. In essence, they achieved "the keys to the kingdom."

Gonzalez's team soon had access to TJX, J. C. Penney, Wet Seal, Hannaford Bros., and Dave & Buster's. They were deep into systems around the country, but they were also in deep trouble. The companies were detecting evidence they were compromised. Owners of the compromised cards were detecting the theft in their monthly payments. Creditors and banks viewed the theft with growing alarm and concern. The FBI and other law enforcement agencies were on the case to hunt down the perpetrators and bring them to justice.

That's because the Secret Service had an agent who was purchasing credit card numbers from one of Gonzalez's associates, Makyam Yastremskiy. Yastremskiy served as Gonzalez's fence, selling the credit card information to the highest bidder. The feds were convinced of the Yastremskiy linkage to Gonzalez, yet they didn't have the evidence to prove it, make it stick, and get a conviction. After months of excruciatingly detailed investigative work, and a lot of good luck, law enforcement officials were able to surreptitiously gain a copy of Yastremskiy's hard drive, from which they were able to trace transaction records back to Gonzalez. They were able to demonstrate the chain-of-custody link of the stolen credit cards from Gonzalez to Yastremskiy. Within days, the feds rounded up Gonzalez and his men.

But the worst was yet to be revealed. While Gonzalez and his team were in custody, Heartland Payment Systems, one of the largest credit card payment companies, discovered their computer systems had been thoroughly invaded. Their computer systems, and thus the essence of the company, were totally under the control of someone else. In the parlance of hackers, Heartland was "owned" by Gonzalez and his henchmen. Heartland, who does business with over 500 banks and is one of the key facilitators of e-commerce, had been breached completely. Over 130 million transactions were known to have been compromised before the hijacking was discovered and corrected. This was a huge heist, but who did it?

The answer came from Gonzalez's jail cell. Perhaps in a bout of remorse or perhaps in an effort to get a reduced sentence, he admitted that he had assisted two Eastern European hackers gain access to Heartland's systems via the SQL injection exploit.

Gonzalez was sentenced to multiple, concurrent 20-year terms and currently is serving his sentence in a federal prison. His actions and that of his associates resulted in the loss of over US $400 million for numerous banks and businesses. Potentially more harmful than the monetary loss is the loss of reputation for many key retailers, many of whom are still remembered for the compromise of their systems by the many people who shop there with cash in hand as opposed to credit cards or checks—or those who don't shop there anymore![29]

This case presents many questions for you and your business. Is your company prepared for Albert Gonzalez? Is your Wi-Fi secured? Are your technical personnel following best practices in cybersecurity? What's your liability if someone accesses your network and uses it to steal from someone else? Is your web page secured from exploits such as SQL injection? Are your networks segmented to prevent someone from hacking in and gaining total control over it? Are your e-commerce procedures properly secured to give you and your customers confidence their information is properly protected? How do you know?

As you and your company assess risk, it is essential that you consider the type of threats hackers like Poulsen and Gonzales present to your company. You need to be

---

[29] In fact, one of the merchants that we used to shop with had their computer system compromised. It was annoying to us but devastating to the retailer. We don't go there anymore and neither do many of their other former customers.

prepared for any contingency as threats come from many directions, even from the company you keep. It is essential you pick your partners well as they too may be the avenue of approach for those who seek to do you harm. An example of this is when hacktivists from the group "Anonymous" attacked Visa, MasterCard, PayPal, and others in late 2010.

2.4.2 Hacktivists

Hacktivists are hackers and other computer-savvy individuals who use their skills to promote social or political agendas. Such groups would like to convince you that they have the welfare of the public at heart when they invade your information systems. In fact, by intruding into your systems without your permission (and often without your knowledge), they are acting unlawfully. Hence, they also are cyber criminals and should be treated accordingly.

While the hacktivist group "Anonymous" is arguably the most famous of these organizations, ad hoc groups emerge every week, fueled by social media to rally against what the hackers view as a fight against injustice. Your business's intelligence unit should keep an eye on social media and websites known to promote hacktivism to alert you if you or your company's name shows up on the Internet as a target.

How did Visa, MasterCard, PayPal, and other firms become the target of Anonymous' ire? It all starts with WikiLeaks.

In 2010, WikiLeaks published classified information from the U.S. government leaked by U.S. Army Private Bradley Manning. Hundreds, and then thousands, of classified State Department cables and military information were made available on the WikiLeaks website. The U.S. government appealed to WikiLeaks and its founder Julian Assange to cease publication of the information, claiming that its public disclosure could put countless lives in jeopardy as military operations and intelligence sources would be revealed. Assange, emboldened by the worldwide publicity the situation generated, refused the appeals and appeared in several news conferences stating that "freedom of the press" would not be impinged and WikiLeaks would continue to publish the documents in parcels.

With the release of the U.S. documents in April 2010, Assange became a controversial public figure, appearing on numerous television shows, at numerous conferences, and in print. Countless people around the world viewed him as a daring protector of freedom of the press, bravely staring down the powerful U.S. government. He was especially popular among the hacker community, who liked how Assange was "giving the finger to the establishment."

Nobody likes being the recipient of being "given the finger," especially the U.S. government, who sought legal means to block the publication of the documents and threw the full weight of the government against Assange and WikiLeaks.

As Assange became more popular, he and his background were subject to increased scrutiny. His personal life was probed and exposed.

Soon, Assange was arrested in the United Kingdom amidst charges of sex crimes in Sweden. To the hacktivist community, this was an example of government overreach and viewed as a petty retaliation for Assange and WikiLeaks "exposing the truth." The hacktivist community, led by the Anonymous group, rallied in support of Assange.

Using the 4chan website, a favorite of the gamer and hacker communities, Anonymous outlined a cyber attack on those businesses and organizations that had turned their backs on WikiLeaks and Assange. They posted directions to those interested in joining the attack and supplied them with cyber weapons, most notably a program called the "Low Orbit Ion Cannon," named after a famous Star Wars rebel weapon. This cyber weapon would enable the user's computer to participate in a global DDoS attack against targets selected by Anonymous.

At the top of the target list were PayPal, Visa, and MasterCard. They removed their support of WikiLeaks, who relied on donations provided through their websites to operate and needed the companies to process the donations. With their support to WikiLeaks severed, like many other businesses subjected to hacktivist attacks, they had some warning they would be targets.

The attacks started in early December 2010 with great media attention. "Operation Payback" was planned and implemented on the Anonymous website against those who stood against WikiLeaks and Assange and gave direction to those willing to join their cause. Postings on 4chan and other websites were used to promote the cause and give instruction to potential recruits. Social media sites, such as Twitter, were used to give play-by-play commentary on the attacks and provide command and control of the hacktivist forces. As the businesses' websites fell victim to the attacks and were rendered ineffective, the "Twitterverse" crowed with victory for the hacktivists.

But the victory was short-lived. The businesses counterattacked. They issued press statements that declared that while the attacks were profound, they had not stopped their customers from being able to execute credit card and other critical transactions. They were able to demonstrate to their customers that the customers were protected against the attack. Further, they implemented plans to counter the DDoS attack and worked with law enforcement authorities, their Internet service providers, and others to neutralize further attacks. Soon, the attacks withered and died.

Does your company know what its exposure to hacktivists is? It is important to understand what activities your organization engages in that may make it subject to hacktivists' ire. Many companies invest in having someone monitor social media for indications that they may be the next target. Are you prepared for an attack by hacktivists? If you are not, now may be a great time to create a plan for responding to media requests if you are attacked. Planning now may be the critical factor in maintaining the confidence of your clients and partners during a crisis.

Hacktivists are out there and can be mobilized quickly. You need to be prepared!

As you consider the hacktivist threat, ask your staff the following questions: What if one of those hacktivists works for us? What if we have someone on the inside working with the hacktivists? How would we know? What do we do? You should be alert that even one of your own employees may side with the hacktivists. You should have a plan to detect and neutralize all threats, even when they come from inside your organization.

2.4.3 The Insider Threat

The insider threat is something very real and troublesome for many businesses. According to the FBI:

> A company can often detect or control when an outsider (non-employee) tries to access company data either physically or electronically, and can mitigate the threat of an outsider

stealing company property. However, the thief who is harder to detect and who could cause the most damage is the insider—the employee with legitimate access. That insider may steal solely for personal gain, or that insider may be a "spy"—someone who is stealing company information or products in order to benefit another organization or country.[30]

Consider the following risk scenarios in your business. Are you and your company equipped to adequately address them? Do you have adequate policies in place? Do your supervisors and workforce have adequate training to observe and report signs of trouble? Do you have the means to detect and act before an issue becomes a problem? See if you can identify which one is the insider threat that may threaten your business or organization by compromising its cybersecurity posture.

Joseph has been working with your company for four years. He is a system administrator in your IT department and is working on his advanced degree through night school. He is an employee of another company (a subcontractor) to whom your company outsourced its IT operations. Although he has only been with that company for a couple of months, he has been working as an IT specialist in your business for many years and has made repeatedly progressive job changes in positions of ever-increasing responsibility. His supervisor reports that Joseph is one of his most talented technicians and your staff expresses satisfaction with his work. He is clean-cut, very articulate, and well traveled. He dresses conservatively and professionally, yet (like many of his fellow technicians) displays his affection for popular cyberpunk culture by decorating his laptop with stickers such as from the Internet Freedom Foundation, the Tor Project, and popular gamer logos. Joseph is well regarded by his peers, who quote Joseph as frequently saying, "Do the right thing!"

Lee is a research chemist in your company. She has been working with other staff chemists to develop a variety of new compounds that show great potential commercial value. She has been with the company for six years and has been exemplary. She shows up early, works hard throughout the day, and has been known to continue her research after work from her home computer by remotely accessing her files on the company's internal databases. Her supervisor calls her "a valuable member of our team—a brilliant scientist!"

Bob is the chief executive officer (CEO) of your company. He founded the company over 20 years ago and presided over its growth and expansion. As the majority shareholder, he is also the chairman of the board and is involved in numerous philanthropic activities, where he freely shares his business card to anyone who asks for it. After all, that's how you generate new contacts and new business! Bob likes to consider himself an active manager of the company and insists on managing his own email, even though he does not have a lot of time. He prides himself on his technical skills and the speed in which he can race through his inbox.

Who among these is a potential insider threat? If you answered all of the above, you are correct!

Each of the employee profiles listed earlier fit profiles identified by the FBI as matching profiles of recent insider threat cases. In fact, it is our conviction that all employees may be susceptible to presenting a risk to your company whether they intend

---

[30] http://www.fbi.gov/about-us/investigate/counterintelligence/the-insider-threat. Accessed on August 29, 2013.

to or not. Let's take a deeper look into the examples given earlier and discuss how you and your business may face an insider threat risk presented by them.

You may recognize the example of Joseph from recent media reports. He is modeled after Edward Joseph Snowden, an IT technician briefly employed by Booz Allen Hamilton in support of the U.S. National Security Agency (NSA) who is accused of pilfering and making public highly classified files taken from the NSA computer systems.[31] In the aftermath of the Snowden incident, many executives now are questioning the reliability and trustworthiness of their workforce and contractors. Do you have a potential Snowden in your workforce? How do you defend against a person like Snowden, who says he is motivated solely by principle and ideology? How do you detect someone acting with potentially hostile intent in your organization? How do you minimize your risk? Do you have a plan?

What about our example of Lee? She is modeled after Yuan Li, a research chemist with a global pharmaceutical company, who pleaded guilty to stealing her employer's trade secrets and making them available for sale through another company that she set up with a partner. For over two years, Li accessed her employer's internal databases, downloaded information to her home computer, and made them available for sale. She was sentenced to 18 months in prison and ordered to make restitution.[32] How do you prevent someone from stealing your company's secrets? Do you have the right policies and procedures in place to deter, detect, and prevent theft of your intellectual property? How do you manage this risk? What is your plan?

What about Bob, your CEO? How can he be a risk? After all, it is his company. According to Rohyt Belani of PhishMe, a Virginia-based security firm, senior executives are a lucrative target for hackers.[33] Because they are high-profile persons, there often is a lot of information about their activities available on the net, making it easier for hackers to craft and send them specially targeted email messages armed with malicious code.

This technique, called "spear phishing" (and pronounced spear fishing), puts you and your company at risk. Often, when you open up the email, the malicious code is launched, where it potentially opens up you and your network to exploitation. According to Belani, senior executives frequently exclude themselves from security training citing lack of time, race through their inboxes, and are most susceptible to click on links in the messages without scrutinizing the link to ensure it is to a trusted site. Interestingly, a new term is emerging to describe the spear phishing directed specifically at the senior executives: "whaling."

Now that you know that some hackers call you a "whale," don't be discouraged. It is not because of your outward appearance. It is because you are a "big fish" in your organization and a big target. You have access to the most sensitive and valuable information in your business. Having access to that information can give someone a great advantage.

[31] Glenn Greenwald, Ewen MacAskill, and Laura Poitras, *Edward Snowden: the whistleblower behind the NSA surveillance revelations*, June 9, 2013, http://www.theguardian.com/world/2013/jun/09/edward-snowden-nsa-whistleblower-surveillance. Accessed on August 29, 2013.

[32] http://www.fbi.gov/about-us/investigate/counterintelligence/insider_threat_brochure. Accessed on August 29, 2013.

[33] Rohyt Belani, *"Why Your CEO Is a Security Risk,"* June 11, 2013, http://blogs.hbr.org/cs/2013/06/why_your_ceo_is_a_security_ris.html?utm_source=Socialflow&utm_medium=Tweet&utm_campaign=Socialflow. Accessed on August 29, 2013.

You have to recognize you are a target for exploitation, follow cybersecurity practices and procedures, keep current through training, and enforce the same discipline throughout your organization. As an executive whose duties include managing risk, recognition that you are a risk is part of "knowing yourself."

Like you, the employees of your business are its greatest asset yet also present risk to your business. In fact, we contend *a poorly trained workforce presents the greatest cybersecurity threat to you and your business.* We are not alone in this view. According to an April 2010 study by the Ponemon Institute, 40% of all data breaches in the United States are the result of negligence.[34] IT departments around the world spend a significant amount of time fixing problems they attribute to "self-induced wounds" caused by employees failing to follow procedures, improperly configuring hardware and software, and "doing some incredibly stupid things." Even the IT specialists fall into the category of "stupid user" from time to time.[35]

Your own people are your greatest risk. We have found that the best way to mitigate this risk is through a combination of training, proper management controls, and strong auditing. We address each in later chapters. Nonetheless, when evaluating the risk that your own staff may adversely affect your information, we advise, "Watch out for the clueless and careless!"

2.4.4 Substandard Products and Services

So far, we've identified human beings as the source of risk for you and your business. Nation-states, criminals, hackers, hacktivists, insider threats, and even the so-called stupid users all present great risk to you and your business. Each is related to direct human involvement. Are these all there are? Regrettably, no.

Another significant risk vector is found in substandard products and services. Poorly designed, poorly crafted, antiquated, or counterfeit products present significant risk to your business. Businesses around the world suffer countless hours of monetary and mission loss due to unexpected equipment and system failures caused by these substandard products. Additionally, similar losses occur when substandard services such as poor or improper maintenance, unqualified and inaccurate advice and counsel (tantamount to "malpractice"), poor performance by contracted personnel, and even inaccurate data from a business partner are accepted on behalf of your business.

Most executives recognize the threats that substandard products and services present to business, yet many do not realize how they may present significant risk to your cybersecurity posture. The same reasons cited earlier that drive monetary and mission loss in businesses also produce risks to your cybersecurity posture. For example, consider counterfeit products. According to the Organisation of Economic Co-operation and Development, in 2005, the counterfeit goods industry was estimated to be worth up

[34] Molly Bernhart Walker, "*Cybersecurity Panel: Federal CISOs Must Focus on Worker Training,* August 12, 2010," http://www.fiercegovernmentit.com/story/cybersecurity-panel-federal-cisos-must-focus-worker-training/2010-08-12. Accessed on August 29, 2013.

[35] Oliver Rist, "*Stupid User Tricks: Eleven IT Horror Stories,*" April 13, 2006, http://www.infoworld.com/d/adventures-in-it/stupid-user-tricks-eleven-it-horror-stories-822?page=0,0. Accessed on August 29, 2013.

to US $200 billion and the goods produced cross every imaginable product sector.[36] It is reasonable to believe that despite best efforts of governments and law enforcement activities to curb the proliferation of counterfeit goods, the counterfeit goods industry has continued to grow significantly.

How is this a cyber threat? The fact is that much commerce, including material and equipment procurement, is transacted using your data and information systems, including the Internet. It is not unusual to never have the benefit of looking a salesman in the eye anymore (even though with some of them that might be a good thing).

The author had a noteworthy experience in dealing with the risks presented by counterfeit products. When serving as a military officer, his unit was deployed overseas and purchased several routers from a channel partner of a trusted and valued manufacturer. The author's supply officer made the purchase directly from the channel partner's local affiliate and struck a deal that seemed too good to be true. It was. It turned out that when the equipment arrived, it looked like what we wanted yet there were telltale signs that something was not quite right. The equipment chassis was not the same as the name-brand routers we already had in our inventory. Additionally, when our technicians opened up the housing of the unit, many of the components were not the name-brand components we expected. The supply officer notified representatives of the manufacturer, who identified the routers as counterfeit. As a result, the manufacturer terminated the channel partner as a business partner, and we made another purchase, this time directly with the manufacturer. Despite the best efforts of the supply officer, our legal staff, and our host nation allies, we never received reimbursement from the disgraced channel partner and lost several thousand dollars of precious operation funds. We lost our money to counterfeits that looked and felt close enough to the real thing yet weren't.

Some may ask why we just didn't use the counterfeit products rather than take the loss and make a repurchase of the genuine equipment directly from the manufacturer. The answer is simple: it wasn't worth the risk.

We could not, and the supplier would not, verify the integrity of the components of the routers, whereas the trusted manufacturer had a strong supply chain management system in place to ensure the integrity of their components and the firmware embedded in them. We were going to use the routers to handle sensitive military command and control information and needed a high level of trust in the integrity of the systems. We could not take the risk that the routers had tampered or altered components that would compromise the security of the devices. Further, we knew the performance parameters guaranteed by the manufacturer. With the counterfeit devices, there would be no warranty in the event the items did not perform to standard. We would not accept those risks, so we dealt with the losses and moved forward to support our mission.

Regrettably, the military is not the only organization to suffer from substandard parts. Examples abound in every business sector with varying effects. In some of the highest profile cases, the failure or underperformance of substandard parts has spelled near disaster for the affected companies, who have to deal with situations

---

[36] The Organization of Economic Co-operation and Development, "*The Economic Impact of Counterfeiting and Piracy*," http://www.oecd.org/industry/ind/38707619.pdf. Accessed on August 29, 2013.

where public safety may be jeopardized, consumer confidence is shaken, and litigation abounds. The risk of substandard parts adversely affecting you and your business may be profound.

Take, for example, the recent case of the Boeing 787 Dreamliner and the challenges it has had with several components, most notably its electrical system.

In designing the 787 Dreamliner, Boeing instituted a revolutionary supply chain model that utilized 50 so-called "tier 1" integrators. Boeing maintained a close relationship with these tier 1 integrators, who would coordinate subordinate suppliers through multiple cascading tiers in the supply chain. This was a radical change from Boeing's previous supply chain efforts, where Boeing maintained direct relationships with their individual suppliers.

Boeing's previous supply chain model is considered a classic or traditional supply chain model. Many executives believe it is the model that gives a business the best visibility and control into the supply chain yet it comes with costs and overhead to develop and nurture the supply relationships and manage the processes to monitor and control the development, production, acquisition, and quality control of materials and products and the requisite inventory control system to sustain operations. Many companies place great value in maintaining control over their supply chain and continue to use such a model.

Boeing executives, on the other hand, decided that the benefits of implementing the new supply chain model would outweigh the risks. Their analysis illustrated that by outsourcing the supply chain through trusted strategic partners who specialized in integration and supply chain management, they could deliver a high-quality product at a lower cost to their customers while garnering a healthy profit margin for their shareholders. It looked great on paper.

How many times have you seen proposals that look great on paper only to see them wilt under stress? Regrettably, in this case, Boeing's new approach did.

Problems resulting from the integration and quality control of the supply chain led to numerous schedule slips and delivery delays. Boeing's tier 1 vendors outsourced subordinate tiers in the supply chain to other vendors who, in turn, outsourced further. Boeing soon found itself unable to accurately track and manage its supply chain, which led to significant integration problems that led to successive production delays, quality control issues, and dramatic increases in costs. Despite a much heralded and celebrated "launch" on July 8, 2007 (deliberately chosen to fall on 7/8/07), that promised first deliveries would enter commercial service in May 2008, the first delivery did not occur until September 25, 2011, with the first commercial flight occurring on October 26, 2011, nearly three and a half years behind schedule.[37]

But delays in deliveries weren't the only problem the beleaguered program would face. As reported by the *Wall Street Journal*, "The delays strained the company's credibility with customers and investors and hurt the bottom line. Boeing initially set a goal of keeping 787 costs to about $5 billion. Barclays Capital conservatively estimates the program ended up costing around $14 billion, not including the penalties

[37] http://www.telegraph.co.uk/travel/travelnews/10207415/Boeing-787-Dreamliner-a-timeline-of-problems. html. Accessed on August 29, 2013.

Boeing has had to pay customers for late deliveries."[38] Could your company absorb that kind of cost creep and stay healthy?

The challenges keep coming for Boeing as electrical system fires attributed to lithium batteries on the aircraft have resulted in several grounding incidents spurring another crisis in confidence for Boeing and their commercial airline customers. Time will tell as to what the root cause is, yet one thing is increasingly evident: substandard parts are a significant problem for Boeing and the Dreamliner. According to Dominic Gates, a *Seattle Times* reporter, Boeing engineers blame the 787's outsourced supply chain, saying that poor-quality components are coming from subcontractors that have operated largely out of Boeing's view.[39]

This case study presents some interesting questions you may want to ask about your supply chain. What risk does your supply chain present? Do you have confidence in your supply chain management processes and oversight to ensure you receive the supplies that meet your specifications for performance and quality? How do you anticipate and prevent the negative effects of substandard parts and materials? Do you have a risk management plan in the event your supplies are tainted or defective?

Boeing is a great company with great employees who will find and fix the problems that plagued the Dreamliner's initial design and production, yet if they are vulnerable to the risks of substandard parts and materials, so are you and your business. You need to be prepared and recognize this risk vector.

Many of our clients don't readily see the linkage between their supply chain management and cybersecurity and risk, but in reality, they are tightly coupled.

You and your business rely heavily on information. You need accurate, timely, and trusted information to make the decisions that drive your business. You need assurance that your information will be there when you need it, delivered in a format you can use, and is "the truth, the whole truth, and nothing but the truth." In essence, it is the way you manage and control.

Nowadays, your information is handled by machines and computers that gather your vital information from numerous sources. It used to be that humans were the principal sources of your actionable information, yet we are living in a period where machine-to-machine interfaces, so-called "smart" systems and sensors, and other automated data feeds are increasingly the prime sources of the information that fuel your awareness of the world around you and drive your decisions. Are you going to risk making those decisions on machines powered by substandard parts that may contribute to feeding you with the wrong information, allow unauthorized people access to your vital information, or deny you the information you need? Ask your team about your supply chain. Do you know where your parts come from? How about your software? Is software a "part" that needs to be managed like other hardware components of your business? We contend that software products pose even greater risks and need even greater oversight. Your questions may expose problems that need to be fixed immediately.

[38] Jon Ostrower and Joann S. Lublin, The two men behind the 787, The Wall Street Journal, January 24, 2013, http://online.wsj.com/article/SB10001424127887324039504578260164279497602.html. Accessed on August 29, 2013.

[39] Dominic Gates, *Boeing 787's Problems Blamed on Outsourcing, Lack of Oversight*, February 2, 2013, http://seattletimes.com/html/businesstechnology/2020275838_boeingoutsourcingxml.html. Accessed on August 29, 2013.

Executives need to recognize the risk presented by substandard products and that extends to how they affect your cybersecurity posture as well. Due care and due diligence need to be taken to ensure that you and your business receive best value in all your procurements and you get what you pay for. In addition to performance and functionality measures, make certain you include warranties, liabilities, and guarantees as part of your risk analysis process when making procurements.

We used the Dreamliner example to show how outsourcing significant portions of a supply chain can negatively impact a company and its commercial airline program. We further showed how you and your business could be susceptible to similar risks because of substandard parts and materials. But what about substandard services? Do they present a risk to you and your company? Absolutely, but particularly when it comes to your cybersecurity posture.

Recall from our earlier discussion on Edward Snowden, the contractor who allegedly stole information from the NSA. He fled the country and publicly disclosed U.S. classified information. He worked there because the NSA, like many government and commercial entities, outsourced its IT operations. Many organizations outsource their IT. Perhaps, you do too. It is essential that the employees your contractor hires are trustworthy. Your board of directors expects you to make sure that the custodians of the machines that contain your company's most valued secrets can be trusted. How do you know that the people who handle the information circulatory system of your business are acting with your company's best interests in mind? How about other services that you might outsource? All of these services are potential risks that warrant your attention.

Take as an example your customer help desk. We submit that your customer help desk is one of your most critical functions yet many businesses and organizations do not invest wisely to ensure that the help desk truly is helpful. As a result, their business is put at risk.

How so? Consider how many times you've called a help desk only to be greeted by someone who hasn't quite mastered your language skills. You have difficulty understanding them and they have difficulty understanding you. As a result, your performance is degraded, introducing risks such as lack of productivity and profit.

Language isn't always the only barrier. You may encounter a poorly trained or inexperienced "level one" help desk agent who passes you on repeatedly through multiple help desk levels in search of the person who can actually help you. Such events waste your time and reduce your productivity as well.

Technology in your help desk may be a risk too. How many times have you spent several minutes struggling to wade through an automated menu of choices when you only wanted to ask a quick question to a knowledgeable technician to resolve your issue? How many times have you found yourself yelling at the phone asking to speak to a human? Services (especially those presented by IT and automated systems) that are tedious and difficult to use introduce risk and need to be addressed immediately.

When confronted by these substandard services, your frustration level rises, and your confidence in the company's ability to service and support their product erodes with every minute you are separated from the successful resolution of the problem you have. The company is at risk of losing you as a repeat customer and risks you telling your friends about the service you received. As such, they risk their reputation and potential to generate new business.

Think this isn't a significant risk factor? Data indicates it may be a bigger problem than many people would expect.

Substandard services produce huge losses for businesses. In 2009, Genesys issued a report that estimated that poor customer service costs businesses in the United States over \$83 billion per year. They found more than 65% of consumers have ended business relationships due to poor customer service and of those who end their relationship, 61% end up with a competitor.[40] We don't believe there are many companies that can survive losing 65% of its business because their help desk is substandard. As a result, help desks and customer service management should be a continual focus for service-based industries. So should cybersecurity and risk management.

You may be asking yourself, "What does my help desk have to do with cybersecurity and risk?"

To answer the question, let's look at two help desks in your business: your external customer help desk and then your internal IT help desk.

Your external customer help desk is a potential treasure trove of information for those who want to better understand the vulnerabilities of your company. Your customer help desk is the first line of feedback from customers who call and email with problems about your products and services. Like all help desks, they keep logs on call volume, problem areas, and who called or emailed whom with records replete with contact information for any follow-ups your company has to make. Help desks often are empowered to offer coupons and discounts, rebates, free returns, and other customer service incentives intended to retain the customer. Imagine if a hacker took control of your customer service software database. What if the hacker took a page out of Kevin Poulsen's playbook and took down your toll-free telephone system? What if the hacker took down your business web page? Your external customer help desk may present a noteworthy risk.

Outsourced help desks frequently do not reside on your corporate network and for good reason. They often are managed on the network of the company that operates your help desk and may not have the same level of cyber-hardening protection as your corporate network. If a hacker is able to penetrate the help desk, they could rapidly gather all of the corporate data generated by the help desk outlined in the previous paragraph and exploit it. They likely will take your customer coupons and rebate information and attempt to sell it like Albert Gonzalez did with credit card information. You can void the coupons and rebates, if you detect the theft, but then, you'll have to deal with the probable bad press that will occur when word leaks of the theft and your action. Most likely, they would offer to sell the information to one or more of your competitors. After all, who wouldn't like to know the weak spots of their competitors? Your competitor may even take to contacting everyone who called your help desk, offering a "better deal" than what they got from you and your business. Is a hack of your customer help desk a cybersecurity risk? We think so. Is it the most tempting target for someone trying to

[40] http://www.slideshare.net/fred.zimny/the-cost-of-poor-customer-service-the-economic-impact-of-customer-experience-in-the-us. Accessed on August 29, 2013.

attack you and your business? Probably not. Nonetheless, it does present a risk and the prudent executive addresses all risks.

Let's look at your internal IT help desk. Like customer service help desks, many businesses outsource these functions, and they are a prime cybersecurity risk to your business. A substandard service performed by your IT help desk could destroy your business.

Help desks are a leading threat vector for hackers and other bad actors who seek to gain access to your networks and the information on them. Using a technique called "social engineering," hackers and other bad actors often are very aggressive, creative, and successful in posing to be someone they are not in order to breach your defenses. Help desks, those same people who reset your password when you forget it or give you permissions to access certain files and directories, are in positions of great power and responsibility yet often are contractors and part of your outsourced IT staff. They also are often the lowest paid employees and least experienced on the IT staff.

Imagine your help desk receives a call at 10 p.m. from you. You sound very anxious and angry. You tell the help desk technician you are traveling on business and have to access some files for a critical presentation the next day and you can't log in. You tell the technician you forgot your password because your secretary (whom you mention by name) always logs you in before you arrive at the office and you have misplaced the slip of paper you always carry with your password. The technician asks you to verify your identity by supplying your date and place of birth, which you do. You tell the technician you are impressed by their customer service and advise the technician you will tell his boss what a great job he is doing. You even mention the task leader of his contract by name to show you indeed know his chain of command. Now, since you have a lot of work to do tonight, you ask for a password reset so you can get working to fix that presentation. The technician, convinced he is going to be getting an attaboy from his boss, resets your password and issues instructions for log-in.

Unfortunately, the call didn't come from you. Rather, it came from a hacker who did his homework. The hacker now has logged in to your network and assumed your identity. He probably downloaded all of your sensitive files and your emails and set up some "backdoors" so he can access your network to gather more information later, even after you have your password reset when the hack is discovered.

Can this really happen? Unfortunately, it happens more often than it should. Because you are a "big fish," there is plenty of information about you on the web that can help a determined foe learn enough about you to overcome simple security procedures. Many corporate leaders have their biographies available online, which give hackers great hints on where to start looking for key personal information. Knowledge of age, hometown, family information, and education is a key data point that leads to information routinely used to verify identities for password resets. Organizational charts, social media sites, and other information readily available on the net help the hacker complete his research to prepare to take your place over the phone. A sharp IT help desk should detect and thwart an attack such as this, but not all IT help desks meet the standards. The takeaway is that substandard services, outsourced or organic, also are a threat vector that exposes you and your company to great risk. Be prepared!

As you read this, don't think that we are against outsourcing. Outsourcing has proven very effective for many businesses, freeing them to concentrate on their core competencies while shedding noncore business activities to partners who specialize in those tasks. Nonetheless, outsourcing presents risks to you and your business. You need to enter into outsourcing relationships with your eyes wide open and conduct a thorough risk analysis before entering into any commitments. Remember though that when contemplating an outsourcing arrangement, you can outsource the work but in the eyes of your customer (and your shareholders), you will always retain the risk and responsibilities—**always!**[41]

2.5 SUMMARY

If you are depressed from reading this chapter, we understand. If you are angry from reading this chapter, we understand that too. There are many cybersecurity risks confronting you and your business. You need to be very concerned.

We firmly believe that if you follow Sun Tzu's advise, "If you know the enemy and know yourself, you need not fear the result of a hundred battles," you will be best prepared to defend yourself and your business in a hotly contested cyberspace environment.

To help you "know your enemy," we introduced the following cybersecurity threat sources facing you and your business:

- *Nation-states*
- *Organized crime and hackers*
- *Hacktivists*
- *Insider threats*
- *Substandard products and services*

We also submit that *a poorly trained workforce presents the greatest cybersecurity threat to you and your business.*

In this chapter, we exposed you to several types of cyber risks. We urge you to relate the examples to your business. Doing so could reveal points of vulnerability that need fixing.

Some people complain that the Cyber Age presents too many risks. We submit that in a perfect world, there would be zero risk but as Yogi Berra said, "If the world were perfect, it wouldn't be."[42]

While one could argue that Yogi is the Sun Tzu of our time, let's not forget that it is critically important that you "know your enemy." If you don't, you'll never properly be prepared. Yet, to prepare, you need to know yourself. Let's now turn our attention to how to manage cyber risk in your business through "knowing yourself."

---

[41] We are reminded of the famous quote of Ronald Reagan after he had a meeting with Mikhail Gorbachev on the SALT Treaty, "Trust, but verify." Another good one is a quote of our grandfather, "Never let somebody else hold your wallet."

[42] http://www.yogiberra.com/yogi-isms.html. Accessed on August 29, 2013.

3.0

MANAGING RISK

*If it is connected to the Internet, your system is exposed to countless risks.
Cybersecurity is primarily about risk management. Minimize your risk
through smart practices. Ensure your systems are properly configured,
patched, and audited. Ensure your workforce is trained and regularly tested.
Make cybersecurity part of your daily business practices.*

Brigadier General Gregory Touhill
Presentation at Southern Illinois University—Edwardsville,
School of Engineering
April 2013

3.1 WHO OWNS RISK IN YOUR BUSINESS?

Risk is the potential of loss resulting from a given action. It is a function of the interaction of threats, vulnerabilities, and likelihood (or probability) of threats acting against you. While there is no universally-agreed-upon prescriptive formula that defines how to measure risk, it is essential that you have a firm understanding of the risk environment (i.e., "know your enemy") as well as your vulnerabilities and the likelihood that deleterious events will occur (i.e., "know yourself").

Cybersecurity for Executives: A Practical Guide, First Edition. Gregory J. Touhill and C. Joseph Touhill.
© 2014 The American Institute of Chemical Engineers, Inc. Published 2014 by John Wiley & Sons, Inc.

Life is full of risk. As an executive, one of your primary responsibilities is to manage risk to protect your business and create an environment for it to grow and thrive.

There are many people who will argue that risk can be transferred like a commodity, brokered away through investments in insurance carefully planned using complicated actuarial tables that calculate probabilities and effects. Insurance is an important investment and mitigation instrument to be sure, yet we contend that you never, ever will remove risk completely from your organization (or life for that matter).

We submit that risk is **managed** at every level of your business, but it is **owned** in the boardroom and C-suite. The responsibility to lead and manage your business is vested there by the owners of your business—your shareholders. While activities are delegated in hierarchical organizations, responsibility never can be. Therefore, we believe it is critically important that you create and maintain a risk management program owned at the most senior levels and designed to cascade throughout the business to where each employee knows and **understands** that they are valued stakeholders in the risk management program.

3.2 WHAT ARE YOUR RISKS?

Understanding where you are vulnerable and to whom or what, as well as the likelihood of someone exploiting those vulnerabilities, is essential to determining your risk posture. In the previous chapter, we identified our top five sources of cyber threats. How do you and your business stack up against them? Where are you most vulnerable? Do you know what your cyber risk is?

If you said no, you aren't alone.

The sad state of affairs today is that most companies do not have a clue as to what their cyber risk profile is nor do they know how to calculate it. There are many who believe that there is no means to calculate your cybersecurity risk. We do not agree. We believe that cybersecurity risk can be calculated using some of the same techniques you use to calculate risk in other sectors. We will show you some examples demonstrating cybersecurity risk calculations, but before we get to the formulas, let's review with you areas that commonly are exploited by the top five sources of cyber threats.

3.2.1 Threats to Your Intellectual Property and Trade Secrets

Next to your treasured workforce, your intellectual property and trade secrets are arguably your most valued assets. These are the most common targets for nation-states, organized crime, and insider threats. Why? For the same reason you retain ownership of your intellectual property and keep secret the special (proprietary) tools of the trade that make your business a success, because possession of intellectual property and trade secrets yields a competitive advantage.

As mentioned in the previous chapter, some nation-states and many criminals actively probe the net, looking to steal intellectual property and trade secrets. This is a lucrative market for the end consumer of the information, be it state-owned businesses or those who purchase such information. It permits them to avoid costly research and development activities, move to production faster, and potentially muscle you out of the market. Recall the Interpol estimate that cyber espionage is responsible for the theft of intellectual property from businesses worldwide worth up to US $1 trillion.[1] This is a serious threat to you and your business.

So, what is the risk that your intellectual property and trade secrets may be exploited? Let's use the following checklist to see if you are vulnerable to cyber espionage, theft, or exploitation:

Vulnerability Checklist (Cyber Espionage, Theft, and Exploitation)

1. Do you have intellectual property and trade secrets you need to protect?
2. Do you currently or in the future have market competitors who would benefit by having access to your intellectual property and trade secrets?
3. Do you store your intellectual property and trade secrets on computer systems?
4. Are your computer systems connected to the Internet?
5. Do your computer systems have Universal Serial Bus (USB) connections that enable thumb drives to be connected?
6. Do your computers have read–write DVD/compact disk drives?
7. Do you have frequent and regularly scheduled backups of your information?
8. Do you store your backup information in an off-site location?
9. Do you use any data feeds from other sources into your network?
10. Do you contract your system administration, maintenance, or software support?

How many "yes" answers did you have? If you had one or more, then you are susceptible to cyber-based risk.

"Wait!" you might ask, "Why do I have cyber-based risk if I answered even one of the questions with a yes?" Here's a quick rundown of how a "yes" to any of the following questions could lead to a cyber-based risk.

1. Intellectual property and trade secrets: If you have them, you need to protect them. Picture the following scenario: you are diligent in protecting your critical information. You do not have it stored on a computer, only maintain hard copies of your classified documents, and limit physical access to the documents. Unfortunately, one of your employees has been recruited by one of your competitors to acquire your information. They gain access to your files, photograph them with their cell phone camera, and upload the images from their phone onto a destination selected by your competitor. Fiction? Regrettably no, as this type of exploitation has occurred multiple times around the world. If you have sensitive information, protect it. We recommend you keep cell phones and similar devices away from it. Don't forget meetings where you discuss sensitive information either. If someone has a phone in the room, your meeting may be broadcast to people and places you don't want to include.

[1] Interpol, *Cybercrime*, http://www.interpol.int/Crime-areas/Cybercrime/Cybercrime. Accessed on September 10, 2013.

2. Competitors: Your competitors want to have a competitive advantage over you. Most are honorable and exercise fair and open competition; however, a rare few employ agents who seek access to your information (unauthorized, of course). Nation-states, organized crime, and unscrupulous businesses all have been known to actively use cyber-based resources to steal or tamper with sensitive intellectual property and trade secrets. Cyber espionage is a growing problem in the marketplace with complaints to law enforcement officials continuing to rise. You and your business are at risk. Additionally, the better you are and the bigger you are, the bigger and more lucrative target you present.

3. Computer storage: If you store your information on a computer, you are like most other entities. Computers and their storage devices have become the preferred storage media for the world's information, far surpassing paper copies. This is because computer-based storage is less expensive, provides much faster search and retrieval capability, and enables near-instantaneous transmission of information to multiple locations. The advantages of computer-based storage are many, yet this mode of storage comes with risks as well. Computers rely on electrical power and therefore must have a reliable, uninterruptible power source. They are machines that require maintenance and have components that sometimes fail. They require software to operate effectively and software requires maintenance, regular updates, and most often licensing fees. There are ample possible points of failure that can deny you access to your critical information or present weaknesses that could be exploited by potential adversaries. This presents risk.

4. Internet access: As mentioned in the author's quote at the beginning of this chapter, if your information is on a computer connected to the Internet, it is potentially exposed to anyone else on the Internet. Certainly you can and should implement prudent security measures such as boundary protection (i.e., firewalls, proxy servers, access control lists, etc.), encryption, and other technical measures, but if your critical intellectual property and trade secrets reside on a system connected to the Internet, there is a risk that someone smarter than your IT team will gain access to that information.

5. USB connections: USB ports add great convenience and transportability for information. You can plug in an inexpensive high-capacity thumb drive to transfer files between the computer and the thumb drive and even launch programs from the thumb drive. How many times have you used a thumb drive to transport a business presentation, sensitive data, or even pictures of your family? Like us, you likely have done so. Regrettably, bad actors have taken note of the proliferation of thumb drives and other devices that connect to USB ports (such as smart phones, digital cameras, and even the author's watch!) and are now using them for malicious purposes. An example is the recent Stuxnet attack, where the destructive code is said to have been inserted into the isolated Iranian nuclear control systems by using an infected thumb drive.[2] Any device connected to your computers via a USB port has the potential to insert or retrieve information. There is a risk.

[2] http://www.f-secure.com/weblog/archives/00002040.html. Accessed on September 10, 2013.

6. DVD/CD read–write drives: These older media devices pose similar risks as do USB devices. They could be the entry point for malicious code or the egress point for your critical information. U.S. Army Private Bradley Manning confessed to having used a compact disk with read–write capabilities to exfiltrate 1.6 gigabytes of classified information that he later uploaded to WikiLeaks.[3] As the U.S. Army painfully discovered, any time you have the ability to download information from your computer or the network it is connected to, you have a risk that the information may leak to unauthorized personnel.

7. Data backups: This is considered a routine maintenance and risk avoidance activity in most professionally run IT departments. Ensuring that you have duplicates of your information helps insulate you from hardware failures like crashed hard drives, software faults that occasionally corrupt files, and even "stupid users" who inadvertently delete critical information. While many backups now are done through automated routines, it is important to find the right frequency and time to execute your backups lest you adversely affect business operations. Because of the volume of data many businesses have, data backup often is done incrementally on a prescribed basis. Many businesses run a risk that a system failure can occur that can erase any data since the last backup. Do you know how often your IT shop backs up your data? You have a risk—do you know what it is?

8. Off-site storage: This is a best practice within the IT community and entails maintaining backup copies of critical information at a location other than the primary location. This is designed to ensure that the data survives in the event of a catastrophe at the primary location. As a result of the terrorist attacks on New York in 2001, many companies recognized the risk to their continuity of operations when their information was inaccessible. Now, most businesses have robust off-site storage and data recovery plans designed to facilitate rapid restoration or capabilities from secured locations.[4] They are reducing their risk by doing this. How are you addressing your storage risk?

9. Data feeds: Many, if not most, businesses rely on data from other sources to execute their operations. Financial institutions exchange transaction information at the speed of light. Similarly, electronic commerce flows through the Internet at ever-increasing volumes every day. Business partners place orders through electronic data interchange (EDI) formats that are standardized around the globe. Data feeds fuel the business world and enable fast transactions at lower cost and greater precision. They also present risk. What happens when your feeds are unavailable? What happens if one of your data feeds is corrupted and is feeding your system with bad information? How would you know? How long would it take to fix? How much would it cost? The integrity of your business depends on the accuracy of your information. You need to address your data feeds in your risk management planning.

[3] Kat Hannaford, "How a Burnt Lady Gaga CD Helped Leak Thousands of Intelligence Files," November 29, 2010, http://gizmodo.com/5701089/how-a-burnt-lady-gaga-cd-helped-leak-thousands-of-intelligence-files. Accessed on September 10, 2013.

[4] Patrick Thibodeau, "How 9–11 Changed Data Centers," September 11, 2011, http://www.computerworld.com/s/article/9219903/How_9_11_changed_data_centers_. Accessed on September 10, 2013.

10. Contracted system administration, maintenance, and software support: Anyone who has access to your information, especially your intellectual property and trade secrets, poses a potential risk to steal or tamper with that information. Your business likely vets each of its employees, but what provisions do you have to ensure that your contracted support is equally trustworthy? What provisions do you have to ensure their competence? As with your own employees, be mindful that your intellectual property and trade secrets are vulnerable to theft, tampering, or destruction by contracted personnel. That is a risk worth protecting against.

We anticipate your intellectual property and trade secrets are potentially vulnerable. You want to protect them from the many cyber-based threats confronting you and your business, but what of other threats? How well do you "know yourself" and your vulnerabilities to other threats?

3.2.2 Technical Risks

Technical risks are those risks presented through the operations and maintenance of the technical systems used by your business, for example, computers, processors, monitors, controllers, timers, alarms, etc. They are plentiful and can be catastrophic to your business. If your chief information officer (CIO) is telling you that the IT staff is a crackerjack team and you don't face a cybersecurity risk, we submit that it is time to begin your search for a new CIO. How do you know you and your business have technical risks? They are there. Do you know what they are and have a plan to address them?

Let's use the following checklist of questions to see if you are vulnerable to some of the most common technical risks found in organizations.

Vulnerability Checklist (Common Technical Risks)

1. Have you or your business ever been hacked?
2. Have you ever found malicious code (such as viruses, trojans, or worms) or unauthorized software on your systems?
3. Is your network being probed by outside entities?
4. Do any of the members of your IT staff fail to maintain current industry certifications in their specialties?
5. Are there more current software versions, including patches, available for your system?
6. Do you store data "in the cloud"?
7. Does your workforce use mobile devices such as smart phones, tablet computers, and laptops to conduct your corporate business?
8. Does your business solely rely on passwords to control access to the network and information?
9. Does your business conduct annual vulnerability scans of your network?
10. Do you allow remote access to your network?

If you answered "yes" to any of these questions, you have technical risks that need to be addressed.

We recognize that most executives have neither time nor inclination to become IT experts (although we have met many executives who mistakenly thought they were already!) Nonetheless, it is important to understand the basics and how they affect you and your business. Let's expand a bit on the aforementioned technical risk assessment (vulnerability checklist) so you can see where you and your business may have cyber-security risks that ought to be addressed:

- Previous incidents of hacking: Organizations that have been hacked before are more likely to face other hacking attempts. Hackers like the challenge of break-ing into systems and often post their results on Internet message boards to show off before their peers. This invites others to try to get into your system as well because you have been identified as vulnerable. Additionally, many hackers who successfully penetrate into systems will create "backdoors" that will permit them to come back whenever they want, undetected by you and your security personnel. They are very careful to cover their tracks and try to leave no trace behind that will lead law enforcement and your security personnel to them or their backdoor capabilities. **If you have been hacked before, you are at great risk of being targeted again!**

- Malicious code: Malicious code includes such things as viruses, trojans, worms, and remote access trojan (RAT) kits. Our glossary of technical terms explains them in greater detail. Suffice to say, however, malicious code can get into your system and cause significant damage to you and your business. There are numer-ous ways malicious code can enter your system. Malicious code can enter through an email message with an attachment or self-extracting file. It can enter your system through a mobile device connecting with a poisoned connection point, such as a Wi-Fi spot, that has been compromised by a hacker. It can enter through contaminated media like the thumb drives cited in the Stuxnet example. It can even enter your system when you visit web sites that have been infected with the malicious code and pass it on to your system. Even if you have the best antivirus detection software on the planet, once the malicious code gets in to your system, eradicating it often is expensive and difficult. If you've been infected before, there is a chance that the malicious code may have opened up your system for the planting of even more insidious and undetectable code. This is a significant cybersecurity risk!

- Probing: If you are being told you aren't being probed, you aren't connected to the Internet or you have an incompetent IT staff. The Internet is chock-full of people scanning the net looking for vulnerabilities. In fact, there is a cottage industry evolving where hackers look for corporate networks that are improperly configured, find the vulnerabilities, and exploit them, leaving behind RAT kits that give them remote access into the corporate networks. They then advertise they have control of the networks and sell their services to the highest bidders, which occasionally includes the affected company, who pays to rid them from their network. The lesson is that you will always be subject to probes looking for vulnerabilities. Ensure your defenses are adequate, properly configured, and technically current to minimize your risk.

- Staff certification: Would you fly on a jet airliner piloted by an individual who only had flown a single-engine propeller airplane a couple of years ago? Who would do that? You expect the pilots to maintain their commercial pilot certifications, which includes the requisite qualification training, physical and mental wellness, continuing education, simulator currency training, and actual flight time, to maintain their proficiency. You should expect the same from your IT staff. The IT industry has numerous professional certification programs to ensure that your IT staff has the current level of expertise and talent to perform at the high levels your business needs and deserves. If you have IT personnel who do not have or do not maintain their professional certifications, they may not be capable of adequately defending your information against increasingly sophisticated threats. As such, you may expose yourself and your company to cybersecurity risks. Moreover, like an airline that has an accident at the hands of a pilot who lacks certification, if your network is managed by technicians who don't have proper certification and qualifications, you may expose yourself and your company to litigation in the event that your network is breached. Our recommendation is that whether your IT staff is comprised of direct employees or contracted personnel, you need to ensure they have the right qualifications and certifications to do their jobs properly. This will reduce your risk of having networks and systems that are not professionally and properly configured and operated. Moreover, it will reduce your liabilities in the event your system or **that of one of your customers** is compromised.

- Software currency: Did you know that Microsoft releases security patches the second Tuesday of every month? Known as "Patch Tuesday," it has been a great help to IT staffs around the world and significantly helps improve the security of Microsoft products. Companies like Microsoft routinely issue patches to their code to improve their products and harden them against vulnerabilities that have been discovered in their code. Unfortunately, it takes time for the software developers to create patches to counter vulnerabilities, so the time between detection of the vulnerability and fielding of the patch is when you are most vulnerable. Therefore, when a certified and tested patch emerges from the vendor, it is in your best interest to patch your system quickly to reduce your risk exposure. Likewise, newer versions of software repeatedly have been found to be better constructed and more secure. Maintaining current software configurations and patches is an IT best practice that minimizes your cybersecurity risk.

- Storage in the cloud: The jury is still out when it comes to cloud storage and security. Cloud storage involves storing data on multiple servers often connected to the Internet and generally is hosted by third parties. Because your data is being handled on devices managed by someone else, likely will traverse across the Internet, and is hosted on "virtual" servers on platforms that host information that belongs to other entities, what could go wrong? We contend that cloud computing presents an attractive and economical means of storing data yet presents a cybersecurity risk worthy of a thorough risk/benefit analysis before making any commitment to put mission critical information into "the cloud."

- Mobile devices: They are everywhere! You likely have a smart phone and a tablet computer to complement the desktop that graces your office. After all, you need to be connected all day, every day, no matter where you are. You need to be connected to your workforce as they execute their duties, no matter when and where they are too. It is intoxicating to see how fast the business community works when it employs mobile computing devices. Choices in devices are exploding too. Employees clamor for the latest and greatest devices, while IT departments struggle to integrate heterogeneous devices powered by disparate operating systems from Apple, Microsoft, Google, and others into the corporate network. Mobile devices often connect to other networks that may not be protected as well as yours and may serve as a means to introduce malicious code into your network when they "return home." Recall Timothy Thomas's observation in Chapter 2.0 about business executives traveling in China observing that their mobile devices were exploited.[5] Mobile devices are great tools yet require the policies, procedures, training, and discipline to minimize your cybersecurity risk.

- Passwords: Passwords are getting easier to crack and exploit. The U.S. Department of Defense recognized this fact years ago and invested in a two-factor authentication system using Public Key Infrastructure (PKI) to verify identities prior to granting network access. The department's PKI system features identification cards with a chip containing electronic tokens associated with the individual. Defense personnel logging into defense networks slide their identification card into a reader that reads the electronic chip to retrieve the token and queries the user for their password. Once that is supplied, the network domain controller polls a trusted server on the network to verify that the password and token indeed are appropriately matched before granting the user access to the network. The commercial sector too is rapidly adopting two-factor authentication in lieu of simple passwords as means to authenticate and grant access to network and information resources. For example, the author's bank offers a similar two-factor authentication system for its electronic banking to reduce its risk of theft. If you do use passwords, there are several best practices you should follow at home as well as in the office.

Password Best Practices

○ Try to make your password something you can and will remember.
○ Don't store your password on a sticky note by your computer, in your wallet, or in your phone. Keep it as secure as the information it protects!

[5] Timothy L. Thomas, *Google Confronts China's "Three Warfares,"* Summer 2010, pp. 101–113, http://strategicstudiesinstitute.army.mil/pubs/parameters/Articles/2010summer/Thomas.pdf. Accessed on September 10, 2013.

○ Don't make your password easy to figure out (e.g., P@$$W0rd), your spouse's or child's name (e.g., M0mm@of2), or your favorite sports team (e.g., $t33LeR$#1). Bad actors run password cracking programs that have thousands of passwords like these already stored in their tables. They also research you and can quickly find the names of your family members and figure out your favorite sports mascots.

○ Passwords of 14 characters or more are statistically most secure. Use the maximum strength password that your system will allow.

○ Never share your password with anyone.

○ Never reuse your username and/or password on other accounts.

○ Make sure your password has two upper case, two lower case, two special characters (e.g., @, #, $, %), and two numbers in it.

○ Avoid using typical character substitution (such as @ for "a," ! or 1 for "l," and 0 for "O") in lieu of letters.

○ Change your passwords often. We recommend you change your passwords every quarter. Now, with automated reminders you can load in your phone, you have no excuse for forgetting to do it.

• Vulnerability scans: Your cybersecurity personnel should be continually scanning your network to detect suspicious behavior and to find and correct vulnerabilities. Scanning is not a once a year event. Your CIO and chief information security officer (CISO) should have the results of vulnerability scans as one of their primary job performance metrics. The scans should show how many vulnerabilities are present. Up-to-date software can categorize the severity of the vulnerability to aid in the risk management process. These are your risks. You own them; they don't just belong to the CIO and CISO. Ask to see the results regularly and incorporate them into your governance and oversight rhythm. It is our experience that when vulnerability information makes its way to the directors and officer level, attention is paid and the number of vulnerabilities quickly drops!

• Remote access: This capability provides increased employee productivity and cost savings when implemented efficiently, effectively, and securely. When it is not properly configured, bad actors may find it to be "the information superhighway" to your corporate secrets. There are several risks that remote access poses to your cybersecurity posture:

○ First, the device you are using at the distant end may be infected or contaminated. You don't know what that device has plugged into before it came to your network asking to be connected. It may have a virus just waiting to infect your network!

○ Second, when you permit that device to connect, you are opening up your security perimeter, making it increasingly difficult to defend against hostile threats.

○ Third, once you open up that hole in your defenses, you need to ensure it is sealed properly after the remote access session is concluded.

We have found a best practice is to implement a policy establishing a limit on the amount of time for the remote connection. When the limit is up, the session is terminated unless the legitimate user on the distant end reverifies their identity to the network. Another best practice enabled by technology is to implement a "comply-to-connect" policy. This means that when a device goes to log in to the network remotely, it is quickly scanned by your network devices to ensure it is properly configured to your standards and is free of malicious codes. This capability is not inexpensive and slows down the log-in process, but it definitely helps prevent contamination from remote access devices. Remote access is a powerful capability for your mobile workforce, yet we advise caution in granting remote access. Not everyone needs it.

An important reminder about remote access is that it not only applies to your administrative and business computing systems but also to your specialized equipment too. Many industrial control systems (ICS) such as your heating, ventilating, and air conditioning controls (HVAC), industrial machinery (e.g., pumps, valves, flow and speed regulators, and fuel systems), water and sewage, and power generation all rely on specialized computer controls to operate. Often referred to as Supervisory Control and Data Acquisition (SCADA—pronounced SKAY-DAH) systems, these embedded computing devices control and regulate the critical systems that support the technology we have grown highly reliant upon. Many SCADA systems are connected to the Internet and have been fielded without adequate cybersecurity controls. Frankly, when many were fielded years ago, the cybersecurity threat was so small that many people did not notice the threat to SCADA systems. As we saw with the recent Stuxnet attack, SCADA systems indeed are vulnerable and cyber attacks on them can have catastrophic effect. Physical security of these systems is important too. Even if the device is not connected to the Internet, if it is accessible to someone physically connecting to it, you are at risk. We recommend you minimize your risk by only granting access to those who truly need it and only during those times when they need to.

The risks identified earlier are just a few of the technical risks that are out there. Fortunately, technical risks can be reduced significantly by professional management of your information technologies, regular independent auditing, and prudent investments to maintain system currency. These are core competencies of your CIO and CISO, yet they need your help and support to ensure that the appropriate mix of plans, policies, and resources is applied to provide the optimum cybersecurity posture to meet your business objectives. **It is a team effort!**

3.2.3 Human Risks

Because cybersecurity is a team effort, as an executive, you need to recognize the strengths and weaknesses of your team. Not everyone on your team is a superstar when it comes to cybersecurity. Recall our key point from Section 2.4.3: "we contend a poorly trained workforce presents the greatest cybersecurity threat to you and your business."

Human risks to your cybersecurity posture are profound. From the top of your organization to the bottom, your workforce presents significant risks that you need to address. Wonder what kinds of human risks you and your company may face in the cybersecurity realm? Here are a few common ones (and they may look familiar in noncybersecurity settings too) that you need to address deliberately before they yield catastrophic results:

- Spear phishing and whaling: We introduced these threats in Section 2.4.3, yet they merit mentioning again as they are recognized as the most favored method of gaining unauthorized access to networks by bad actors. In a spear-phishing attack, a target receives a carefully crafted email that looks like it came from a legitimate source. It has the right look and feel to make the recipient think it is an ordinary email. The recipient is lured to either download a seemingly harmless file attachment or to click a link to a malware- or an exploit-laden site. The file, often a vulnerability exploit, installs malware in a compromised computer. The malware then accesses a malicious command and control server to await instructions from a remote user. At the same time, it usually drops a decoy document that will open when the malware or exploit runs to hide malicious activity.[6] Are you and your workforce susceptible to these email-based attacks? Absolutely! We all are. How can you reduce your risk? We suggest you educate your workforce to follow my colleague Mike Jenkins' **READ** the message technique.[7] Before opening any email, look at the message information in your inbox and ask the following questions.

  ### Email Queries

  ○ Relevant: Is this message relevant to me and what I am doing?
  ○ Expected: Did I expect this message?
  ○ Authenticated: Did this really come from the person that it says it came from? Is it from a different email address than I am used to?
  ○ Digitally signed: Is this digitally signed? Digital signatures are increasing in use and help verify the identity of the sender. Look to see if the sender signed it to verify their identity.

      If you answer "no" to any of these questions, you need to be on alert that the email may be tainted. Never click on an embedded link without knowing for sure where it is going! Never click to open an attachment that comes from a suspicious source! **READ your mail carefully!**

- Social media: Social media is a great means of communicating quickly and effectively to a wide variety of people. When used as part of a well-managed business strategy, it can be a boon to your market presence and give you a decisive advantage over your competitors. It can also be a huge cybersecurity risk that can sink your reputation and open your business to attack. Don't believe that your

---

[6] Trend Micro Inc., "Spear-Phishing Email: Most Favored APT Attack Bait," 2012, http://www.trendmicro.com/cloud-content/us/pdfs/security-intelligence/white-papers/wp-spear-phishing-email-most-favored-apt-attack-bait.pdf. Accessed on September 10, 2013.

[7] James M. Jenkins is a Certified Information Systems Security Professional (CISSP), Information Systems Security Engineering Professional, Information Systems Security Management Professional, and Certified Chief Information Security Officer in the Greater Saint Louis area.

Facebook or Twitter account could open you to attack when not used properly? Think again. Look up "Koobface" on the Internet (yes, it is an anagram of Facebook.) It is a computer worm that appeared on social media sites including Facebook, MySpace, and Twitter. It was designed to gather log-in information, set up botnets to do the bidding of the bad actor behind the malicious code, and open the user's computer up to further exploitation. It originally spread quickly through friend requests on the social network. When the user clicked a link, it sent them to a poisoned site where the malicious payload was delivered and installed on the user's system. Despite the strengthening of security at Facebook and other social media sites, Koobface versions still abound in 2013. Koobface is an example of how malicious code promulgated through social media presents risk.

What about other known cybersecurity risks of using social media? Bad actors have been known to use social media to map organizations by making hierarchical associations using the friends feature of the social media tool. It is not unusual for people to "friend" their boss and subordinates on social media sites. Bad actors know that and with a little work are able to ascertain from the social media site, web searches, and other sleuthing who does what in organizations. They then take that information and invest it into their spear-phishing efforts. Aren't you more likely to respond when you get an email from your boss correctly referencing his boss as well as members of your work group? Most people would and bad actors seek to leverage this fact to use a variety of technical and social engineering techniques to gain access to your information. What about instances where employees in your company go onto their social media site and bad-mouth you and your company? In some instances, employee disclosures of corporate impropriety and trade secrets have occurred over social media outlets, resulting in great embarrassment to the business, dismissals, and temporary loss of value in the marketplace. Our advice to reduce your social media cybersecurity risk is to regularly and thoroughly train your workforce on how to use the tools safely and responsibly. Consider conducting internal exercises such as seeing if they are able to identify a potentially malicious email or malicious social media activity. This will help you fine-tune your training program as you discover where your weaknesses are. Also, don't be afraid of using social media just because there are threats. You and your business should not be strangers to social media. Social media enables business growth through market presence and visibility, rapid communication to prospective clients and yields valuable feedback from your customers. Ensure someone on your team has responsibility for posting your message and monitoring social media sites to ensure your valued brand remains in good stead.

• Inadvertent disclosure: Your employees may inadvertently disclose sensitive information without even realizing it. Numerous examples abound where unwitting employees post information to web sites, send out letters and emails, and even conduct press conferences revealing sensitive material that senior leaders in the organization want protected and withheld. Such sensitive material is not limited to just trade secrets. It can just as easily be personally identifiable information protected under the Privacy Act, or it could be copyrighted material you do not have rights to use. Just the other day, my college-aged son received a note from

Netflix informing him that the next season of "Fringe" would have to be pulled from their site as they did not yet have rights to show it. We already watched the first episode but will have to wait another month to resume the series. Imagine what happened behind the scenes at Netflix when they found they had a problem. Imagine what the liability implications are behind such an inadvertent disclosure. Training is essential to reduce the likelihood you will have inadvertent disclosures and thus reduce your risk.

- Ignorance: Some may argue that inadvertent disclosure and ignorance are one and the same. We disagree. While there is some overlap and they often share common results, ignorance is the result of not knowing something, while inadvertent disclosure is the result of a mistake made contrary to a known policy or procedure. People often are ignorant of rules, procedures, concepts, and even of the effects of their actions, yet we believe that the vast majority of people try to do the right thing. Take the following cybersecurity incident into account and see if ignorance had a hand in how the situation developed:

> In April 2013, the administrative assistant to a vice president at a French-based multinational company received an email referencing an invoice hosted on a popular file sharing service. A few minutes later, the same administrative assistant received a phone call from another vice president within the company, instructing her to examine and process the invoice. The vice president spoke with authority and used perfect French. However, the invoice was a fake and the vice president who called her was an attacker. The supposed invoice actually was a Remote Access Trojan (RAT) that was configured to contact a C2 server located in Ukraine. Using the RAT, the attacker immediately took control of the administrative assistant's infected computer. They logged keystrokes, viewed the desktop, and browsed and exfiltrated files.[8]

Would you think that the administrative assistant was ignorant of policy and procedures? Should the administrative assistant have confirmed the call prior to processing the invoice? Was it unusual for the administrative assistant to receive a phone call from another vice president in the company instructing her to process the invoice? One certainly can make the case that there were warning signs of a potential cybersecurity threat that a well-trained employee could have caught. Ensuring your employees are well trained, understand and employ policy and procedures, and act as fully empowered members of the team are core attributes of executive leadership. Look within your own organization with this type of cyber attack in mind. What should you do to train your workforce to ensure something like this never happens to you? How will you change the ignorant to the informed and thus reduce your risk?

- Negligence. Many lawyers will tell you that negligence and liability are often spoken in the same sentence in courtrooms. Here is an important definition to remember: "A person has acted negligently if he or she has departed from the

---

[8] Symantec, "Francophoned – A Sophisticated Social Engineering Attack," August 28, 2013, http://www.symantec.com/connect/blogs/francophoned-sophisticated-social-engineering-attack?goback=%2Egde_1765567_member_269221828#%21. Accessed on September 10, 2013.

conduct expected of a reasonably prudent person acting under similar circumstances."[9] Increasingly, lawsuits are emerging in the courts as plaintiffs allege negligence against organizations that fail to protect their personally identifiable information such as social security numbers. Other lawsuits allege negligence to properly follow their own policies to maintain their cybersecurity posture. Consider the following case:

> In Baidu, Inc. v. Register.com, Inc., a search-engine operator, Baidu, Inc., sued Register.com, its traffic-routing services provider, after a hacker gained access to Baidu's account and directed its web traffic elsewhere. Imagine the business next door diverting all of your phone calls to it. Baidu sued.
>
> Baidu asserted breach of contract, negligence and gross negligence claims. Register.com moved to dismiss, arguing that its security policy contained a broad limitation of liability provision. And it did. But it also contained statements about how Register.com protected its customers' information and employed security measures to guard against data breaches.
>
> Baidu argued that Register.com's failure to follow its own policies constituted a breach of contract and gross negligence. The Southern Distinct of New York agreed. The court held that the limitation of liability provision barred an ordinary negligence claim, but not the breach of contract and gross negligence claims. The court stated that if Baidu proved what it had alleged, "then Register failed to follow its own security protocols and essentially handed over control of Baidu's account to an unauthorized intruder, who engaged in cyber vandalism. On these facts, a jury surely could find that Register acted in a grossly negligent or reckless manner."
>
> A few months later, the case settled for an undisclosed sum.[10]

Can you and your business afford to be negligent when it comes to cybersecurity? What is your liability risk if the information in your care is compromised through the negligence of your employees? What mechanisms do you have to detect and mitigate negligent behavior?

- Apathy: Apathy is a dangerous condition under any circumstance but especially when it comes to cybersecurity. When people have been trained, informed of the threat, understand the impacts, but don't care, then you have a recipe for cyber disaster. Apathy is a leading (and frustrating) cause of cybersecurity incidents. For example, hackers and identity thieves increasingly target small businesses, yet only 28% of small businesses consider cybersecurity a priority, according to an AT&T report. The National Cyber Security Alliance (NCSA) warns that this "cyber apathy" can be costly to both small businesses and consumers.[11] We agree. The best cure for apathy is prevention and strong positive leadership is essential.

---

[9] http://legal-dictionary.thefreedictionary.com/negligence. Accessed on September 10, 2013.

[10] Jordan M. Rand, "Cashing in Cyber Security Checks," April 10, 2013, http://www.business-law-counsel.com/2013/04/cashing-in-on-cyber-security-checks.html. Accessed on September 10, 2013.

[11] http://www.techjournal.org/2007/04/%E2%80%9Ccyber-apathy%E2%80%9D-costly-to-small-businesses/. Accessed on September 10, 2013.

Look for signs of apathy such as failure to follow policy and procedures, resistance and failure to complete cybersecurity training, and other behaviors that point to lack of support of your cybersecurity program. If you make cybersecurity a priority, reinforce its importance with your words and deeds, and hold employees accountable, apathy likely will fade away.

- Stupidity: This is a controversial topic. Calling someone stupid is politically incorrect. Nobody likes to be accused of being stupid, but people do stupid things. Even intelligent people make mistakes, especially in the cybersecurity realm. Nonetheless, this is a discussion of risk and the threat of stupidity is real, making you and your business vulnerable. You **have** to address stupidity. Don't ignore the possibility that you or your people may do stupid things! Penetration testers (the folks who specialize in testing your cyber defenses, also known as Pen-testers) find that stupidity is a HUGE threat vector they can exploit to gain access to systems. Take for example a recent exercise conducted by the DHS. They deliberately planted several USB thumb drives and data disks in the parking lots of federal agencies and their contractors. Despite the requirement for comprehensive cybersecurity training among the workforce at those agencies and their contractors and the known possibility that the drives and disks could be infected, 60% of those drives and disks ended up loaded on government computers in contravention of existing policy and training. DHS found that if the drive or disk had "official" government markings, the "success rate" for it being inserted in the computer rose to 90%.[12] In the aftermath of the test results' public release, the usual sniping of the government briefly rose, yet criticism was oddly muted as corporate America found they too were susceptible to similar tests. We imagine that many who read the stories of the testing were uncomfortable as they thought about how they and their colleagues would react if they were part of the test. How should a business executive address stupidity to reduce their risk? We think John Verry, principal enterprise consultant of Pivot Point Security, says it best: "You can't fix stupid. You can only try to make people more aware."[13]

- Curiosity: Curiosity is essential for creativity and is the type of trait we seek in our employees. The curious are the people who find new and better ways of doing things and who develop the new products and services that yield the best profit and growth in your business. They also are the most susceptible to social engineering by cyber criminals. Cyber criminals can use the simplest of methods and maximum yield by simply exploiting human curiosity.[14] How? The most common method is

[12] Henry Kenyon, *"Found Thumb Drives: Another Way Employees Are a Security Menace,"* June 30, 2011, http://gcn.com/articles/2011/06/30/dhs-test-found-thumb-drives-disks-network.aspx. Accessed on September 10, 2013.

[13] Jenny Hirsch, "Consultant: Stupidity Threatens Cyber Security," August 10, 2010, http://fordhamnotes.blogspot.com/2010/08/consultant-stupidity-threatens-cyber.html. Accessed on September 10, 2013.

[14] Alice Decker, "Curiosity Is the Nourishment of Social Engineering," April 21, 2008, http://blog.trendmicro.com/trendlabs-security-intelligence/curiosity-is-the-nourishment-of-social-engineering/. Accessed on September 10, 2013.

via email. It doesn't matter if the email is part of a widespread spam mailing or a targeted spear-phishing message as long as it is well-crafted and interesting.

People tend to click on links that promise to lead them to appealing locations. Techniques successfully used by cyber criminals include alarming the recipient about problems with their credit or banking information and providing them with a link that alleges to take them to a location where they can learn more about what the problems are and how to resolve them. When the link is clicked, a remote access toolkit or other malicious code is downloaded onto the recipient's computer and the criminal now has control. Other appeals that sucker even the most discerning of users include links that promise imagery of recent catastrophes or sporting events, political controversies, or business insider information. Emails containing attachments are among the most dangerous to the curious. Recently, after Mandiant Corporation had released its report on Chinese computer espionage,[15] emails containing an attachment alleging to contain a copy of the report made the rounds on the Internet. Everyone wanted to read the Mandiant report, and here, someone presents it for recipients to open and read without having to search for it. How convenient! While many people opened the attachments and eagerly read the report, they also exposed themselves and their businesses to danger as the attachment contained hidden malicious code that allowed bad actors to access the recipient's computer and its information. The lesson? If you are curious about a topic, get your information directly from the trusted source. How do executives reduce risk by addressing curiosity? Set your policies, explain them, train your employees, test your employees, and stay on message. Mark Rasch, director of network security and privacy consulting for Computer Sciences Corporation (CSC), advises, "Rule No. 1 is, don't open suspicious links." Rasch continues, "Rule No. 2 is, see Rule No. 1. Rule No. 3 is, see Rules 1 and 2."[16] We agree. Curiosity killed the cat. It can also kill your business. While we strongly encourage and foster curiosity in our business, you need to channel it away from activities proven to be deleterious.

- Lack of leadership: Have you ever noticed how leadership sets the tone for an organization? I once had a boss who came to work every morning angry, and that anger spawned fear and angst that rippled throughout the organization. Fortunately, his boss saw it too and replaced him with a positive leader who rejuvenated and inspired our organization to do great things. Your leadership makes a difference, both positively and negatively. When it comes to your cybersecurity risk management program, if you aren't leading it, it will fail. Why? Because if you don't make it a corporate priority and delegate it to your technical staff, others in the company will see that it is not one of your priorities and will not support it either. Many executives exclude themselves from cybersecurity training, citing they don't have time. Don't fool yourself. Word gets around when the boss does

---

[15] Mandiant, op cit.

[16] Cliff Edwards, Olga Kharif, and Michael Riley, "Human Errors Fuel Hacking as Test Shows Nothing Stops Idiocy," June 27, 2011, http://www.bloomberg.com/news/2011-06-27/human-errors-fuel-hacking-as-test-shows-nothing-prevents-idiocy.html. Accessed on September 10, 2013.

that. Every time you order an exception to policies for yourself, the word gets out that the boss is not serious about cybersecurity. As a result, your risk goes up as your cybersecurity posture erodes. Our recommendation is that you make it clear throughout your organization that you feel strong personal ownership in your cybersecurity risk management program. Lead by example. Put it on agendas. Include cybersecurity messages in your interactions and correspondence with your employees. Take the same training as your employees to ensure it is up-to-snuff and meeting your corporate objectives. It is expected that you will delegate the administration of your cybersecurity risk management program to subordinates, but you never delegate responsibility and ownership. The moment you delegate responsibility and ownership, you fail—every time.

• Lack of accountability: Lack of accountability is one reason why organizations fail. When things go wrong, what happens if nobody is responsible? If nobody is responsible, then the wrong things keep happening. How do you handle situations where things go wrong? Do you have guidelines that outline consequences for certain actions? Are they well known by all employees? Are they published? Are they followed? Like other critical business functions, cybersecurity must be viewed with the same rigor as traditional profit-generating activities. People need to know what their responsibilities are and be held accountable to deliver upon them. When they fail, there have to be consequences; otherwise, you risk that others in the organization will see there is no incentive to uphold their own responsibilities. When this happens, morale wanes, discipline erodes, and you find yourself the captain of a sinking ship. You already know good people make mistakes. Nonetheless, there have to be consequences for improper conduct. The consequences ought to be commensurate with the conduct and the impacts. When it comes to cybersecurity, the stakes are high as the average cost to clean up a cyber incident in 2013 is reportedly US $616,000.[17] If someone clicks on a link in an email that brings a virus into your network that costs significant amounts of money to remedy, what do you do? Your directors and officers, your employees, and shareholders expect you to provide decisive leadership and hold people accountable. Cybersecurity has evolved to a critical business imperative. You must hold people accountable to manage and control your risk.

3.3 CALCULATING YOUR RISK

In preceding discussion, we raised some questions you should ask as you evaluate your cybersecurity risk. Exposure of your intellectual property and trade secrets as well as technical and human risks are all critical items of interest you should factor into your risk analysis. You should ask your staff tough questions and verify their answers. Your business is at stake!

---

[17] Average cyber-attack clean-up totals $616K, *Infosecurity Magazine*, June 28, 2013, http://www.info security-magazine.com/view/33177/average-cyberatttack-cleanup-totals-616k/. Accessed on September 10, 2013. Author's note: there are various reports that show a range of costs on cyber incidents that range from a low of near US $300,000 to a high of US $8.9 million per incident. We selected a contemporary figure from a trusted source that is backed by solid data to represent the costs associated with loss and cleanup generated by cyber incidents.

As mentioned earlier in this chapter, there is no universally agreed-upon prescriptive formula to calculate risk. Actuarial science has evolved to where several risk specialists are available to help you using some well-researched complex proprietary formulas backed by empirical data. You are well advised to investigate insurance and actuarial advice when selecting options to address the risks you possess and the costs they might entail. But where do you start when calculating your risk? Do you call in one of the expensive actuary experts? Perhaps. But before you do, you can begin framing your analysis yourself by doing your own calculations based on "knowing your enemy and knowing yourself."

Hopefully, previous sections got you thinking about your vulnerabilities and threat sources we've discussed thus far. Understanding the threats, from whom and where they may come, and your vulnerabilities is essential to calculating your cybersecurity risk. The next step is determining what is actually at risk.

There are two popular techniques in calculating risk. The first is quantitative risk analysis, which is based on assigning real and meaningful numbers to all elements in your risk analysis. The second, qualitative risk analysis, does not use calculations. It is based on scenarios. We'll demonstrate how to use both by citing examples.

3.3.1 Quantitative Risk Assessment

Quantitative risk analysis is a mathematically complex subject that is the hallmark of insurance companies and financial institutions, but it is rarely used in the context of information technologies and cybersecurity because of the difficulty in assigning value to information and even greater difficulty in determining the likelihood of loss. Both areas, value assessment and probability of loss, tend to be approached subjectively and do not lend themselves to objective and quantitative analysis. Nevertheless, we believe that with prudent judgment and management oversight, reasonable estimates on the valuation of information are feasible. Thus, it is possible to carefully analyze threat stream and statistical information to make informed estimates on the likelihood of events. When these conditions exist, we believe quantitative risk analysis methodology can be used to assess cybersecurity risk. We believe you should incorporate quantitative risk assessments into your corporate business processes, wherever possible.

Let's walk through a high-level example calculation to illustrate how you can use the quantitative technique to assess your risk to a cybersecurity threat.

We submit that your intellectual property and trade secrets are your principal valued assets **at risk to cyber incidents**. Further, we submit that like hard assets, your intellectual property and trade secrets have value that can be calculated and factored into your risk equations. Think about it for a moment. If you were contemplating the sale of your company, you would have to estimate the value of all of your assets, and we are certain that you would come up with a number that would be credible both to you and to the potential buyer. Moreover, in all likelihood, the buyer would require the segmentation of asset valuations in order to turn his "due diligence" accountants loose. They in turn would use accepted accounting techniques to validate (or invalidate) your estimates. What we are suggesting here is that you use the same methodology to establish a value for your intellectual property.

How much is your intellectual property worth to you? How much is that secret family recipe worth? Often, you'll hear executives touting that their secrets are priceless, but nobody really believes that. Everything, *including information*, has value and value is the principal concern when calculating risk and making investment decisions.

We submit that one way to establish the value of your intellectual property and trade secrets is a summation of the following costs:

1. <u>Profit value</u>: Your intellectual property and trade secrets give you a competitive advantage that translates to increased profits. Do you know the impact that your intellectual property and trade secrets have on your bottom line? Do you have statistics that indicate before and after effects? Can you put a value on what they mean to your business?

2. <u>Cost to acquire or develop</u>: How much did the acquisition or development of the information cost? Whether you did an outright purchase or developed it from in-house resources, your information represents an investment with a tangible value. You should know how much you have invested.

3. <u>Cost to maintain</u>: Maintenance costs for information often are camouflaged in budget sheets yet they are noteworthy. First, you have to store the information you already have. Hardware to host it, software to manage and read it, and staff to maintain it are all costs. Information itself often is perishable and needs to be maintained. An example is financial data that is continually updated and added to models that calculate opportunities and trends used by investment specialists. The addition and integration of that data, maintenance of the data feeds, and the periodic addition of additional storage as the volume of information increases all ought to be factored into your cost to maintain figures. Similarly, the expenses associated with securing the information and providing adequate system redundancy to keep it available should be included in your cost to maintain calculations.

4. <u>Cost to replace</u>: This is not as straightforward as it may seem. In calculating this cost, don't forget you need to factor in all the costs to replace your information. Cost items to consider include the loss you incur while the information is being replaced, the cost to acquire or develop the replacement, and costs associated with any substitutes or proxies used in lieu of the lost information. For example, suppose that all of the data from your quality control analyses for your main chemical product were lost or completely compromised. How much would it cost to repeat all of the chemical analyses?

5. <u>Cost if unavailable</u>: This represents the cost to you and your business if the information is unavailable. For example, if you rely on your information to create or generate business, not having it available through theft, alteration, or other malicious or unintentional activity deprives you and your business of revenue. This is a cost that ought to be factored into your calculations.

6. <u>Liabilities if compromised</u>: You and your company may find yourself open to liability if your information is compromised. For example, as a director or senior executive, you likely are familiar with indemnification insurance that protects officers of the corporation against lawsuits from shareholders. If your intellectual

property is stolen by a cyber criminal, it is not unreasonable to expect that a lawsuit may be filed, alleging lack of adequate management controls to protect the business' vital information. Other potential liabilities come from lawsuits filed by partners with whom you may share portions of the intellectual property or even clients for whom you were developing the intellectual property.

Once you have summed all of these costs, the result represents what it would cost for you to replace the intellectual asset. It does not represent the true value of the asset. The true value is the cost plus what a buyer would be willing to pay over and above the replacement cost. Nevertheless, in order to focus on the value of information for purposes of this book, let us assume that the summation suggested earlier is a good approximation of value, at least as a starting point for the subsequent discussion.

In this example, let's assume you are the CEO of Plieno Corporation, a specialty steel corporation in Western Pennsylvania. You produce a special alloy that is renowned for its durability, light weight, and incredible strength. Your metallurgists refined your formula over several years and consider it a trade secret along with the rolling techniques you use to process the finished products. Prior to production of this alloy, you were a niche company with total gross revenue of US $100 million with an annual net profit of U$ $10 million (in a good year!) Now, after three years of offering the new alloy, your business generated US $500 million in total gross revenue with a record net profit of US $100 million. You have vaulted to the top of a US $1 billion market for this type of specialty steel. Your shareholders are thrilled with the returns and prospects for the future. Your customers are delighted with your product, and word-of-mouth advertising is causing orders to soar. You have three shifts working around the clock and have expanded your facilities twice to keep up with demand. You now are contemplating opening another facility to handle further increases in demand.

Such exceptional growth has caught the eye of investors and competitors alike. Industry associations are praising Plieno for revitalizing the regional steel industry. You are getting a lot of positive media attention too with numerous requests for interviews. University professors have contacted you asking for permission to do case studies analyzing your success. Competitors have made it known quietly they intend to soon offer special alloys to compete against your product. You aren't overly concerned as they haven't been able to duplicate your formula and manufacturing process in three years. You have a big head start and momentum. Life is good!

But you realize the good times may not always be there. You recognize there are many risks facing your business and you need contingency plans. You mentally walk down some of the risks you face: reduced market for your steel, someone introduces a better product or undercuts your prices, labor and material shortages or interruptions, and flooding on the Monongahela River (your facility sits along the river). Are these realistic threats? Absolutely! You already have plans in place that analyze the risks in each of these areas and how your company would respond. Moreover, you feel comfortable you have the right kind and amount of insurance to cover the greatest threats to your business.

Later that evening, you are sitting at home sipping your Iron City beer watching the KDKA evening news when you see a report of a cyber attack on a local retailer where thousands of credit card numbers were stolen. You know the CEO of the company from the local chamber of commerce meeting but haven't shopped at one of his stores, so you blow a sigh of relief at not having your credit card compromised. You are astonished at the estimated liability and direct loss associated with the incident. Good thing you are not in retail and don't face the risk of someone using a computer stealing credit cards from you.

But wait, the local reporter interviews a professor from Carnegie Mellon University who is a cybersecurity specialist. The professor says that retail establishments aren't the only businesses vulnerable to cyber attack. In fact, she advises, all businesses are vulnerable in one way or another and should take proactive measures to protect themselves. You finish off your beer, planning to look at your cybersecurity risk the next day.

You start by meeting with the head of your IT department. He serves as your *de facto* chief information officer (CIO). You consider him your resident "geek." You haven't considered him part of your senior executive team, although he has been very effective integrating new technologies to help accommodate the growing business. He oversees the operation of business unit networks, the telephone system, web presence, and mobile devices. He manages the electronic data exchange between your procurement department and your suppliers. He even works with the manufacturers of your milling equipment to ensure your business unit networks get reliable data directly from the milling equipment's smart controls. You tell him you are concerned about cybersecurity risks and want to know where you are most vulnerable.

"Boss," he tells you, "You are vulnerable everywhere."

You bring in your chief financial officer (CFO), your chief operating officer (COO), and your general counsel to continue the discussion. Together, the five of you determine the most damaging threat to your business is someone getting your alloy's formula and using it in direct competition against you. You decide you want to know what the cybersecurity risk is of someone taking your alloy's formula. Because you've used it to calculate other risks, such as calculating potential loss of the mill to fire, you order a quantitative risk assessment.

Quantitative risk can be expressed as annualized loss expectancy (ALE), which is the expected monetary loss for an asset due to a risk being realized over a one-year period. ALE is the product of the impact of the loss (expressed as the single loss expectancy or SLE) and the likelihood or how often the loss occurs (expressed as annualized rate of occurrence or ARO):

$$ALE = SLE \times ARO$$

Your team follows a disciplined process to evaluate the cybersecurity risk scenario.

Let's begin with Step 1: assigning value to assets.

3.3.1.1 Assigning Value to Assets. Recall our discussion earlier that everything has a value, including information. It is important that you have a thorough understanding of the value of your assets. The key valuation figures your team will look at in this scenario are:

PV = profit value

CAD = cost to acquire or develop

CM = cost to maintain

CR = cost to replace

CU = cost if unavailable

L = liabilities if compromised

The team starts with a look at the value of the alloy formulation. It is at the heart of your process and is largely responsible for your profits rising tenfold. There is great debate among the team members as to its value. Your COO believes it is priceless. After hours of fruitless discussions, your CFO proposes that the value is equal to the difference between your profit before implementing the formula and the current profit level (US $100 million – US $10 million = US $90 million). You and your team accept his proposal and assign PV = US $90 million.

Your CAD is a sunk cost. Your team knows that it took five years of research to develop the formula for your alloy and kept meticulous records. Factoring in materials, equipment, salaries, and other direct and indirect costs, your CFO validates CAD = US $20 million.

Compared to CAD, the cost to maintain your formula is relatively low. It is stored on the company's central server with off-site storage at a commercial vendor facility in West Virginia. A tertiary copy is kept on a drive stored in a safe deposit box in a local bank. Because the IT staff spreads their time across multiple systems and your software is licensed as part of a corporate licensing agreement, your CFO and the IT chief base their estimates on staff costs, software licensing, and network and computer hardware on a pro rata basis supporting the storage and use of the formula in the manufacturing process. Based on their analysis, CM = US $2 million per year.

Cost to replace (CR) is hotly debated by your team. Given that your team has carefully provisioned for both on- and off-site storage of the formula, it is relatively easy to get a backup copy and reload. Your CFO argues that auditors could make the case that the cost to replace is zero. Your COO disagrees. He was part of the technical team that developed it and knows all that went into developing it. He believes that once your formula is revealed, its value plummets to zero as your competitors adopt it and you lose your competitive advantage. He makes the compelling case that it will need to be replaced with a better formula. Based on his experience and knowledge of the technology, he estimates it will take three years to develop at a CR = US $50 million. The team agrees to accept the figure in this calculation.

Determining cost if unavailable (CU) is easier for your team. Your business produces US $500 million over the year. You have 300 production days per year with other days being consumed by holidays and Sundays (you believe the only steel production in Pittsburgh on Sundays should be football at Heinz Field). Given this, your team determines a daily cost of US $1.67 million if the alloy formula is unavailable.

Finally, your team addresses the liabilities (L) issue. Your general counsel and the marketing director advise that you have numerous contracts in place that specify on-time deliveries to customers with significant penalties for delays. Most contracts have a cushion of a mere three days before monetary penalties kick in. Additionally, your contracts for transportation of finished products are firm fixed price regardless if you have a delay in production. Your logistics director advises that delays will affect the supply chain of raw materials used to manufacture the steel; you do not have on-site storage to absorb an interruption of more than six days. You determine if the interruption is less than three days, L = US $0. If it between three and six days, L = US $1 million per day. If it is over six days, L = US $5 million per day.

With values assigned to assets, the team turns its attention to Step 2: estimate the potential loss.

3.3.1.2 Estimate the Potential Loss.
Estimating your loss is difficult and has to be predicated on making some key assumptions. In this case, the key assumptions you make that drive your next steps include:

- Your formula is stolen and gets in the hands of a competitor.
- The competitor uses your formula to create an identical product to compete against you.
- It takes the competitor one year from receipt of the formula to bring their copycat alloy to market.

The type and degree of analysis to estimate the potential loss due to an event depends on the types of threats encountered. Analysis of the threat of a massive fire at your facility varies greatly from that of a zombie epidemic, meteor strike, mudslides, or even a cyber-based attack. Each asset faces potential loss based upon the threats that they face.

This is where the mathematics get complicated very quickly, involving complex statistical modeling as you want to evaluate each and every possible scenario. Regardless of what threat you confront in your risk analysis, remember that it is important to address all the potential courses of actions and impacts. For the purpose of brevity and to illustrate the methodology, we will only look at one specific line of threat.

To analyze the potential loss from a cybersecurity incident, your team makes additional assumptions to bound the analysis:

- You are evaluating the loss associated with a hacker gaining access to your formula.
- You are still able to operate using your formula and meet your production and delivery objectives.

Your team continues its analysis to calculate the potential loss (SLE). The SLE is expressed as a dollar amount representing the potential loss if a specific threat takes place. It is calculated as

$$SLE = \text{asset value}\,(AV) \times \text{risk exposure}\,(RE)$$

Your team assumes that in the event a hacker accesses their data to steal the formula, they quickly will use backup software to restore full operations within one day. They determine that in this scenario the asset value is the sum of the aforementioned values:

$$AV = PV + CAD + CM + CR + CU + L$$

where:
PV = US $ 90M
CAD = US $ 20M
CM = US $ 2M
CR = US $ 50M
CU = US $ 1.67M
L = US $ 0

Therefore, they determine the value of the company's secret formula to be

$$AV = US \$163.67 \, million$$

Risk exposure (RE) represents the percentage of loss the threat will have on a certain asset if it occurs. With many assets, this is fairly straightforward to calculate. For example, if you are a car dealer with a million dollar inventory of vehicles and a damaging hailstorm hits but 30% of the vehicles are protected by being inside your facility, then you face an exposure factor of 0.70 because 70% of your assets are exposed to the risk of hail damage.

Calculating risk exposure for information regrettably often is a binary data point. There doesn't readily appear to be any such thing as partial loss when it comes to information; either you have a total loss or no loss. If someone has destroyed your data and you have no backup, then you have total loss and your RE = 1. If you do have a backup, then you have no loss and your RE = 0.

What's your risk exposure in this scenario? While you do have solid and reliable backup procedures, because of the impact of the formula ending up in the hands of a competitor, your team believes it would be a total loss where RE = 1.0.

But wait! If your competitor can't bring their hijacked product to market until a year after receipt of the formula and it will take you three years to replace it with your new formula, can you make the case that your RE actually is loss of two out of the next three years? Absolutely. After all, you do not have an expected loss for the first year after the incident but do for the two subsequent years until your expected replacement arrives where you anticipate you will resume your market dominance. Given that assumption, RE = 0.67.

The team accepts that assumption and calculates the expected loss of an incident (SLE):

$$SLE = asset \; value \, (AV) \times risk \; exposure \; (RE)$$

where:
AV = US $163.67M
RE = 0.67

Therefore, they determine the expected loss to be:

$$SLE = US\$109.66 \text{ million}$$

3.3.1.3 Estimate Threat Likelihood. The team next evaluated the ARO, which
is a measure of how likely the threat will take place in a 12-month period.

In calculating the ARO, the team started by looking at availability rates on their
computer systems. They had been prudent in their planning and implementation of
their computer systems and invested well in their production equipment and software. As a result, they maintained a 99.95 operationally ready rate over the last three
years.

They also consulted with a cybersecurity intelligence specialist, who conducted a thorough anonymous search of Internet resources to see if there were any indications that
an entity was expressing interest in your company on hacker forums or other potentially
dangerous venues. The results surprised you and your team as the specialist found that
indeed there had been discussions in a popular forum referring to a company that was a near
dead ringer for your business. Upon deeper digging, your specialist found that the query
came from a country where one of your overseas competitors has their headquarters. You
are suspicious.

The specialist's search could have been chocked up to coincidence, but your IT
chief comes back with some disturbing news. After seeing the initial report about the
hacker forum, he had his boundary protection team check the firewall and router logs to
see if there was any unusual traffic hitting your network. There was! In fact, over the last
five months, there had been a growing number of probes and scans against your network
with two failed log-on attempts in the last month. Many of those scans and probes originated in the country where your competitor is headquartered. Whoever was behind it is
executing a "low and slow" strategy. Had you not been looking for the specific evidence,
it would have been very difficult to find them. Now, you had evidence that someone was
indeed trying to access your network.

Your IT chief advises you that this month's vulnerability scans indicate there are
several software and configuration vulnerabilities that exist on your network. They've
been there for a couple of months but have been low priorities for correction. Now, given
the increased threat, he recommends they be remedied as soon as possible. He says he
needs additional resources to complete the task and will come to you with details the
next day after he consults further with his staff.

You have your IT chief contact the professor at Carnegie Mellon to help estimate
how many times the threat can take place in a 12-month period. She is very helpful
and points out that data collected by the government and insurance companies indicates that a company like yours with comparable defenses has only been successfully
attacked once every two years. She also refers you to DHS and FBI programs that can
help identify threats and tactics bad actors use. You authorize your IT chief to sign your
company up for the next FBI InfraGard meeting in Pittsburgh as well as to join the
manufacturing sector's Information Sharing and Analysis Center that partners with
the DHS.

Given the information you have, you know that you are at risk, yet the data indicates the estimated frequency of a successful cyber attack is one every two years. Therefore, you and your team calculate the ARO = $1/2 = 0.50$.

3.3.1.4 Calculate the Annual Loss Potential. Now that you have your SLE and predicted ARO, you calculate the entire equation:

$$ALE = SLE \times ARO = US\$109.67\,million \times 0.50 = \$54.83\,million$$

This means that you can expect an annual loss of US $54.83 million in the event of a cyber attack that compromises your intellectual property and trade secrets. This is your risk in this scenario.

It is important to note that the quantitative risk assessment method is the standard method of measuring risk in many fields such as insurance and manufacturing, but is not commonly used to measure risk in IT.

As you measure your cybersecurity risks, this method may prove challenging. It is very difficult to measure the value of information, but we submit that it is possible. Moreover, valuation of shared assets such as networked systems, virtual devices, and software used across an enterprise poses a challenge to actuarial computations. Additionally, while you can use statistics to determine the anticipated failure rate of an information system, it is nearly impossible to accurately predict the likelihood, frequency, or severity of cyber attacks against your organization. We believe the vagueness surrounding calculation of the likelihood of a cyber attack drives many to use an alternative method of measuring risk: the qualitative risk assessment.

Nonetheless, you look over the quantitative risk assessment again. You may think, "Holy Smokes! Our cybersecurity risk is huge! What do we do next?"

Before we advance to a discussion of the next steps such as risk mitigation, avoidance, acceptance, and other postanalysis decisions, let's turn our attention to this other method of determining cybersecurity risk: qualitative risk assessment.

3.3.2 Qualitative Risk Assessment

When it comes to IT and cybersecurity risk assessment, the qualitative risk assessment model may be more attractive and useful for you and your business.

Qualitative risk assessments do not utilize detailed calculations to assign monetary values to assets and losses like the quantitative method. Rather, the qualitative risk assessment method recognizes the difficulty present in assigning realistic values to information and the likelihood of risk. As such, this method provides relative measures of risk and asset value based on ranking specific items into categories such as high, medium, or low or on a numeric scale.

Qualitative risk assessments are a popular method of calculating cybersecurity risk. While not as precise as the quantitative method, they generally are faster, easier, and less expensive to produce and give senior decision-makers actionable information in a more timely manner.

We'll use the example of another fictitious Western Pennsylvania company to illustrate the qualitative risk assessment methodology of calculating cybersecurity risk to a business.

You are the CEO of BigRX, a large (US $10+ billion) regional medical enterprise with over 20 major hospitals and 400 operating locations. Your business is an industry leader and has a good reputation. You carefully guard your brand.

Cybersecurity is on your agenda. Reports from across the medical sector indicate an increase in violations of the Health Insurance Portability and Accountability Act (HIPAA) as systems fall out of compliance with HIPAA standards and disclosures of sensitive patient records have spawned litigation that have cost other similar businesses tens of millions to repair and litigate. You are concerned about hackers penetrating your systems, which would expose your business to potential disclosures and/or corruption of data that could cost your business tens of millions of dollars and potentially sully your sterling reputation.

BigRX has a large medical information management system called BigMIMS that is the heart of its business operations. BigMIMS has approximately ten million sensitive records in its database. Medical providers at your remote and contracted facilities love BigMIMS as they can access the records through a convenient web interface that your IT department delivered through a contract with a major software vendor. BigMIMS cost you US $20 million to develop and field and costs you US $5 million to operate and maintain. Your accounting team recently conducted an analysis of BigMIMS's information and determined that the replacement cost of each record is US $100.

Yesterday, you attended a chamber of commerce luncheon celebrating the Pittsburgh Pirates' winning season where you sat next to the CEO of Plieno Corporation. As you shared lunch and conversation together, he pointed out the CEO of a major Pittsburgh retailer across the room. The Plieno CEO asked if you had heard about the cyber attack against the retailer that resulted in the loss of thousands of credit card numbers and threats of litigation. When you said you hadn't, he advised he was conducting a risk assessment of his cybersecurity posture and recommended you consider doing the same at BigRX, "...if you hadn't already." Good advice. Perhaps, these luncheons do have value after all!

When you return to your office, you have your regularly scheduled senior leadership meeting with your COO, CFO, chief medical officer, CIO, and chief risk officer (CRO). You tell them that you are concerned about reports of cybersecurity incidents and the major retailer incident is hitting "too close to home." You want a cybersecurity risk assessment conducted, starting with BigMIMS.

Based on their experience with qualitative risk assessments, your staff recommends using this methodology to assess your risk.

3.3.2.1 Threat Identification. The first step of the qualitative risk assessment is to identify your threats and threat sources (know your enemy!)

Your team categorizes the threats and threat sources into the following table,[18] which they present as part of their report to you.

---

[18] This is not an all-inclusive list. Your team brings you over 100 threats and threat sources, but you tell them you want to focus on the most likely and ask them to focus on the top nine.

Threat Source	Threat	Description
Human	Improper data entry	This is an improper entry of data into BigMIMS, either intentional or deliberate, that compromises the integrity of the data in the database
Human	Virus infection	This is the insertion of malicious code into the computer network that compromises the security and integrity of your network and jeopardizes the information residing on it
Human	Unauthorized access	This is the access of patient information in the BigMIMS database to individuals not authorized to view or handle it
Human	Hacker attack	This is an action where a hacker gains access to BigRX networks and information. It may or may not result in malicious activity yet will drive costly remedial activities and notifications in accordance with the HIPAA
Natural disaster	Earthquake	This takes into account the possibility that an earthquake could strike, disrupting BigMIMS operations
Natural disaster	Flood	This takes into account the possibility that a flood could affect the BigMIMS facility and interrupt operations
Natural disaster	Tornado	This takes into account the possibility that a tornado could affect the BigMIMS facility and interrupt operations
Physical/environment	Power failure	This takes into account the possibility that a power failure in the BigMIMS facility could damage the system or otherwise interrupt operations
Physical/environment	HVAC failure	This takes into account the possibility that a HVAC failure in the BigMIMS facility could damage the system or otherwise interrupt operations
Physical/environment	Fire	This takes into account the possibility that a fire in the BigMIMS facility could damage the system or otherwise interrupt operations

You like this format and are comfortable with it as BigRX uses this format across the organization. You standardized the format for risk assessments to improve management oversight, consistency, reliability, and repeatability. Employees across all operating units are trained to use this format, which was developed as a result of a previous risk management exercise. Having a standard and repeatable risk assessment process across the organization reduces variance and confusion while enhancing accuracy.

You agree that these are reasonable threats but you still want to see your vulnerabilities. You turn to the next page.

3.3.2.2 Vulnerability Identification. Your team produces numerous tables identifying hundreds of vulnerabilities. Because you are focusing on cybersecurity vulnerabilities to BigMIMS and its data, your team consults with those most familiar with the system: the system developers, the system and database administrators, program managers, and cybersecurity personnel.

Technical teams are a treasure trove of information in identifying potential vulnerabilities. Based on their technical knowledge and their daily interaction with the systems, they know the strengths and weaknesses of the system. If you want to know where your greatest cybersecurity risks are, they are the best people to ask. They will either have the information you need or know how to get it for you.

Vulnerability scanning results are a prime source of information to identify your cybersecurity vulnerabilities. Good technical teams routinely run vulnerability scanning software to examine operating systems, network devices, applications, databases, and other critical infrastructure for known flaws by comparing the systems and their responses against databases of known flaws or signature files. Internal scans are standard procedure for professionally managed networks. Great technical teams not only do regular internal scanning but also do external scanning of your network boundaries as well.

Great technical teams also ask for help and do regular independent penetration testing to find out where their security is weakest and can be exploited. Penetration testing (also known as "Pen-testing") features specialized security analysts who exercise threats against the system under controlled nonmalicious circumstances. The best ones don't just challenge your technical team, they also use social engineering, on-site physical security probes, and other techniques to find ways to penetrate your defenses. In essence, Pen-testers figure out ways to hack into your system so you can find your weakest links. We highly recommend you include Pen-testing on a regular basis with vulnerability scanning to provide you with the vulnerability information you need to make informed decisions.

Vulnerability scanning and penetration testing are not the only sources of vulnerability analysis. Your organizational and management control program also ought to be used to identify areas of vulnerabilities. Internal audits and control procedures are used to ensure that your policies and procedures are routinely and accurately adhered to. We believe this is an essential part of your internal control program and a rich source of vulnerability information.

As an example, several years ago, the author found a vulnerability through his internal audit and management control program that is worth sharing. The author was responsible for the network operations, maintenance, and security supporting a 160,000-person organization with 20 major operating locations around the world. In order to maintain effective, efficient, and secure network operations, the author ordered standardized procedures to be followed in the installation of software and patches. Nothing was to be installed on the network or devices until it had been properly tested in the organization's central cyber test facility. Once software and patches had been cleared by the lab, we used technical means for designated system administrators to automatically push software updates to devices across our network enterprise. This process saved significant resources by reducing the need for touch labor, reduced the time to patch and install from weeks to minutes, and significantly improved reliability and security.

Key to the process was the system administrators following the process in a disciplined manner. We routinely ran scans looking for unauthorized software appearing on the network as part of our cybersecurity program and saw an alarming rise in the appearance of unauthorized software. We were concerned because the unauthorized software not only could contain malicious code that could jeopardize our operations, but also it could be unlicensed software that could open us to litigation for using copyrighted material without proper permission.

Only a system administrator could install software and the entire technical team had received thorough training; they knew the process and swore they were following it. We had to find who was installing the unauthorized software, why they did it, and what caused them to do so. Only then could we resolve our problem.

I directed my deputy to lead an internal control audit of the system administrator process and procedures to see if he could find the root cause. Sure enough, he did. The internal control audit discovered that indeed the system administrators on our technical team were well versed on the policies and procedures. They were regularly tested and followed the procedures with discipline and rigor. What my deputy discovered through the internal audit, however, surprised us. Not everybody with system administration privileges were on the technical team.

The internal control audit revealed that business unit administrative staff members at one of our operating locations had asked for and been given system administrator level privileges to enable them to assist members of their business units with routine computer problems. There was no evidence that they had received the requisite training on our software and patching policy nor were they formally trained as system administrators. Several of them had violated corporate policy and had installed untested and unlicensed software. We quickly moved to remedy the situation by removing the software, implemented very tight access control procedures to centrally manage privileges, and alerted management at the operating location of the issue. Fortunately, we detected and fixed the problem before damage occurred, but it highlights the positive impact internal control and management programs have in helping you find your weaknesses. Do not rely solely on your technology to reveal your problems!

BigRX uses all the techniques cited in the preceding text to expose their list of vulnerabilities to BigMIMS. Their internal and external security scans reveal a list of software and configuration weaknesses that are common with many of the vendor products. In fact, the technical team tells you that these vulnerabilities are well known and available for anyone to see on the Internet.[19] Your staff identifies hundred of vulnerabilities, but you zero in on the one below; the same one you heard was used to exploit the Pittsburgh retailer. You have the same vulnerability!

Vulnerability	Description
Web page software is vulnerable to SQL injection	SQL injection is a code injection technique, used to attack applications, in which malicious SQL statements are inserted into an entry field for execution (e.g., to dump the database contents to the attacker)*

*This is one of the most common and dangerous vulnerabilities around. For more detailed information on SQL, we recommend you view the Microsoft SQL server web site at http://technet.microsoft.com/en-us/library/ms161953%28v=SQL.105%29.aspx. Accessed on March 1, 2014.

[19] Vendors often are forthcoming with the vulnerabilities to their systems and software, perhaps out of liability avoidance but more likely due to honesty. They frequently disclose vulnerabilities on their web sites, post patches, and inform consumers of their schedules to mitigate the vulnerabilities. Additionally, public sources such as the National Vulnerability Database (http://nvd.nist.gov) and the Common Vulnerabilities and Exposures (http://cve.mire.org) are available for staff to review known information about vulnerabilities.

Your attention level just went from concern to alarm. How bad is this? Let's take the next step and determine if you have a threat related to the vulnerability.

3.3.2.3 Validating Threat and Vulnerability Matching. Matching threats to vulnerabilities is an important part of your risk management process. The reasoning is straightforward. A threat without a vulnerability does not produce a risk. Similarly, a vulnerability without a threat does not produce a risk. However, a threat from a legitimate threat source directed toward a vulnerability generates risk, risk that you need to address.

In the case of BigRX, the SQL injection vulnerability has been identified. It can enable an attacker to gain access to the BigMIMS database potentially revealing, altering, or destroying sensitive patient records and opening BigRX up embarrassing litigation, regulatory fines, and damage to its valued brand. The vulnerability is serious.

But is there a threat? How do you know?

There are several methods to determine whether you have a threat directed against a cyber vulnerability. Let's introduce you to some of the most common:

- The threat source identifies you as a target: Strange as it seems, some threat sources clearly identify their targets, giving them a heads-up they are the subject of future attack. The previously cited anonymous DDoS attacks on PayPal, MasterCard, and Visa are examples of this type of threat and vulnerability match.

- The threat source performs reconnaissance against you: Potentially hostile threat sources are continually scanning the Internet looking for vulnerabilities to exploit. Your IT team should be continually reviewing their security logs to see who is scanning you. If there are a lot of repeat visits from the same Internet address, be concerned and block them.

- The threat source has a pattern of misconduct indicating "you are next": Cyber crime statistics indicate when cyber criminals find a technique that works, they continue to tap it until it runs dry or they are apprehended. Albert Gonzalez had his acolytes execute successful attacks against retailers by hacking in through their Wi-Fi and later web pages (using a SQL injection technique!) to steal credit card numbers. Do you think if the other retailers in the area knew about the exploits they would have made the linkage between the threat and their own vulnerabilities? We would have!

BigRX suspects there is a problem. Nobody has directly communicated a specific threat to the company's information systems, but the network is constantly being bombarded with scans and probes. You are not sure that it is part of widespread scanning or is directed toward BigRX, but conclude that regardless of the source it is reconnaissance of your network. Moreover, your neighbor in retail was just burglarized through a SQL injection exploit that is buffeting their reputation and driving embarrassing litigation and potential losses due to the theft of sensitive customer data.

Does BigRX think there is a threat and vulnerability match? Absolutely!

So what's next? How likely it is that someone will attack you?

3.3.2.4 Estimate Incident Likelihood. Before we continue with the BigRX example, let's use a cyber-related example to highlight how some people look at how to decide that an event is likely (event likelihood).

Some people like to think that it is unlikely Apple products will be hacked. They point out that Microsoft often patches their software to remedy vulnerabilities and most hacking activity is directed against Microsoft products. They point to Apple as an example of a company that "doesn't have to do that" and use the software patch metric as a measure of relative quality. Is that true? Not entirely.

The fact of the matter is that Microsoft has become the world's single largest source of software, making their product set the largest target for hackers. Why? To quote the famous bank robber Willie Sutton, "Because that's where the money is."[20]

Because businesses predominantly use software based on the Microsoft architecture, hackers pay great attention to Microsoft products, relentlessly searching for vulnerabilities they can exploit. Cybercrime is big business and it is logical the widespread use of Microsoft products by businesses, governments, and the public at large would make Microsoft products the huge target it is for hackers.

But just because Microsoft gets a lot of attention from the hacker community doesn't mean you are safe with your iPad, iPhone or Macbook. In fact, Apple's resurgence and increase in market share have made it an increasingly inviting target for hackers. Don't believe it? Even Apple itself was recently hacked and had to temporarily shut down its application developer web site.[21] The lesson is that you have to be careful when you are deciding "event likelihood" to not succumb to bias and tradition. Rather, be strategic in your view and look to multiple diverse sources of trusted information in making your judgments.

BigRX uses their standard corporate model to characterize the likelihood or probability that the threat will be acted upon in the next 12-month period. Like numerous other companies, they use a format familiar to those who have graduated from business schools and other executive development programs.

Likelihood	Definition
Low	0–33% chance that the event will occur in a 12-month period
Medium	34–66% chance that the event will occur in a 12-month period
High	67–100% chance that the event will occur in a 12-month period

This is the method used by BigRX but there are many other ways you can categorize the likelihood of an event. Some people prefer more categories (e.g., very low, low, medium, high, and very high). Others prefer different ranges for their categories (e.g., high = 90–100%, medium = 60–90%, and low = <60%).

We have found that regardless which characterization is selected, there is great benefit in consistency. When your organization and its employees are trained to employ a standardized methodology, are comfortable with it, and use it as designed, the resulting analysis is consistent, reliable, and trusted across the organization.

[20] http://www.fbi.gov/about-us/history/famous-cases/willie-sutton. Accessed on March 1, 2014.

[21] Mark Knapp, "Has Apple Solved Its Hacking Problem?," August 5, 2013, http://wallstcheatsheet.com/stocks/has-apple-solved-its-hacking-problem.html/?a=viewall. Accessed on September 10, 2013.

When considering which likelihood category to select, there are many methods you can use. They include but are not limited to:

- Leadership selection: The boss or delegate picks.
- Nominative group decision: Everyone involved in the process votes and you (the boss) select the average.
- Delphi group technique: Everyone involved in the process presents their recommendation, and the group debates options until consensus is reached.
- Plurality rules! Everyone votes. Whichever category gets the most votes is selected.

Which one your organization selects depends on the culture of the organization and the decision to be made. Those that are time sensitive are more likely to use either the "leadership selection" or "plurality rules!" techniques. Where the decision is potentially very contentious, the "nominative group decision" or "Delphi group technique" often are preferred.

So what did you do at BigRX? You followed your established corporate risk management process. You gathered experts from your IT and financial departments and business operations and even some cybersecurity consultants. They used the Delphi group technique to make a recommendation to management that the likelihood was HIGH that BigRX would face a successful hacking incident using the SQL vulnerability in a 12-month period.

Based on the reports you are seeing in the news about cyber attacks at home and abroad, you are not surprised.

3.3.2.5 Define Incident Impact.
The next step for BigRX is to define how you measure the impact the expected cyber attack would have.

BigRX uses the same process to determine impact as they did to assess incident likelihood; they use their corporate risk model.

An incident can have multiple impacts, and it is appropriate for organizations to analyze each as part of their deliberate risk analysis process. As stated before, it is beneficial to maintain consistency in methodology and approach throughout the process.

BigRX leadership (you, directors, and principal officers) is most concerned about the impact on patient safety and the economic implications.

In estimating the impact, the team uses the following definitions.

Impact	Patient Impacts	Economic Damages	Damage to Business Operations	Potential for Litigation
Low	No harm to any patients	<US $10K	BigMIMS unavailable for less than an hour	Low
Medium	Patient records damaged but no direct physical patient effects	US $10K < damage < US $100K	BigMIMS unavailable between 1 and 4 hours	Medium
High	Patient records damaged causing adverse effects on patient care	Damage > US $100K	BigMIMS unavailable over 4 hours	High

Now, ready with estimates of likelihood and impacts, BigRX is ready to move to their risk assessment.

3.3.2.6 Risk Assessment. Risk assessment is a process. Regardless of whether you are measuring risk from natural disasters, new product launches, or even cybersecurity incidents, you use the process to determine the likelihood (or probability) of a threat occurring against a vulnerability resulting in an impact.

Using the qualitative risk assessment method, you create a matrix to determine the relationship between the likelihood of an event occurring and the impact it will have if it does. You've already analyzed likelihood and impact in previous steps, so you can compare them in your matrix to portray the relative risk you have calculated.

Given BigRX's three-tier measurement techniques, the following risk evaluation matrix is created.

↓ Likelihood/Impact →	Low	Medium	High
Low	Low	Low	Medium
Medium	Low	Medium	High
High	Medium	High	High

Remember that in the qualitative risk assessment, you normally do not use numbers in your risk measurement. Since in this cybersecurity-related example you do not have accurate numbers to estimate the likelihood of the event, using this construct adequately conveys the range of risk to focus management attention to matters of gravest concern.

As the CEO of BigRX, you review the team's work and conclude you most likely face high risk of a significant cybersecurity event in the next 12 months. You want options on what to do next.

3.3.3 Risk Decisions

Life is full of risk. Recall that as an executive, one of your primary responsibilities is to manage risk to protect your business and create an environment for it to grow and thrive.

In our opinion, you have four basic options when confronted by risk: mitigate, transfer, accept, and avoid. Note that the four options also hold for any type of risk encountered in life. Each one should be supported by the facts and a thoughtful review. During your evaluation of options, remember that you can choose one or more in making your decision. The options are:

- Mitigate: This is one of the most common techniques used to address cybersecurity risks as part of your risk management strategy. Mitigation focuses on fixing the deficiency that creates vulnerabilities and/or leveraging some other form of compensation that controls your vulnerability. For example, mitigation techniques we have used include patching software to close security vulnerabilities, training personnel, installing and configuring new and/or better security apparatus like firewalls and encryption devices, and adding improved physical security controls such as special access control devices. We cannot overemphasize the importance of the business case analysis as part of your mitigation process. If you agree it does not make sense for you to spend 10 dollars on a lock to protect a five-cent pencil, you'll probably also agree that it doesn't make sense to spend a million dollars on an IT system to protect information valued at US $500,000. Mitigation is a business decision enabled by technology to support business objectives. Make sure you have a good business case before you invest in any mitigation technique! The right investment should jump out at you as a result of your business case analysis! As a reminder, after you implement your mitigation steps, make sure you reevaluate your residual risk in light of the new controls and configurations you may have placed into effect. Whenever you confront a risk, some of your first questions to your subordinates should be, "How can I mitigate this risk?" "How much will it cost?" "How long will it take?"

- Transference: While you can never transfer responsibilities, you can transfer risk. You do it all the time. You likely have car, property, and life insurance policies in effect right now. You pay premiums to the insurance company who in turn underwrites your liability based on how much coverage you are willing to pay for. Can you underwrite cybersecurity risk? Absolutely! In fact, there are several insurance companies around the world that now offer insurance for cybersecurity events. It is estimated that the cyber insurance market already has surpassed US $1 billion.[22] Beazley and Chubb were among the first firms to offer cybersecurity risk packages for businesses, and many insurance companies have followed suit. Packages range from comprehensive coverage for large multinational businesses to offerings tailored for small local firms. Should you pay for cybersecurity coverage? Probably. As the risk of significant financial losses due to cybersecurity incidents continues to climb, it appears that insurance to cover these losses is rapidly becoming less and less optional as part of risk management strategies. Essential cyber coverage includes third-party liability for damages associated with a data or network security breach, typically bundled with related first-party crisis-management costs—forensics, notification, call-center staffing, credit monitoring, and legal guidance.[23] First-party insurance may prove to be critical in incidents involving cyber extortion,[24] data breaches, or business interruptions. Regardless of your current insurance posture, it is well worth looking into your options regarding cybersecurity incidents and insurance coverage, just as you would in addressing other risk assessments. While insurance won't eliminate your risk, it will help reduce the overall impact if disaster strikes.

- Acceptance: Often, the cost of fixing a vulnerability is more than the asset you are trying to protect. Sometimes, you don't have the resources to fix the

---

[22] Michael Voelker, "As Cyber Coverage Soars, Opportunity Clicks," January 21, 2013, http://www.property-casualty360.com/2013/01/21/as-cyber-coverage-soars-opportunity-clicks. Accessed on September 10, 2013.

[23] Ibid.

[24] Cyber extortion is a relatively new term. An example is when criminals threaten a firm with a cyber-based action, such as deleting a critical database, if they are not paid a tidy sum. Like any other case of extortion, this is illegal activity.

vulnerability. Other times, you may decide that the high costs associated with mitigation are too much to pay based on the likelihood of an event and its potential impact. In cases like this, many people decide to accept the risk and allow their systems to operate with the known risk. Acceptance of risk is a decision reserved for senior leadership and management. As an executive, insist on a formal risk acceptance process for each and every risk acceptance decision. Ensure that all documentation regarding the risk assessment and decision-making process is complete and accurate. Also, make sure the risk acceptance decision is in writing and accepted by the senior leader making the decision. Remember, with great power comes great responsibility. Acceptance of cybersecurity risks is a business decision senior executives will be called to make. Be ready. Know your enemy. Know yourself. Know what mitigation and transference options you may have. When you know all of these, your decision will be much easier to make and be auditable and defendable.

- Avoidance: Avoidance happens when you stop doing that which exposes you to risk. We exercise the avoidance technique all the time in the cyber environment. An example of cyber avoidance is the practice of removing or disconnecting the vulnerable component or system to avoid risk. Let's say you have a faulty old web server configured with antique software that has numerous vulnerabilities. Rather than spending valuable staff time trying to resurrect the antique equipment and load contemporary software on it (which may or may not work on the older gear), you find it is cheaper and more effective to replace the server and software completely to avoid the risk. Another simple example addresses information itself. Many senior executives post their biographies online. Many post information about their spouse, children, and homes in their biographies. (For example, President X is married to the lovely Y of Trenton and they have four lovely children, A, B, C, and D. They currently reside in Palm Beach.) What a treasure trove of information for criminals! While your business may harden your cyber defenses at work, does your family have the same cyber protection? Could a criminal or hacktivist use that information to threaten your family? Avoiding placement of personal and other potentially exploitable information on your web site is an important risk management technique. Don't forget to check your official biography today! Finally, few companies operate alone. Your organization likely shares information with one or more organizations, often with so-called trust relationships, that permit transparent information sharing with the other organizations. Opening your network to a less secure partner may impose undue risk to your organization. Since a risk taken by one is a risk taken by all, make sure you choose your partners well. You may very well find that you need to avoid entering into a business relationship because your proposed partner does not maintain an effective cybersecurity program.

So, what about Plieno Corporation and their risk? What risk decisions does their CEO face? What are his options? What strategy does he adopt?

The CEO and his senior leadership team know they are at risk of losing their intellectual property and trade secret (the alloy formula) to a cybersecurity incident.

Their analysis of threat sources, potential threats, vulnerabilities, and exposure indicates they are at high risk and the estimated loss is over US $50 million. Their estimates based on available data indicate it is likely they will face an incident soon. They have a new sense of urgency to address this risk.

Based on the scenario provided, we have several risk management strategy suggestions for the Plieno CEO and his senior executive team. Perhaps you will find these helpful considerations as you look at your own organization:

- Mitigation: Here are our top ten recommended mitigation actions for Plieno Corporation. You too can significantly reduce your risk by accomplishing the following mitigation actions:

  1. Ensure your cybersecurity policies are well documented, that all personnel are trained on them, and that they are regularly tested.

  2. Ensure your software configurations and patches are all up to date. This applies to your antimalware software, applications, and operating systems. Only use approved and tested secure software, especially operating systems. This hardens your network against attack.

  3. Implement strong boundary connections and intrusion detection systems. Test them regularly through independent third-party penetration testing.

  4. Implement a policy of "Deny All, Permit by Exception," which filters all network traffic and denies all traffic not explicitly allowed. This can stop someone from "walking out the door" with your information.

  5. Implement a policy of "least privilege" where users only get the privileges and access to information and services they need. This significantly reduces the risk of someone hijacking the identity of one of your employees and elevating their privileges to gain access to your most sensitive information.

  6. Encrypt your data. All of it. Encrypt while it is at rest and while it is in transit. Encrypt your hard drives on your desktops, laptops, and other devices whenever possible. Make sure you have a key management system to assure you retain positive control of the keys to unlock your data.

  7. Implement a robust vulnerability management program including internal and external scans. Install and use an intrusion detection system on your network. This will provide the ability to deploy threat-specific detection signatures that will trigger immediate alarms for traffic of interest. Don't you want to catch insider threats or external penetrations red-handed and stop them?

  8. Make cybersecurity a corporate priority. Disable CD/DVD readers and USB drives by policy, and only provide that capability by exception under controlled conditions. Make importing and exporting of data a conscious decision. Implement comply-to-connect policies to reduce threats of contamination. Tightly control remote access.

  9. Invest in your IT staff commensurate with the value of the information you want to protect. Make sure you have the right team, properly trained and certified, and in the right amount to do the work you need them to do.

10. Disconnect Internet access to all critical and sensitive information that doesn't need an outside connection. Segment your mission critical business data from the outside world (who doesn't need to see it) as well as from administrative functions. This limits and/or contains the effects of compromises and also speeds recovery. Does your key intellectual property and trade secrets need to reside in the same place as your general correspondence? Generally, no. Prioritize, segment, and secure your information based on risk.

- <u>Transference</u>: We recommend you investigate your options to insure your business against loss from a cybersecurity incident. We recommend your discussions should include first- and third-party liability discussions. Additionally, we believe you should have conversations with multiple insurance firms before you make any decisions on risk transference as the cyber risk insurance market is still developing and wide variances in coverage, premiums, deductibles, and other factors exist. Be a discerning shopper when it comes to insurance. Ask for quotes. Ask for referrals. Ask a lot of questions and do your business case analysis before you sign up for anything.

- <u>Acceptance</u>: Your network readiness rate is at a very high level, indicating your staff is effective at meeting business operation needs, but removing the vulnerabilities indicated by the internal scanning may best be addressed by temporary technically qualified reinforcements rather than hiring additional full-time staff. Consider accepting risk of temporary hires of certified professionals to bring your vulnerability posture to an acceptable baseline, look to lean processes to better utilize existing staff, and defer the request for additional manpower for two months after the posture meets objective and processes are controlled.

- <u>Avoidance</u>: There is an avoidance option to consider for Plieno. Does their production system need to be connected to the Internet? What happens if they pull the plug? They still would have to address an insider threat and external attacks a la Stuxnet, but they can avoid the threat of hackers if they disconnect external connections. They still can maintain a connection for their administrative functions but can keep their core intellectual property and business functions insulated from external cyber attack. This is an option worth exploring.

What about BigRX? What recommendations do we have for their CEO?

BigRX appears in pretty good shape. They have disciplined processes for managing risk, and the employees seem to be well trained. Nonetheless, software testing procedures appear to be lacking as the vulnerability analysis indicates the SQL vulnerability. This should have been caught in the software testing process and fixed before it was put on line. This is a significant problem that needs immediate remedial action.

Here are our recommendations to the BigRX CEO as he contemplates his risk decisions:

- <u>Mitigation</u>: Frankly, if BigRX had not already implemented the top ten recommendations we gave to Plieno, we'd urge them to implement the same controls. We'd also recommend the additional following specific mitigation measures:

1. Fix the SQL injection vulnerability immediately. Test the fix before putting it on the production system.

2. Reinforce your defenses while the new code is being written. Be on the lookout for someone attempting access through a SQL injection technique.
3. Prevent further instances of putting deficient code on your system by implementing disciplined software acceptance and testing protocols. Never let bad code get on your system again.
4. Implement regular external and internal vulnerability scans to better expose your risk.

- Transference: We definitely recommend that BigRX consult with insurance brokers to discuss their options for risk transference through insurance. Unlike Plieno Corporation, who operates in a product-based environment, BigRX operates in what many consider a service-based environment. BigRX operates in a market sector where litigation is plentiful. They likely have very robust insurance packages addressing risks like medical malpractice. They ought to investigate adding insurance for cyber malpractice as well. Clearly, there is a risk. They owe it to their stakeholders to protect the business.
- Acceptance: We recommend BigRX fix their SQL injection issue immediately. Unfortunately, it isn't like flipping a switch and the problem goes away. The code will have to be written, verified, and thoroughly tested before being loaded on to the active production system. In the meantime, we recommend BigRX consider accepting the risk of keeping the existing configuration online until the new code can make its way through the appropriate repair, testing, and delivery process. Given the urgency to fix the code, we do not believe it would take an inordinate amount of time to receive the fix.
- Avoidance: It may be possible to remove the flawed code from BigMIMS and still be able to maintain effective operations until the new code is ready for deployment. This is an option that may be viable but would have to be explored in greater detail before making a recommendation to implement it. A business case analysis taking into account the technical and business operations considerations is warranted.

We also would recommend to both Plieno Corporation and BigRX that they consider investing in a cybersecurity business intelligence capability. Back in the "old days" before computers, such services used to be provided by people who clipped articles from newspapers and magazines. Now, many companies maintain technically enabled sophisticated in-house business intelligence functions to maintain situational awareness over key items of interest in their business sector, supply chain, and other areas that possibly could affect their business. Others subscribe to services that provide them tailored information to heighten their awareness of key market trends, threat warnings, etc. Both companies need cybersecurity business intelligence as part of their "know your enemy" early warning capability.

Cybersecurity has become a key business component, and both companies need to have the type of information a cybersecurity business intelligence function provides. It can provide information to let you know when you may be targeted for cyber attack, who is doing the targeting, and why. Business intelligence professionals specializing in cybersecurity issues can provide you with analysis of current threats that can prove to be invaluable in preparing your risk assessments. Executives need solid actionable information to make operational and strategic decisions. We recommend both companies secure a cybersecurity business intelligence capability.

3.4 COMMUNICATING RISK

Risk must be communicated to be properly managed. Ask any manager whether they understand how to manage risk, and they will tell you they know how to manage what they understand. If they understand the risk, they can manage it. Therefore, it is important to clearly communicate the risks and risk management strategies, policies, and procedures in a manner that is readily understood by key stakeholders throughout the organization.

It is easy to frighten people when it comes to cybersecurity risk. There are so many vulnerabilities and threats that it can quickly overwhelm even the stoutest heart. Not everyone understands the lingo that has evolved in the cyber ecosystem, and some people are offended when they believe the technical community is deliberately trying to obfuscate by "speaking in technical tongues." Likewise, the technical community is offended when they try to communicate highly complex technical topics in the simplest terms only to be derided for "dumbing down" the conversation. Barriers to effective communication only increase your risk!

We are reminded of the scene in the movie "Ghostbusters" where Dan Ackroyd's and Harold Ramis' characters are arguing over the dangers of using their tools to catch a ghost. The characters played by Ackroyd and Ramis are the highly technical "geeks," while Bill Murray's character is not. Murray's character is told that crossing the powerful streams of electricity from their tools would be bad and not to do it. He does not understand what "bad" means and asks for clarification. Ramis says, "Try to imagine all life as you know it stopping instantaneously and every molecule in your body exploding at the speed of light." Murray responds, "Right. That's bad. Okay. All right. Important safety tip. Thanks."[25] Failure to communicate actually increases your risk if that particular risk is not understood properly and acted upon. It is imperative for executives to ensure that risk is effectively communicated. Let your people know what the impacts of "crossing the streams" are in your business.

Risk needs to be communicated to several constituencies. First, it has to be communicated internally. Every employee has a stake in the business's risk. It has been said that risk management is a team effort. Therefore, the team needs to clearly communicate as a team. Second, there are some communications of risk based upon regulatory guidance that have to be considered. While such communications make many executives uncomfortable, they have become a fact of life. Precision, honesty, and brevity are our three watchwords for this communication requirement. Finally, you have to communicate with your shareholders. They are the owners of your company and expect to know what risks their company faces.

[25]*Ghostbusters*, Columbia Pictures, Culver City, CA, 1984. Written by Dan Ackroyd and Harold Ramis and directed by Ivan Reitman, the film is a comedy about three unemployed parapsychology professors who start a business exterminating ghosts.

3.4.1 Communicating Risk Internally

We submit that communicating cybersecurity risk is best done when everyone uses the same language. Communicating risk focuses on sharing information about threats, vulnerabilities, and impacts. Management can set the tone by establishing a risk management program that includes the following:

- Establish a standardized risk management process: A disciplined process yields rich dividends as you are more likely to identify threats, threat sources, and vulnerabilities and, thus, predict the likelihood of events with greater precision. Perhaps more importantly, you will have common understanding of risk among all team members. Define key terms and procedures on how you identify, characterize, and manage risk. Make it part of your culture and train personnel throughout the organization to follow the process.

- Reinforce that while senior management owns the risk in the business, everyone has a stake in it: This bears emphasis. As an executive, your leadership is essential to ensure that each employee understands their responsibilities in managing the risk your company faces. Everyone has a stake, if for no other reason than it will have an impact on their wallets and pocketbooks.

- Ensure your team is well informed regarding the risk you face and your program to manage it: Clearly communicate your "Five W's": (1) What risk you face, (2) Who has responsibility to manage risk, (3) Where the risk is, (4) When to look for it, and (5) hoW to avoid it.

- Establish and document a Critical Information Reporting process to maximize leadership's risk visibility: Key components of the process include:

  ○ Identify your key information: It is essential senior management let subordinates know what information they require. Don't keep your information needs a secret.[26]

  ○ Identify who needs the information: If the right people don't know, the right actions will not occur. Management needs to let subordinates know who needs the information. You get bonus points when you tell them what decisions you make from the information from various sources because everyone wants to know "why." People who know why information is needed are more likely to act with greater vigor and precision that those who feel they are just "passing on another report."

  ○ Define the timeline for reporting: You need to define what is a **wake me up in the middle of the night** situation versus **it can wait until morning** event. Don't torture your staff by making them guess what you want and when you need to know things. Tell them! Give them the leadership and clear direction they deserve!

  ○ Define the process for reporting: Does the key information go right to the top? Does it go to a central control center that filters and feeds it out? Does it make

---

[26] We are reminded of the quote (perhaps apocryphal) attributed to Yogi Berra, "If you don't know where you are going, you probably won't get there!"

its way through the hierarchy to its destination? You need to define the flow of information from detection all the way through to receipt by the person who needs the information.

○ Define the reporting format: Clearly defining how the reporting message is conveyed beforehand will save time, money, and angst.

• Listen well: Your personnel often will detect threats and vulnerabilities that yield risk well before the C-suite even imagines it. Empower your employees to sound the alarm and incorporate procedures in your risk management process that employees can use to identify risks. Most companies already have safety programs to minimize risk of industrial accidents where employees can identify threats and vulnerabilities to management. They have been very successful in reducing accidents. Note that when employees identify safety risks and report such risks expeditiously, most employers make a big deal out of it and frequently give awards, sometimes in cash. Do you have a similar program to minimize cybersecurity risk? If not, why not?

• Control and monitor your risk with metrics: Let your personnel know how they are doing in managing risk through visible metrics that are shared throughout the organization.

• Celebrate risk management successes: Praise and award star performers and teams. "Success breeds success" and positive messages about risk management will encourage your team to perform at high levels.

We submit that your people are your most valued and treasured resource. Successful businesses have great processes. Great businesses have great people who manage great processes. To make sure you have a great risk management process, ensure that you invest in your workforce and clearly communicate, communicate, communicate, and listen.

3.4.2 Regulatory Communications

On October 13, 2011, the Securities and Exchange Commission (SEC) Division of Corporate Finance issued "CF Disclosure Guidance: Topic 2, Cybersecurity" (CF DG 2), which substantively changed the way businesses communicate cybersecurity risks.[27]

As an executive, you are well advised to be aware of the content of this guidance and understand how it affects you and your business.

3.4.2.1 Disclosure Obligations. The mission of the U.S. SEC is to protect investors; maintain fair, orderly, and efficient markets; and facilitate capital formation.[28] Created in 1934 during the height of the Great Depression, the SEC has a long history of interaction with American business and those foreign firms who do business in the

---

[27] Securities and Exchange Commission, Division of Corporate Finance, CF DG 2, October 13, 2011, http://www.sec.gov/divisions/corpfin/guidance/cfguidance-topic2.htm. Accessed on September 10, 2013.

[28] http://www.sec.gov/about/whatwedo.shtml. Accessed on September 10, 2013.

United States. The SEC seeks to foster a competitive climate that will prevent another Great Depression from occurring again.

The SEC is led by five commissioners, appointed by the president and approved by the Senate, who oversee the commission. Its responsibilities include:

- Interpret and enforce federal security laws
- Issue new rules and amend existing rules
- Oversee the inspection of security firms, brokers, investment advisers, and ratings agencies
- Oversee private regulatory organizations in the security, accounting, and auditing fields
- Coordinate U.S. security regulation with federal, state, and foreign authorities

The SEC interprets U. S. law and issues rules and regulations to implement those laws. These rules and regulations go through a deliberate process that often starts with public release of a rule proposal with a 30–60-day comment period for the public to provide comments. After the comment period, the commissioners consider the public comments and, after any requisite editing, vote on the proposed rule. Upon agreement of the commissioners, the rule goes into effect and has the force of law.

The SEC CF DG 2 guidance emerged during a period of great national debate regarding cybersecurity and the government's role in developing a series of laws and policies to address the growing cybersecurity risk environment. At the time of issuance of the SEC guidelines (and even through today's writing), there were no national level regulations backed by the force of law that applied to business reporting of cybersecurity risk to shareholders and potential investors. While CF DG 2 explicitly states that "This guidance is not a rule, regulation, or statement of the Securities and Exchange Commission…," evidence points to it being applied as if it were.

CF DG 2 calls for public companies to disclose cybersecurity risks and cyber incidents in the following six areas[29]:

- Risk factors: If your company is registered with the SEC, the guidance calls for you to disclose "the risk of cyber incidents if these issues are among the most significant factors that make an investment in the company speculative or risky." How do you determine whether these incidents are significant enough to disclose? The SEC believes you should "consider the probability of cyber incidents occurring and the quantitative and qualitative magnitude of those risks, including the potential costs and consequences resulting from misappropriation of assets or sensitive information, corruption of data or operational disruption." You should also "consider the adequacy of preventative actions."

    This rhetoric is considered highly controversial. Couldn't the public disclosure of such detailed information serve as an invitation to hackers to visit? We think so, and even the SEC acknowledges that. They state in the guideline, "We are mindful of potential concerns that detailed disclosures could compromise

---

[29] CF DG 2, op cit.

cybersecurity efforts…by providing a 'roadmap' for those who seek to infiltrate a registrant's network security…and we emphasize that disclosures of that nature are not required under federal securities laws" [emphasis added]. Nonetheless, the guideline calls for specific information to be disclosed. The guideline goes on further to state that your disclosure should "avoid generic risk factor disclosure" and you need to be prepared to discuss specific attacks and their "known and potential costs and other consequences." They conclude their discussion on risk factor disclosures by stating, "…registrants should provide sufficient disclosure to allow investors to appreciate the nature of the risks faced by the particular registrant in a manner that would not have that consequence."

While the SEC is noble in their objective of informing shareholders and potential investors through the disclosure process, we advise great caution when addressing these cyber-related disclosure guidelines. Advertising your vulnerabilities to potential foes is dangerous and may invite bad actors to see just how vulnerable you are. We believe the SEC recognizes this and is using its administrative leverage to spur businesses to invest prudently in cybersecurity so that they in fact do not have significant vulnerabilities that can be exploited. Regardless of the underlying intent, public disclosures of your cybersecurity risk can have profound influence on your brand reputation, consumer confidence, and (ultimately) your bottom line. Craft your risk factor disclosure carefully when disclosing cybersecurity risk information. Make sure you have your **best** lawyers drafting it and have your technical staff review it to advise whether your disclosures present additional risk. **Pay close attention to this disclosure!**

• Management's Discussion and Analysis (MD&A) of financial condition and results of operations: The MD&A is an essential part of your annual report that allows you to provide a narrative explanation of your company's financial statements. It often is referred to as telling the story "through the eyes of management." Your MD&A can improve your overall financial disclosure by providing context to the financial information presented in the rest of the report and presents a venue where you can provide information about your company's earnings and cash flow (among other important financial disclosures). Shareholders and potential investors alike use this information to make judgments about the likelihood that past performance is indicative of future performance. The SEC recognizes the importance of the MD&A in the disclosure process and advises in the guidelines that registrants should address cybersecurity risks and cyber incidents if "the costs associated with one or more incidents or the risk of potential incidents represent a material event, trend, or uncertainty that is reasonably likely to have a material effect on the registrant's results of operations, liquidity, or financial condition or would cause reported financial information not to be necessarily indicative of future operating results or financial conditions." Recall our discussion in Section 3.2.2, which states that companies who have been hacked before are at greater risk of being hacked again. With the CF DG 2 guideline, you are being strongly encouraged to disclose incidents and risks with the strong suggestion that past performance is indicative of future performance. Fear not, though, as, in your MD&A, you can use and take advantage of the narrative in this disclosure to inform shareholders and potential investors what bold and strong management controls you have taken to reduce your cybersecurity risk and eliminate vulnerabilities. Remember, though, that you can only reduce your risk, not entirely eradicate it. We hope that investors recognize this fact of life.

- Description of business: CF DG 2 states, "If one or more cyber incidents materially affect a registrant's products, services, relationships with customers or suppliers, or competitive conditions, the registrant should provide disclosure in the registrant's Description of Business." This is tricky and consultation with legal counsel is warranted when you discuss this during your disclosure deliberations. What if you have a business partner who has a cyber incident and that incident is a consideration contributing to you not renewing your contract with that partner? One could make the argument that you should disclose that under this provision. Yet, what if you want to leave the door open for a future relationship with that company? Disclosing that you dumped a business partner because of a cyber incident or risk can have positive and negative consequences. Be careful in how you characterize your business and its relationships.

- Legal proceedings: This part of the guideline is fairly straightforward. If you are engaged in litigation due to a cyber incident, you are instructed to disclose it in your "legal proceedings" disclosure.

- Financial statement disclosures: Cybersecurity risk management drives financial decisions, and SEC guidelines call for you to appropriately characterize your expenditures associated with cybersecurity in your financial statements. You make investments to prevent cyber incidents. You may have a cyber incident that results in a loss, diminishes cash flows, or drives you to further investment in response. The guidelines call for you to ensure that your financial statements disclose "the nature of" cyber incidents and an estimate of their financial effects. Further, registrants must explain any "risk or uncertainty of a reasonably possible change in its estimates in the near-term that would be material to the financial statements." As with all financial statements, precision and **brevity** are imperative, but you may find it difficult to accurately characterize all cyber-related costs. For example, how would you determine what loss of cash flow can be directly attributed to a cyber event? The inquisitive minds at the SEC want to know.

- Disclosure controls and procedures: In the aftermath of major corporate and accounting scandals such as at Enron and Tyco International, the U. S. Congress passed the Sarbanes–Oxley (SOX) Act of 2002. The law was intended to provide greater accountability and oversight over corporate finances to better protect shareholders interests as well as the greater economic health and well-being of the market. Other nations have adopted similar legislation, and executives have become abundantly familiar with the internal control processes and procedures the act promotes. Internal control processes often are reliant on automated reporting mechanisms, information contained in computer databases, and other cyber-reliant sources. Management is required under law to certify the integrity of their internal controls process. Yet, what do you do if a cyber incident affects one of your data sources or internal control processes? CF DG 2 addresses that and reminds registrants that you must consider any effects of cyber incidents that may cause deficiencies in your disclosure controls and procedures and make appropriate disclosures. This begs the question, "How do you issue certification of your disclosure controls and procedures if you had a cyber incident that potentially tainted your information or processes?" While we recommend you consult with your general counsel, our visceral default response is "always be honest."

3.4.2.2 Why Disclose? We believe the SEC's Division of Corporation Finance is strongly encouraging cybersecurity disclosure to accomplish two objectives:

- Bring cyber threats to light and prompt companies to invest in adequate cybersecurity controls.
- Provide a mechanism to inform shareholders and potential investors about the cyber risk to which companies are exposed.

Are these objectives appropriate? Should government guidelines steer you to disclose your cybersecurity risk information or should you do so voluntarily?

Answers to these questions depend on who you are and where you sit.

Some people will argue that the objectives are appropriate based upon the SEC's charter from Congress to act "as necessary or appropriate in the public interest or for the protection of investors."[30] There certainly is a good case to be made that the public at large and potential investors would want to know whether a business has a significant cyber risk. After all, who wants to put their money into a bank that is likely to fall victim to cyber theft?

On the other hand, some argue that Congress has not enacted laws directing these actions, so the guidelines are presumptive and perhaps representative of government overreach. They argue that the commission is shaping public policy without warrant from the people's representatives in Congress. They contend that if the people really want disclosure requirements, they will direct so through the law. In the meantime, these people believe the SEC should limit their actions to what the law explicitly dictates.

Should you disclose in accordance with the guidelines? We leave that determination to you and your advisors.

3.4.2.3 Reasons to Not Disclose. There are three primary reasons why you may not disclose cybersecurity risk in accordance with the DG CF 2 guidelines:

First, you don't know what your cybersecurity risks are.

Second, if you do disclose your cybersecurity risk, you may attract hostile bad actors who will try to exploit your vulnerabilities and damage your business.

---

[30] Securities and Exchange Act of 1934, http://www.sec.gov/about/laws/sea34.pdf. Accessed on September 10, 2013.

Third, if you do disclose your cybersecurity risks, you may face multiple negative effects, which could include but are not limited to:

- Loss of investor confidence
- Increased risk of liability lawsuits
- Loss of brand reputation
- Loss of share value

These three are all legitimate reasons cited by companies as to why they are reticent to publicly disclose cybersecurity risks and incidents in great detail. Do you and your company share in these concerns?

3.4.2.4 How to Disclose. SEC regulations direct several reporting mechanisms that should be used to report cybersecurity risk:

- Annual report, Form 10-K
- Quarterly report, Form 10-Q
- Current report, Form 8-K

Both the annual report and the quarterly report are well-established report formats with which companies and their staffs are very familiar. Under provisions of the CF DG 2 guidelines, specific information regarding cybersecurity risk and incidents is now expected to be included in the reports as spelled out in the CF DG 2.

The current report, Form 8-K, is used when any "material events" arise inside the timelines directed for the quarterly and annual reports. Examples include bankruptcies, "material definitive" agreements, amendments to articles of incorporation, and "other events."

What are "material events"? Lawyers have argued over the definition and interpretation of those words for years and likely will do so for years to come. Let's use the definition the Supreme Court used, a "…fact is material if there is a substantial likelihood that a reasonable shareholder would consider it important in deciding how to vote."[31]

Given the Supreme Court definition of "material," is it reasonable to assume the expectation of the shareholder is that they want to know if you have cybersecurity incidents and risk as key information as they consider their votes? We believe such an association indeed would be made by a reasonable person. However, as we discussed earlier, incidents or events can have a broad range of impacts, some inconsequential and others devastating. Notwithstanding the opinion of the Supreme Court, shareholders expect company management to exercise good judgment in assessing and reporting upon "material" occurrences. It does not appear that the SEC has considered or provided insightful guidance on that subject.

---

[31] TSC Industries, Inc. Et. Al. v. Northway Inc., http://scholar.google.com/scholar_case?case=898547504021 2340102&hl=en&as_sdt=2&as_vis=1&oi=scholarr. Accessed on September 10, 2013.

What would we do? We would convene our management and legal counselors and together decide what in our considered judgment serves the best interests of our shareholders. Readers should not lose sight of the fact that the management and the board frequently represent a substantial percentage of ownership.

3.4.2.5 What If You Don't Disclose? Given the reasons not to disclose cited earlier and the CF DG 2 statement that its guidance "is not a rule, regulation, or statement of the Securities and Exchange Commission" and that "the Commission has neither approved nor disapproved of its content," there doesn't appear to be any statutory or regulatory requirement that demands that you have to disclose your cybersecurity risk information through the annual, quarterly, or current reports. So why do it? What's the worst that can happen if you don't include it in your reporting?

Preparing disclosure reports is not a trivial task and involves noteworthy analysis and production costs, including the use of high cost outside professional services. As we've learned in our discussion of quantitative and qualitative risk assessments, determining cybersecurity risks can be difficult to quantify and characterize. Moreover, communicating your cybersecurity risk to potentially hostile bad actors may invite further trouble as hackers and other threat sources attempt to exploit your vulnerabilities. At first blush, withholding **detailed** cybersecurity risk information from public disclosure may be in your shareholder's best interest.

The SEC doesn't see it that way. While they officially maintain a voluntary disclosure program, their staff repeatedly has pushed companies to disclose cyber attacks. There are several reports of SEC using aggressive tactics to encourage companies to disclose cyber attack information. This is not surprising. According to Peter Henning, a former SEC lawyer, the SEC can force disclosure without making rules because companies need to stay on good terms with the regulator, which reviews their financial filings and can "make things difficult." Resisting a letter from the agency can be costly, amounting to US $250,000 in legal fees, according to Henning, even if the company is found to be fully compliant. "If it's complex, your lawyers write drafts in response, you have conference calls with them," he says. "The SEC knows that's their power. If you want to litigate with them, it costs millions."[32]

What if you want to take on the SEC and their aggressive tactics? What is the worst that could happen?

According to the SEC, they are first and foremost a law enforcement agency. They investigate violations of securities and exchange laws and initiate civil and administrative actions to address them.[33]

Common violations that may lead to SEC investigations include:

- Misrepresentation or omission of important information about securities
- Manipulating the market prices of securities

[32] Linda Sandler, "The SEC Says Speak up about Hack Attacks," September 6, 2013, http://www.businessweek.com/articles/2012-09-06/the-sec-says-speak-up-about-hack-attacks. Accessed on September 10, 2013.

[33] SEC, How Investigations Work, http://www.sec.gov/News/Article/Detail/Article/1356125787012#.Uio_fzashcY. Accessed on September 10, 2013.

- Stealing customers' funds or securities
- Violating broker–dealers' responsibility to treat customers fairly
- Insider trading (violating a trust relationship by trading on material, nonpublic information about a security)
- Selling unregistered securities

If the SEC initiates a **civil action** against your company, you face the possibility of:

- An injunction that will prohibit you from taking further legal action
- A monetary penalty
- The return of any profits that were deemed to be acquired through illegal means
- Barment or suspension of directors and officers of the corporation
- In the event you violate the judgment of the court in the civil action, you face contempt charges, with accompanying fines and possible imprisonment.

If the SEC initiates an **administrative action** against your company, you face the possibility of:

- Sanctions including a cease and desist order that freezes your activities
- Suspensions or revocation of registrations
- Censures
- Barment from associations

Is "the juice worth the squeeze" to contest the SEC's CF DG 2 cybersecurity disclosure guidelines? Major companies such as Google and Amazon concluded it wasn't. After repeated volleys of requests for further cybersecurity information, Google and Amazon relented and edited their disclosures to satisfy SEC staff demands.[34]

Sooner or later, you will need to make decisions on how and what cybersecurity information to publicly disclose. Choose wisely.

3.4.3 Communicating with Shareholders

While the mandatory SEC disclosure requirements mentioned previously arguably could suffice as a means of communicating cybersecurity risk and incidents to shareholder, we don't believe SEC reports should be the primary means of communicating with your shareholders.

Your shareholders are increasingly sophisticated and appreciative of the risks presented in a cyber-enabled marketplace. While they may not understand the technical underpinnings behind them, the vast majority of your shareholders understand that cybersecurity risks exist and they expect you, the executive, to properly set conditions to protect their investment by mitigating that risk.

---

[34] Sandler, op cit.

The severity of a cybersecurity incident could range from a very minor event to an existential threat to your business. You need to have a plan on how to communicate regularly with your shareholders so they retain confidence in their business, that their investment is in good hands, and that you are in control.

Here are some suggestions on how to best communicate with your shareholders.

- Ask them how they want to hear from you: Surprisingly, many companies do not even ask their shareholders what their preferred means of communications are. I prefer to receive emails and electronic reports yet still get piles of paper-based prospectus information in the mail that ends up shredded and recycled.[35] Disposing of the paper products is time-consuming and just increases my frustration. Do you ask your shareholders how they want you to communicate with them? Do they prefer letters? Email? Phone calls? Web chats? Videos on your web site? There are many options. We suggest you let your shareholders pick their preference.

- Ask your shareholders what kind of information they want: Nobody looks forward to receiving spam, even when it is from your company. Don't waste company resources sending out unsolicited and undesired information. Do shareholders want to know when you have cybersecurity risk? Do they only want to know when you are attacked? Or do they not care to know at all as long as the company stays safe, under control, well managed, and profitable? You'll never know unless you ask them!

- Solicit communications from your shareholder to you: Not every shareholder has the means of attending stockholder meetings where they can give you direct feedback. For most shareholders, sending an email or a letter is the primary means they use to communicate with you and your staff. When you receive a letter or email from a shareholder, make sure you answer it completely and quickly, and by all means, make it warmer and more pertinent than the responses you get back from Congressmen and Senators!

- Have a plan to communicate in crisis: A cyber attack could pose an existential threat to your business, placing your shareholder's equity at risk. You ought to have a plan on how to communicate during a crisis. Here are some best practices for communication during a crisis:

  i. Get yourself a world-class Public Relations consultant: Believe us, it's no fun sitting in front of a bank of twelve microphones while TV cameras grind and facing questions from a bunch of hungry reporters who want you to tell their listeners and viewers why you were so damned stupid to let this mess happen to begin with. Based upon hard experience, immediately consult with an accomplished public relations specialist. We did and defused a couple of situations that, although they were difficult, could have been terminal financially.

---

[35] On the other hand, my father, the junior author of this work, prefers "hard copy." He likes to relax in his recliner and drive into the details at his leisure. To him, reading annual reports online is a pain in the neck, only he doesn't say neck.

ii. <u>Communicate early and often</u>: It is essential to have open lines of communication with your shareholders and to remember that communication goes both ways. When confronted by a crisis such as a cyber attack, the early hours after the event are critical and set the tone for the duration of the crisis. When communicating with your shareholders, be prepared to answer these questions:
- What happened?
- Where did it happen?
- When did you find out?
- What you are going to do about it?
- Who's to blame?
- Were there warning signs?
- How will you prevent it from happening again?
- What does this mean for us?

iii. <u>Take responsibility</u>: Don't beat around the bush. Take responsibility, express regret, apologize as appropriate, and decisively inform your shareholders what your next steps are to address the problem.

iv. <u>Speak with one voice</u>: Ensure your message is consistent throughout the organization. Centralized control of information and talking points have proven to enhance accuracy and timeliness of information.

v. <u>Establish a crisis team</u>: Create and train a crisis team (including your PR specialist) as part of your business continuity planning effort. Operate a command post to coordinate and synchronize response efforts. Establish a scheduled rhythm to share information with your shareholders and other key stakeholders.

vi. <u>Plan for the worst</u>: While you hope for the best, you need to plan for the worst. Anticipate having to deliver and respond to bad news. Have your script ready. Don't "wing it" in delivering your message, and by all means, do not deviate from the central message or ad lib or try to be the least bit humorous. You must convey that this is a serious situation and you are acting accordingly.

vii. <u>Get your message out</u>: Communicate, communicate, and communicate. These are the watchwords of crisis communications. There are plenty of ways to get your message out to your shareholders:
- Emails
- Letters/mailings
- Television and radio commercials
- Web site postings (including video messages)

Your shareholders trust you to protect their investments. They expect you to professionally manage their company and deliver success. They also expect you to keep them informed. Do so in a manner that retains their trust and confidence in your abilities.

3.5 ORGANIZING FOR SUCCESS

Many companies have come to realize that they need disciplined processes and procedures to forecast, measure, and control risk. Great companies do something about it, and more often than not, they implement organizational structures specifically to address risk.

3.5.1 Risk Management Committee

An example of specific organizational structures to manage and control risk is found in the proliferation of risk management committees at the corporate level. It is not unusual for corporate boards of directors to establish committees to address auditing, compensation, and governance. Now, many companies have added committees to focus on risk management. We believe this is a terrific concept, particularly in regard to improving cybersecurity risk management.

Committees are formally chartered by corporate boards of directors to provide oversight and governance over key functions of the business. Charters include direction regarding board purpose, membership, organization and operations, duties and responsibilities, reporting requirements, resources and authorities, meetings, and other needs of the board. It is important that the board chart the course for the committee yet recognize that the charter should often be reviewed for any changes or improvements.

Risk management committees usually consist of nonmanagement directors. This is important as nonmanagement directors are more likely to be unfettered by the organizational bias that often accompanies management positions. Likewise, because it is highly unlikely that the nonmanagement committee members were participants in the detailed decision-making that led up to any emerging risk, they are more likely to focus on the risk rather than the daily running of the business. Many companies have discovered this alignment to be powerful and one that delivers excellent results.

Risk management committees monitor and control the "material enterprise risk" of the organization. Typically, they are the approval authority for and provide oversight of management proposals, leading to the creation and subsequent assessment of a risk management framework submitted for approval by the board. The framework includes the definition of the categories of risk, standards in relation to each category, and an approach to risk tolerances adopted by the company. These standards will be reviewed periodically (and at least annually) to take into account changes in the internal and external environment as well as reports and findings of the audit committee as it relates to performance of controls.

Cybersecurity is increasingly at the top of the agenda for risk management committees. Because it is, the risk management committee must have the resources it needs to posture itself to make informed decisions.

One of these resources is quality information and insight into the business. Quality information yields quality decisions, which yield quality results. When it comes to cybersecurity risk, in addition to close communication with business unit directors and officers, the risk management committee should have very close communication with the CIO, who should provide the committee with information regarding architectures, performance and expenses, and other information regarding the information systems that drive the business. Similarly, the committee should be in close communication with the chief information security officer (CISO) who should provide the committee with information on cyber vulnerabilities and threats.

Another critical resource the risk management committee needs is threat awareness. If the organization has a business intelligence function that focuses on external threats, they should be tightly coupled with the committee. If the organization does not have a function such as this, the committee is urged to recommend that the company should contract or subscribe for such a service to aid in maintaining comprehensive threat awareness.

A third and critical resource for the board is the technical awareness to understand the challenging and complex cyber environment. While members of the risk management committee do not need to be certified technical experts, they need to have a basic understanding of both the business and the technology that supports it in order to make the best decisions. We are familiar with many boards of directors that have invested in continuing education to ensure their directors, including members of the risk committee, have the requisite contemporary knowledge to stay "on top of their games." As an example, a colleague of ours who has been in corporate America for over 50 years and currently serves as a director on several boards says about his continuing education, "I may not be able to see or hear as well as I did, but I can still smell when crap is being shoveled my way!" The takeaway for you regarding technical awareness is that the committee has to understand cybersecurity to understand its risks. You need to plan to invest in your committee's continuing education to keep them current!

Corporate risk management committees largely have been very successful in helping to highlight, manage, and control risk. Board-level oversight of management's efforts to manage and control risk is appropriate and fosters more disciplined, professional, and complete risk identification, accounting, and control. If your business does not have a risk management committee, we highly recommend you consider creating one.

A closing thought is in order based upon the foregoing discussion of the risk management committee. *It is the job of the board of directors to* **direct**, *and the job of managers to* **manage**. *To forget this aphorism is an invitation to trouble.*

3.5.2 Chief Risk Officers

Many companies invest in chief risk officers who support directors and officers in the strategic management of the corporate risk program. For some firms, the investiture of a senior executive designated with strategic responsibilities over the corporate risk program yields improvements in compliance, strategic planning, and governance.

Because the chief risk officer (CRO) is a relatively new position, many organizations who have appointed one haven't yet mastered how to integrate them into their management structure. The successful (and satisfied) ones typically report to the CEO for their daily duties. They oversee the strategic risk management program, its processes, and its metrics. The CRO ensures that processes are maintained and current and that personnel are trained in accordance with corporate objectives.

The CRO often also oversees compliance programs, working with the general counsel and across business units to ensure that compliance actions and reporting are accomplished. Frequently, the CRO interacts with outside professional, legal, accounting, and public relations (PR) consultants.

While corporate CROs are now emerging as powerful and important senior executives, some boards wisely are asking whether they need both a risk management committee and a CRO.

Many companies who have both report they intend to keep them. An example is KeyCorp, an US $87 billion asset regional bank headquartered in Cleveland. They have had a board-level risk management committee and a CRO for several years. According to the bank's senior executive vice president and CRO, the risk committee's primary role is to establish the "risk appetite level" for the bank's various business lines, expressed in the form of measurable data like nonperforming loans or customer service complaints, and "it's part of my job to translate that appetite into a risk control structure for the company." That control structure includes a risk reporting process where the CRO regularly provides the committee with a variety of forward-looking metrics that will not only tell the committee what the bank's risk profile is today but also where it might be trending in the future. KeyCorp's risk committee meets six times a year, "so every other month we're also having face time with the [committee members]," says the CRO.[36]

How you organize to manage your risk successfully depends on your company, its goals, and the threats it faces. If you haven't already done so, we strongly urge you to consider establishing a risk management committee as part of your corporate board structure. The committee should be chartered to provide strategic oversight and governance over management's risk management program and should determine a "risk appetite" measure for the board's consideration and approval.

If your company is complex, faces significant risks across many business functions, and has the means, you may want to consider investing in a CRO to provide the strategic and operational management of your company's formal risk management program. Those companies that have made that type of investment generally reap positive rewards.

3.6 SUMMARY

Life is full of risk.[37] As an executive, one of your primary responsibilities is to manage risk to protect your business and create an environment for it to grow and thrive.

Risk is managed at every level of your business, yet it is owned in the boardroom and C-suite. Responsibility to lead and manage your business is vested in you by the owners of the business: your shareholders. While activities are delegated in hierarchical organizations, responsibility never can be.

It is critically important that you create and maintain a risk management program owned at the most senior levels and designed to cascade throughout the business to

---

[36] Jack Milligan, "Do Risk Committees and Chief Risk Officers Make Banks Safer?," November 2, 2012, http://www.bankdirector.com/index.php/magazine/archives/4th-quarter-2012/what-s-the-risk/. Accessed on September 10, 2013.

[37] Touhill, C. J., The environmental challenge and the risk of living, *Washington State Library Association Annual Meeting*, Richland, WA, May 9, 1970.

where each employee knows they are valued and essential stakeholders in the risk management program. A formal and disciplined risk management program best postures you for successful identification of risk, management, and control over risk factors and sustained risk awareness. The best risk management programs have well-defined processes, well-trained and motivated employees who understand and implement the program, and active leadership who maintain ownership over the risk management program.

Your need to "know your enemy" and "know yourself" in order to have a successful risk management program. When addressing your cybersecurity risk, it is imperative that you understand your threats, threat sources, and vulnerabilities and have as accurate a measure of the likelihood of an incident as possible. You must consider all vulnerabilities including those presented by technical means, procedural or material defects, or human failures or deficiencies.

It is possible to measure and estimate cybersecurity risk. While cybersecurity risk estimation processes generally are not as mature as traditional risk estimations used by most corporations, cybersecurity risk can be quantified in monetary terms using the **quantitative risk assessment** technique. This technique is difficult to employ due to the difficulty in assessing precise value to information and even greater difficulty in determining the likelihood of loss. We believe that with prudent analysis and management judgment and oversight, reasonable estimates on the valuation of information are possible. Moreover, it is feasible to carefully analyze threat stream and statistical information to make informed estimates on the likelihood of events. When these conditions exist, we believe quantitative risk analysis methodology can be used to assess cybersecurity risk. (We have cited examples to amplify upon our contention.) We believe you should incorporate quantitative risk assessments into your corporate business processes, wherever possible.

Qualitative risk assessments are a popular method of calculating cybersecurity risk and present potentially preferable means of determining cybersecurity risk for businesses, in contrast with quantitative risk assessments. Qualitative risk assessments do not utilize detailed calculations to assign monetary values to assets and losses like the quantitative method. Rather, the qualitative risk assessment method recognizes the difficulty present in assigning realistic values to information and the likelihood of risk. As such, this qualitative method provides relative measures of risk and asset value based on ranking specific items into categories such as high, medium, or low or on a numeric scale. While not as precise as the quantitative method, they generally are faster, easier, and less expensive to produce and give senior decision-makers actionable information in a timelier manner. Moreover, in most respects, results are easier to understand.

We recommend you consider investing in a cybersecurity business intelligence capability. Many companies maintain in-house business intelligence functions to maintain situational awareness over key items of interest in their business sector, supply chain, and other areas that possibly could affect their business. Others subscribe to services that provide them tailored information to heighten their awareness of key market trends, threat warnings, etc. Your business needs cybersecurity business intelligence as part of your "know your enemy" early warning capability.

You can manage risk through mitigation, transference, acceptance, or avoidance. Whatever technique you decide to implement to manage risk ought to be influenced by a business case analysis. If you do your business case analysis well, the right decision should jump out to you!

Risk must be communicated to be properly managed. It is important to clearly communicate the risks and risk management strategies, policies, and procedures in a manner that is readily understood by key stakeholders throughout the organization. You must communicate risk internally within the company to its employees and those who manage and control the risk. You must consider disclosing risk through channels identified by regulatory rules and guidelines. You also must regularly communicate risk to your shareholders, both in times of calm and times of crisis.

Organizing well can lead you to success when addressing cybersecurity risk. We recommend the charter of a risk management committee at the corporate board level to produce strategic governance and oversight over your corporate risk management program. We believe it is imperative that your risk management committee establish the "risk appetite" level for your business and work with senior management to ensure that the requisite processes and controls are in place and used to minimize your corporate risk.

If your company is complex, faces significant risks across many business functions, and has the means, you also may want to consider investing in a CRO to provide the strategic and operational management of your company's formal risk management program. Those companies that have made that type of investment generally reap positive rewards.

4.0

BUILD YOUR STRATEGY

He who defends everything defends nothing.
Frederick the Great[1]

4.1 HOW MUCH "CYBERSECURITY" DO I NEED?

Can you defend against every cybersecurity threat that exists? We bet that you've met at least one smooth talking salesperson who claims to have a product suite designed to protect you from every cyber threat in existence. Don't believe it and keep your hand on your wallet! If you try to defend against everything, you will drain your precious resources and still face gaps in your coverage. The key to a great cybersecurity program is having the right strategy to manage your risk.

Earlier chapters of this book may have concerned or perhaps even frightened you regarding the prospect of a cyber attack or incident; as well, they should. After all, there are a seemingly endless number of cybersecurity threats, threat sources, and vulnerabilities that can cause devastating impacts upon your business. Bad actors seeking to exploit you and your business potentially include nation-states, hacktivists, hackers, and even

[1] Michael Keane, *Dictionary of Modern Strategy and Tactics*, Naval Institute Press, Annapolis, MD, 2005, p. 57.

Cybersecurity for Executives: A Practical Guide, First Edition. Gregory J. Touhill and C. Joseph Touhill.
© 2014 The American Institute of Chemical Engineers, Inc. Published 2014 by John Wiley & Sons, Inc.

your own employees. Attack venues include emails, web pages, thumb drives, laptop Wi-Fi connections, and even your own cell phone. You and your company are at risk. How do you make you and your information bulletproof against all the potential threats?

You can't.

Why? Because you likely don't need to nor have the resources to do so. Instead, you have to identify your risks and manage them consistent with your corporate strategy and the associated risk appetite that drives the strategy.

To be effective, cybersecurity needs to be an integral part of your corporate strategy. It needs to be a consideration in your corporate plans, policies, and procedures. It needs to be "baked in" to everything you and your company do. Unfortunately, many potential cybersecurity solutions can be costly and may not be affordable to many businesses, especially small ones. You will have to make some tough decisions to find the right fit for your particular business.

We have a client who is just starting their business and rightfully asks, "How much do I need to invest in cybersecurity?" That's a terrific question and one you probably are asking yourself right now. In fact, that is a question that every executive ought to ask when evaluating their potential cybersecurity investments and solutions.

Our client is a small business but with demonstrated potential for rapid growth. They have almost all of their capital invested in their personnel (i.e., salaries and benefits), overhead (e.g., rent, facilities maintenance, administrative costs, etc.), and IT (e.g., computers, network hardware, software, and telecommunications). Because they are a start-up, they don't have a "fist full of dollars" to invest heavily in cybersecurity, yet they want to invest well. What are their next steps?

The answer lies in their corporate strategy.

Strategy is about making the right choices to meet your objective. Strategy is derived from the Greek word referring to a military general—"strategos," which combines the words "stratos" (the army) and "ago" (to lead).[2] Generals know the strategy of their organization inside and out and communicate it to their troops in unmistakable terms that are easy to understand. Generals ensure that plans are developed and they lead the execution of those plans in support of the strategy. Leadership is at the heart of every successfully executed strategy. Remember the old aphorism in the military, "lead, follow, or get out of the way."

There are numerous examples of great generals whose strategy was instrumental in leading to success. For example, General Ulysses S. Grant arguably turned the tide of the American Civil War upon his appointment as commander of all Union armies. Grant abandoned the previous strategy of trying to capture the rebel capital of Richmond in an effort to get the rebels to capitulate. He recognized the rebel center of gravity to be their military forces, not their capital (which had already moved from Montgomery, Alabama, to Richmond, Virginia, and could just as easily move somewhere else), and focused upon destroying the South's military capability to fight in order to force them to surrender. Grant, with the full support of President Lincoln, implemented his strategy and clearly communicated it to his subordinates. His approach was epitomized by his phrase: "The art

---

[2] http://www.oxforddictionaries.com/us/definition/american_english/stratagem?q=Stratagem. Accessed September 17, 2013.

of war is simple enough. Find out where your enemy is. Get at him as soon as you can. Strike at him as hard as you can and as often as you can, and keep moving on."[3] Grant's leadership and vision was decisive and led to Union victory. Is your corporate strategy that clear? Do your employees understand it? Do your employees know how cybersecurity supports your corporate strategy?

There is a difference between strategy and strategic planning. We believe strategy drives planning and plans implement strategy. We are disappointed when we hear fellow executives talk about their marketing strategies, their investment strategies, their engagement strategies, their cybersecurity strategies, etc. We contend there is only *one* strategy in your business, not an amalgamation of disparate strategies acting in concert or potentially in competition. Everything else, including marketing, investments, engagements, and even cybersecurity, is done in support of the overall strategy. We are not alone in this belief. For example, Roger Martin, writing in the *Harvard Business Review Blog Network*, says, "…strategy is a singular thing; there is one strategy for a given business—not a set of strategies."[4]

Determining "how much cybersecurity" you need is indeed reliant upon your strategy. Your strategy should provide a clear picture of what you want to do and how to get there. According to Jack Welch, former CEO of General Electric, "In real life, strategy is actually very straightforward. You pick a general direction and implement like hell."[5]

4.2 THE MECHANICS OF BUILDING YOUR STRATEGY

Strategies assist organizations in establishing priorities. The great ones are simple, well documented, and easy to understand; are based on the current situation, with an eye to the future; and are given enough time to bear fruit. Remember, nobody has confidence in a "Five-Year Plan" that changes every six months!

You don't have to be fancy in building your strategy. Just remember to focus on asking the following key questions:

- Where are we now?
- What do we have to work with?
- Where do we want to be?
- How do we get there?

The process of creating your strategy can be very expensive and time-consuming if you don't manage and control it well. We contend that crafting a strategy is a human process, not an automated one. Unfortunately, it is too easy to spend an inordinate amount of precious time and resources chasing possibilities rather than probabilities. Do not fall victim to paralysis by analysis. The time to build your strategy should take days, not

---

[3] http://www.u-s-history.com/pages/h102.html. Accessed September 17, 2013.

[4] Roger Martin, *"Don't Let Strategy Become Planning,"* February 5, 2013, www.blogs.hbr.org/2013/02/dont-let-strategy-become-plann/. Accessed on September 17, 2013.

[5] Jack Welch, Winning, Harper Business, April 5, 2005, p. 165.

months. Your leadership is essential to keep the team focused on the strategic view, answering the key questions above, and to avoid wandering too far "into the weeds" thus losing sight of the "big" picture.

Incorporating cybersecurity into your strategy is essential in today's information-enabled society but often is an afterthought in many strategies. You can't afford to ignore cybersecurity as you build or update your next strategy. As you go through the process to build your strategy, we recommend you get input from a diverse variety of sources. Gather the most creative and innovative team you can. When you charter your strategy team, explicitly tell them that you want them to consider cybersecurity as an essential part of their thought process; tell them you want to protect the business's vital information. Share your thoughts with them so they have a clear idea of your initial direction. Otherwise, what they come up with may drive you and them back to the drawing board.

Building your strategy is a team effort, and in addition to your senior executives, you will benefit adding high-performing employees to your strategy team. "Disruptive thinkers," that is, those who look at things from a different direction, are invaluable and ought to be part of your team. Consider also including technical experts such as those who design and/or maintain your software and those who sustain and defend your networks. We recommend you consider including a trusted technical expert who maintains credentials such as a "Certified Ethical Hacker"[6]. This individual may prove valuable in evaluating cybersecurity strengths and weaknesses in your strategy. When creating a strategy, you are making choices about the direction your business will go. Don't limit yourself by asking the same folks who got you where you are now to recommend where you should go next. Broaden your aperture and include those who don't fit your standard mold.

As you launch your strategy building, focus on the four key questions we provided earlier. It is easy to stray away from strategic thought (i.e., "why" do things) and dive into tactical matters (i.e., "how" to do things). Don't drift away from the reason you create a strategy: that is, to chart the course for your business to follow. There will be plenty of time to focus on the details of implementation after you have settled on what you want to accomplish.

Your organization needs to know where it is going and your strategy sets the compass. Don't assume that everyone in your business knows what the company strategy is and what direction you are going. Unless you clearly define your strategy, keep it simple and easy to understand, and communicate it well, your workforce may find itself doing things they believe are productive but are not moving your business in the direction you want. This is akin to when Yogi Berra and his wife Carmen were out driving one day when Carmen told him she thought they were lost. "Yeah," he supposedly said, "but we're making good time!"

---

[6] *Certified Ethical Hacker* is a professional certification provided by the International Council of E-Commerce Consultants (EC-Council). Organizations often employ certified ethical hackers to attempt to penetrate networks and/or computer systems, using the same methods as a hacker, for the purpose of finding and fixing computer security vulnerabilities. Such penetration testing helps identify weaknesses in an organization's security posture so they can be fixed before they can be exploited.

Do you know where you are? Do you know what you have to work with? Do you know where you are going? How are you going to get there? Unlike the stereotypical male driver, who refuses to ask questions when lost on a drive, you need to ask the right questions in order to build the best strategy possible.

Our intent in this book is not to focus upon overall strategy development. There are many excellent references that deal with that subject. As we said earlier, cybersecurity must be incorporated within the overall strategy. Hence, the sections that follow help to guide strategy development from a cybersecurity perspective.

4.2.1 Where are We Now?

Ask yourself where your business is right now. Are you a market leader? Are you profitable? Do you have any patents or copyrights that give you an advantage over competitors? Is your brand respected? Is your company on the rise or falling? Have hackers penetrated your defenses? Is it possible they have and you don't know? Are bad actors spear phishing you or your employees? Are there people who would benefit from having access to your information? What is your cybersecurity posture? Your organization's strengths are the resources and capabilities that you leverage to generate a competitive advantage.

Many organizations answer the question, "Where Are We Now?" through the lens of a technique called the SWOT analysis. SWOT analysis takes a look at the organizations' *strengths, weaknesses, opportunities, and threats.* The first two are primarily internally focused, while the latter are primarily externally focused. You may find SWOT analysis helpful in characterizing where your company stands in regard to its cybersecurity posture as you create your broader corporate strategy.

4.2.1.1 SWOT Analysis: Strengths. Strengths define what the organization is good at and is doing well. We've found that taking a focused view of your strengths can be very helpful in setting the stage for determining next steps to take in the future. In fact, you may find that as you assess your strengths, a perceived strength may in fact be a weakness or liability, or vice versa.

Cybersecurity traditionally is not a focus area for many businesses conducting SWOT analysis, but in today's information-enabled environment, it ought to be. Here are some cyber-related topics you ought to consider when evaluating your strengths:

- Presence: What is your Internet presence? Are you properly presented to the public to maintain a competitive advantage? Is your web page clear and understandable? Does your web presence generate business or enhance your reputation? Is the information you present on your web page valued, timely, and relevant? How many people view your web page? Of those who view your web page, how many use it to conduct business with you? Do you have positive control over all access points to your information?
- Information: Have you identified your critical information? Do you understand its value? Is it protected with effective controls, procedures, personnel, and technology? Is it insured against loss, damage, or denial?

- Plans: Do your plans produce the cybersecurity results you desire? Are your plans feasible, acceptable, suitable, and affordable (FASA) in support of your corporate strategy? Do your plans adequately address the cybersecurity risks that you've identified? Do your employees understand and implement your plans well? Do you have an effective metrics program that measures the effectiveness of your plans?

- Policies: Do you have the appropriate policies to support your strategy and plans? Are your policies realistic and enforceable? Are they enforced? Are your policies followed? Are employees held accountable when they fail to comply with policies?

- Processes: Do you have well-documented and followed processes that yield the results you want? Are your processes efficient and effective? Are your processes easy to understand and add value to the business? Do your employees follow the procedures? Do you have a formal change management process that encourages innovation and is responsive to changes in technology and/or risk profiles?

- Talent: Do you have the right people in the right jobs? Are your people qualified in all the tasks you demand of them? Do they have appropriate training? Are they properly credentialed and certified in their specialties? Do they have the aptitude to learn new skills as technology advances?

- Technology: Do you have the right technology to be competitive in your market? Is it state of the art? Is it properly configured and operating at optimal capability? Does it complement your processes? Is it effective? Is it efficient? Is it secure?

- Resources: Do you have the resources to accomplish all your critical tasks? Do you have any unfunded requirements? Do you have the resources to recapitalize your information systems on an industry best practice cycle?[7] Do you have the resources to implement new technologies to enhance your business? Have you budgeted for unforeseen technologies that can improve your competitive advantage?

When evaluating your strengths, look at what you do best. Do you do it better than anyone else? Are you more efficient? How do you stack up with your competitors? Ask others what they think are your strengths. What do your competitors think your strengths are? What about your clients and customers? A thorough and honest approach will yield the most complete picture of your strengths.

4.2.1.2 SWOT Analysis: Weaknesses.

4.2.1.2 SWOT Analysis: Weaknesses. Weaknesses define what the organization is not good at and what is not going well. After reading Chapters 2.0 and 3.0, you most likely have been thinking about your vulnerabilities as weaknesses. We hope so. You also may have reviewed your strengths and found that what you formerly perceived

---

[7] Recapitalization of hardware and software varies per device and program. For example, desktop computers are considered for replacement after three to four years of use. Printers have a longer life cycle and are routinely replaced after six years or earlier if significant vulnerabilities are found in their software. Software, such as operating systems or applications, generally is recapitalized when the next version of the code is available and has been successfully tested.

as strengths may indeed be weaknesses. That happens. Be glad that you've discovered them before catastrophe strikes.

When assessing your weaknesses, we've found that going through the list of topics in your strength analysis is worthwhile to assess and compare your weaknesses as well. One can argue that when you answer "No" to one of the questions in the strength analysis, it is a candidate for consideration as a weakness. For example, the absence of a certain strength could be a weakness.

In addition to looking at the topics from the strength analysis, the following are some cyber-related topics you ought to consider when evaluating your weaknesses:

- <u>Priorities</u>: Do you have your priorities right? Are you doing the right things at the right times? Are your resources aligned to the highest priorities?
- <u>Risk analysis</u>: Is your risk analysis complete? What cybersecurity weaknesses did your risk analysis yield? Do you have a complete list of your threats, threat sources, and vulnerabilities? Do you know your risk profile?
- <u>Third-party assessment</u>: Have you conducted a third-party analysis of the weaknesses of your cybersecurity posture? Have you had a penetration test of your network and systems? If so, what were the findings? (Note: this may yield a strength if the findings are favorable.)
- <u>Resources</u>: Are your cybersecurity controls a good value? Do your cybersecurity controls have a high cost compared to the value of information they are protecting?

Your analysis of your weaknesses should help you identify what you should improve and what you should avoid. Be honest when looking at your weaknesses. Consider asking others what they perceive to be your weaknesses. Benchmark against competitors and those who exhibit best industry practices. You may just find your Achilles heel.

4.2.1.3 SWOT Analysis: Opportunities.
Opportunities come in many flavors. They may come from changes in technology or the marketplace. Government policies may change thus opening up new opportunities, such as those introduced by changes in tariffs or export controls. Potential partnerships may present opportunities to expand capabilities and open new markets. Trends, social norms, population changes, resource scarcity, and other factors all may present opportunities that you and your business may leverage to improve your business posture.

Your analysis of your strengths and weaknesses may present the avenue to opportunity. For example, your analysis of your strengths may expose you to opportunities you may not have thought possible. Similarly, you may be able to exploit some opportunities if you eliminate the weaknesses you've identified.

Here are some cyber-related topics you ought to consider when evaluating your potential opportunities:

- <u>Presence</u>: Are you reaching the right people and markets? Can you improve your market presence by expanding your web presence? Can you improve your web site by incorporating videos, widgets, and other technologies to better present your capabilities and products?

- Partnerships: Do other entities have access to technologies or capabilities that could jump start your business and open opportunities? Are you ready, willing, and able to partner (or maybe even acquire or merge) with them to secure access to these technologies or capabilities?
- Data sharing: As we've discussed, information is a valuable commodity, and gaining access to valued information may present great opportunity for your company. Numerous companies already recognize this and often subscribe to services such as those whose predictive analytics expose shopping and Internet search patterns, cluster buying (e.g., "Customers who bought this also looked at…"), and market trends. Other companies find that when they institute peer-to-peer data sharing relationships, they can electronically share information through electronic data interchanges (EDIs) that speed transactions, reduce manpower and overhead costs, and increase the accuracy and precision of their business together. Are there opportunities to reduce your costs and expand your markets if you improve your access and management of information and do so securely?
- Payments: Are there opportunities to improve your accounts receivable capabilities by leveraging the latest in electronic payment capabilities? Can you reduce overhead costs and improve precision and accuracy? Can you better identify delinquent accounts through automation? Can you implement positive changes while maintaining a satisfactory security posture?

Opportunities should be positive but if you don't account for cyber-based threats, they may turn from potential victories to probable disasters. You should evaluate every anticipated opportunity through the lens of a hacker. How is your potential opportunity exposed to cyber risk? Can that risk be mitigated, accepted, avoided, or transferred so that it remains a viable opportunity? Opportunities that appear to be too good to be true probably are. Do your homework and ensure your opportunity evaluation is done with cyber-based risk in mind.

4.2.1.4 SWOT Analysis: Threats.
We spent considerable time discussing threats in earlier chapters, so we believe that you understand and appreciate the cyber-based threats you and your business face. Although we won't rehash the previous discussion, we believe the following will be helpful as you analyze threats as part of your SWOT analysis:

- Threats occur at home too: Some people believe their business's cyber protections are terrific and adequately protect them against cyber threats. That often is the case. Unfortunately, many people bring work home with them and do that work on home systems that do not share the same level of protection as business environments. People who do their work at home and bring it back to the business often introduce malicious code and other threats that expose you and your business to risk. The good news is that you can mitigate this through educating your workforce, offering antimalware programs to employees to install on their home systems, or even providing the employee a company-owned and configured device from which they can work securely from home. The bad news is that there always will be a risk that one of your employees will bring something in from home (i.e., a viruslike influenza or a computer virus such as the Conficker worm) that threatens your business. Plan for that and defend appropriately.

- <u>Technology changes</u>: Is there a change in technology that may render your current capabilities obsolete and drive a recapitalization effort? If you upgrade to that new operating system, do you need to buy new hardware to support it? If you don't keep current with technology, is your market position threatened? What are your competitors doing?

- <u>Standards</u>: Are there any new or anticipated standards that you need to comply with? Does compliance drive changes in your risk or fiscal posture?

- <u>Benchmarking</u>: Here's where checking to see what your competitors are doing is noteworthy. Are your competitors exposed to the same threats? Have they taken proactive measures to defend against cybersecurity risks? Are you leading, following, or preparing to get out of the way?

Threats can seriously threaten you and your business. Identifying and assessing your threats as part of your risk assessment is essential. In this information-enabled market, cybersecurity risks likely present the greatest number and most serious threats to your business. Invest the necessary time and treasure to find and address your threats properly.

After you have completed your SWOT analysis, you will be well postured to evaluate next steps. Some organizations are tempted to jump toward those opportunities that look like they *may* make the most money. Be careful. While some paths initially may appear attractive, those strategies with the greatest chance of enduring success and profit are those which best align the organization's strengths and opportunities while controlling and managing threats and weaknesses.

4.2.2 What Do We Have to Work With?

After you've determined where you are, it is time to assess what you have to work with. Some people argue that this ought to be part of the discussion of "Where Are We Now?" and we acknowledge there is merit in that argument but we contend that focus on the resources available is very helpful in setting the stage for determining your strategy and its implementation.

Determining what you have to work with includes your technical, human, and financial resources. It also includes your information. Don't forget that your intellectual property and trade secrets represent powerful assets that give you a competitive advantage over your peers. They should be included in your calculus.

As you survey your resources, we suggest you address the following cyber-related topics:

- <u>Information</u>: Do you have intellectual property or trade secrets that give you a competitive advantage in the marketplace? If so, will this information retain its value over time? When or under what conditions will it lose its value? How does your intellectual property or trade secrets contribute to your organization's success?

- Plans: Are your existing plans adequate? Have they been followed? Are they producing the success you anticipated? Do you know what your critical information is, understand its value, and have a plan to protect it? Do you have policies and procedures that accompany your plans? Do they follow industry best practices? Do your employees follow them? Do you have accurate and timely metrics that give you visibility into the effectiveness of your plans, policies, and procedures?

- Technology: Is your technology current (up to date)? Does it meet the needs of your anticipated future? Do you expect you'll need major technology upgrades soon? If so, do you have reasonable cost estimates and the funding already secured? Do you have antivirus and other software to thwart common threats? Is it kept current? Do you employ boundary protection capabilities such as firewalls and intrusion detection systems? Do you protect your information through encryption? Do you have the means to detect and thwart both insider and external threats to your information?

- Personnel: Do you have the right type and amount of personnel? Do your technical personnel have the proper certifications and skills to accomplish their assigned work? What type of keep-it-current training is expected to maintain their skills or meet industry standards? Are your people "world class," average, or "not up-to-snuff"? Do you need to upgrade your workforce? Is there a ready talent pool of people with the skills you need available? Are universities and other means of training and educating your workforce available?

- Finances: What is your current financial situation? What does next year look like? If you don't change anything, what is your forecast for the next five years? What are your "must-pay" bills? What are things you believe you really need to invest in but cannot due to priorities and limited resources. Are there any issues that may spur changes to your forecasts or available revenue (e.g., tax changes, environmental concerns, market changes, etc.)? Do you have sufficient liquidity and cash on hand to address "surprises" that may drive unexpected expenditures?

Defining what you have to work with is critical as you evaluate potential next steps in building your strategy. A strategy without the resources to implement it is not feasible and destined to fail. Remember, it is important to "know yourself," even when building your strategy.

4.2.3 Where Do We Want to Be?

Where do you want your business to be? Perhaps more importantly, where do your employees think you want to be? Are your visions of the future and theirs congruent or divergent? How do you articulate it to be meaningful, measurable, and understood by all? How do you convince your employees to "buy in" to the vision?

We think Bill Gates did a great job identifying where he wanted Microsoft to be when he expressed Microsoft's vision as "A computer on every desk and in every home, all running Microsoft software."[8] Such a succinct yet clear vision of the future led Microsoft

---

[8] Farhad Manjoo, *"Steve Ballmer Needs To Be Replaced as Microsoft CEO...by Bill Gates"*, www.fastcompany.com/1647014/Microsoft-needs-bill-gates-back. Accessed on September 17, 2013.

to phenomenal success and market domination (and Gates' ultimate recognition as the world's richest man!). Do you think that Microsoft employees understood where Gates wanted the company to go? Do you think his shareholders did? What about potential investors and partners? We contend that the success of Microsoft was directly attributed to the vision of putting a computer on every desk and in every home, all running Microsoft software. Gates and his employees relentlessly pursued this vision, and more than 30 years after its founding, it remains the largest software company in the world.

Similar visions of the future abound in successful businesses. We especially like the Amazon team's statement of where they want to be: "Our vision is to be the earth's most customer-centric company; to build a place where people can come to find and discover anything they might want to buy online."[9] Do you think that Amazon is following this vision as the core of their strategy? Is it clear and simple? Is it easy to remember? Do you think they can measure their success based on this vision?

We believe the Microsoft and Amazon statements indeed are powerful declarations of where the company wants to be and are valuable examples to emulate. They share three traits that we believe are essential in determining where you want to be: they are *simple*, they are *measurable*, and they are *memorable*.

Simplicity is a trait intrinsically linked with success. No doubt, you have heard the phrase Keep It Simple, Stupid (aka K-I-S-S). Simplicity is the ability to distill complex concepts into something that appears plain, natural, and easy to understand. Simplicity in articulating your vision of where you want to be has a multiplying effect. If you confuse your employees and customers with flowery prose and lengthy vision and mission statements, they are unlikely to understand what you want and what you do. Keep it simple.

Bill Gates is able to measure how successful his vision of the future is by looking at how many computers and Microsoft software licenses were sold. Good visions have the enviable ability to measure their effects. Great visions have simple measures that quickly and clearly tell you whether your strategy is yielding the results you want or is it a dud.

Declaring where you want to go is meaningless unless your people act upon your declaration. If you make your vision so long and complex that your employees can't easily remember it, they are less inclined to efficiently and effectively go in the direction you want.

You can and should consider cybersecurity in charting your course for the future. The Microsoft vision did not originally emphasize security, yet after significant feedback from clients as well as employees, Mr. Gates recognized the need to incorporate it in the company's ethos as part of his vision of the future. In January, 2002, he released a memo to all Microsoft employees, stating that "Trustworthy computing is the highest priority for all the work we are doing."[10]

Your future depends upon trustworthy computing too. Regardless of your business sector, reliance on IT is a fact of life and exposes you and your organization to countless risks. Unless you envision a future that does not include information technologies

---

[9] Frequently Asked Questions, amazon.com. Accessed on September 17, 2013.

[10] Bill Gates, "Trustworthy Computing," January 15, 2002, http://www.computerbytesman.com/security/bills-memo.htm. Accessed on September 17, 2013.

(which is highly unlikely in today's business environment), incorporating cybersecurity into your decisions on where you want to be now and in the future is essential. Attainment of a cybersecurity vision that melds seamlessly with your strategic plan enhances value, reduces risk, and sets the stage for success.

As you contemplate where you want to be, we submit that the following are questions you should be asking during your deliberations:

- Value: Is there value associated with our organization being one that invests in cybersecurity industry best practices? Do we have a good handle on the value of our information? Is our information our most valued asset? Is our brand enhanced when we invest in cybersecurity? Are our intellectual property and trade secrets better protected? Are we keeping up with our competitors, falling behind, or sprinting forward when it comes to cybersecurity? Do we get a good return on our investment?

- Risk management: How does cybersecurity fit into our risk management program? Is it a prime driver or a subordinate task? What are the consequences of not emphasizing cybersecurity? What will be the impact of a breach? What will be the impact of a loss, damage, or denial associated with our critical information? How much risk will we accept when it comes to cybersecurity?

- Effectiveness: Will we be effective in our desired future without a cybersecurity program that manages and controls cyber-based risks? Is secure information an asset valued by our shareholders, potential investors, and business partners? Do we need cybersecurity to survive and thrive?

- Competencies: Is cybersecurity something we want to execute in-house or do we want to outsource it? Is it one of our core competencies? Should it be? What risks are associated with outsourcing our cybersecurity program elements?

Where you want to go depends on many factors, including the sector you are in, the resources you have, the opportunities you seek to explore, and how much risk you want to assume. We contend that the future of your business will continue to be reliant on information technologies (and probably even more so than it is today). Further, we contend that your information has an intrinsic value that enables you to maintain your competitive advantage. This information must be secure from theft, tampering, unauthorized access, denial, or destruction. Failure to do so jeopardizes your business's ability to achieve its goals and objectives.

Great strategies start with a great vision; are simple, well documented, and easy to understand and remember; are based on the current situation with an eye to the future; and are given enough time to bear fruit. When the vision of the company is easy to remember, you and your employees are more likely to focus on the vision and incorporate it in to everything you do. It almost becomes second nature to include your vision into daily activities. Practicing sound cybersecurity principles should be second nature as well. In today's information-enabled age, your vision of the future needs to include cybersecurity as a critical component of your vision.

4.2.4 How Do We Get There?

By now, you have a good idea of where you are, what you have, and where you want to be. You may have a terrific "vision statement" that is simple, well documented, easy to understand and remember, and is based on the current situation with an eye to the future. That's the easy part. You have set the direction for the future. Now, you have to figure out how to get to that future, a challenge that (to paraphrase Sun Tzu) is the "acme of all skill." You need to have a plan.

Many organizations use a vision statement, a mission statement, and core values as touchstones of their strategy. Vision statements give the high-level "strategic" view of what you want to achieve. Mission statements are general statements of how you will achieve your vision. They generally are short, concise statements that start with the word "To" and capture the essence of the vision. One of the most famous of all is, "To boldly go where no one has gone before."[11] Core values define the principles you will follow as you conduct the activities while carrying out the vision and mission.

Each one of these documents can and should reinforce your commitment to cybersecurity as part of your strategy. For example, do you believe Bill Gates' vision statement is improved if it is articulated as "A computer on every desk and in every home, all **securely** running **trusted** Microsoft software"?

What about mission statements? Let's take a look at Bristol-Myers Squibb, a great Fortune 500 company that manufactures pharmaceuticals and health-care products. Their mission is, "To discover, develop and deliver innovative medicines that help patients prevail over serious diseases." What happens if a hacker gets into their databases and alters the formula for a pharmaceutical or tampers with the manufacturing process that turns a helpful product to something potentially harmful? Can we reinforce our cybersecurity vision in this mission statement? Absolutely! We submit the simple addition of as little as a single word can provide the emphasis that can inspire the desired outcomes your vision seeks, "To discover, develop and **securely** deliver innovative medicines that help patients prevail over serious diseases."

Core values too ought to incorporate cybersecurity principles. Core values reveal a lot about the senior executives at an organization and how they manage their business relationships and functions. Employees, prospective investors, and potential clients often use the core values of an organization as a strong discriminating factor that shapes their decisions.

We believe one of the best core values we've seen is Google's "Don't Be Evil."[12] It is extremely simple and is extremely relevant considering the vast amount of information they have access to and control. Not only is it an out-of-the-box declaration of a core value, it can even be measured. For example, how many negative articles are

---

[11] Star Trek, launched in 1966, the franchise originally chronicled the five-year mission of the Starship Enterprise to explore strange new worlds, to seek out new life and new civilizations, and to boldly go where no "man" has gone before. In 1987, when the franchise returned to television with the Star Trek: The Next Generation series, the word man was replaced by the word "one."

[12] Ten Things We Know To Be True, http://www.google.com/about/company/philosophy/. Accessed on September 17, 2013.

written about Google's privacy and customer policies every year? They are the world's largest search engine, so they can clearly search for and mine that information and compare it over time, creating a metric to measure the perception of "evil" activities.

Do your core values state that your organization is committed to protecting client and corporate information through cyber best practices? Do your core values include a commitment to the ethical use of the Internet? Core values are powerful symbols of what the organization represents and where it is committed.

4.2.5 Goals and Objectives

Once you've defined your vision, mission, and core values, your next step is to use them as the foundation to create the goals and objectives to achieve your vision.

Goals and objectives are at the heart of planning. There are numerous definitions regarding goals and objectives. Some of these have goals subordinate to objectives and vice versa. For our purposes, we define them as:

- Goal: Statements of what you want to accomplish. For example, "Score a touchdown."
- Objectives: Specific, measurable, and time-relevant statements of what is going to be achieved and when. For example, "Get a first down every offensive series."

While there are numerous ways to define goals and objectives, we've found there are also many criteria to gauge their merit. We've found that **goals** are most effective when they are *feasible, acceptable, suitable, and affordable*. Likewise, **objectives** are strongest when they are *specific, measurable, achievable, realistic, and timely* (SMART).

Goals need to be feasible, especially when it comes to cybersecurity. If it isn't possible to achieve, why make it a top priority and dilute resources investing in something that isn't possible when other potentially better returns are available elsewhere? The senior author has considerable experience related to the topic of multilevel security. Multilevel security (i.e., being able to operate securely in unclassified, secret, and top secret modes on the same computer) has eluded the military for decades despite it being an operational requirement. Several technical organizations have made its creation a top goal. We know of a three-star Air Force general who rose from an Airman Basic (the lowest enlisted rank) to the senior officer in his field who commented about multilevel security, "Over 40 years ago, when I was a young Airman with no stripes, multi-level security was right around the corner. As I conclude my career it remains there, right around the corner." Multilevel security remains a goal, but only now with the advent of advanced processors and software does it appear to be feasible for implementation in the near future. Are your goals feasible? Is achievement of the goal possible or a possible source of frustration for your team? We recommend you set goals and objectives that will challenge your team and advance your organization. Don't lower the bar by making your goals easy to attain, but don't make them impossible to achieve either. Ensure they are feasible.

Establishing goals that are acceptable is essential in order to be effective and efficient. Determine whether a goal is acceptable by comparing it to your core values and vision. If the goal is in conflict with either your vision or values, it is not worthy of your organization and should not be a goal. For example, if your organization is committed to progressive labor practices that pay a fair living wage to all employees and a goal is introduced to increase foreign outsourcing to reduce overhead costs because they have significantly cheaper labor, you may find yourself digging deep to ensure that there is no collision between your values and your proposed goal.

Similarly, your goals need to be suitable, that is, does the goal fit with the vision and mission of the organization? One of our favorite examples of a goal not fitting the mission of the organization is found in a case study of behavioral economics the author was introduced to during postgraduate studies at Harvard University. In the study conducted by Camerer, Babcock, Loewenstein, and Thaler,[13] the researchers investigated why it was so hard to get a cab in New York City on a rainy day. Initially, they thought the problem was one of demand, that is, because it was raining, more people wanted cabs, but their investigation produced unexpected results. Instead of just being a demand issue, it was also a supply issue as there actually were fewer cabs available during rainy days. Why? The researchers found that the cab drivers had specific goals to earn double the amount it cost them to operate their cabs during a 12 hour period. Once they had achieved their goal, they pulled in to the garage and ended their days. So because they carried more fares during rainy days and earned money faster, the number of available cabs diminished the longer it rained! Just think what the impact would be if the goal was to make as much money as you could during the 12 hour period rather than the "double the amount it costs to operate the cab" goal. We know a lot of soggy people in New York City who'd appreciate more cabs on the streets at the end of the business day. If you were the CEO of the cab company, would you think the cab driver's goal was congruent with the vision and mission of your company? Was it suitable? How about your own organization goals? Do they fit with your vision and mission?

Finally, your goals need to be affordable. Setting goals that aren't affordable can be disastrous. We know of a company that established the goal to expand their production and increase their volume. Sounds great, right? It is until you take into account that their manufacturing facility was already running at 80% capacity. Pursuing the goal required them to invest either in expansion, recapitalization, or acquisition of additional capabilities. Executives at the company anticipated they could get financing to expand their existing production facility and open a second to increase their capacity and pursued their goal with vigor. Their marketing department aggressively moved forward and secured orders for the additional products, offering discounted prices as incentives for future business. All looked good until unforeseen events occurred that made the goal unaffordable. First, the expected financing failed to materialize as a downturn in the economy caused many of the prospective investors to back out of the planned expansion. Second, the planned expansion required a highly skilled workforce, and there were not enough of the skilled technicians available to meet the demand. The costs and time to train new employees to the

[13] Camerer C., L. Babcock, G. Loewenstein, and R. Thaler, Labor supply of New York City cabdrivers: One day at a time. *The Quarterly Journal of Economics* 112(2, In Memory of Amos Tversky (1937–1996)): 407–441, 1997. Cited in "Behavioral Economics" by Colin Camerer, https://www.google.com/url?sa=t&rct=j&q=&esrc=s&source=web&cd=5&ved=0CEYQFjAE&url=http%3A%2F%2Fwww3.nd.edu%2F~pweithma%2Fjustice_seminar%2FBehavioral%2520Economics%2FCamerer%2520(Behavioral%2520Economics).docx&ei=sEE8UpbOOan-2gWh2IGIBg&usg=AFQjCNGNlg_hAcuEFkVPG8bA15gE5KBNtA. Accessed on September 17, 2013.

necessary level were too expensive and would take too long to honor the contracts that were quickly entering the company. Sadly, the once prosperous and respected company was unable to meet its commitments, went into reorganization, and executed a complete change of leadership. Are your goals affordable? Do they have the resiliency to maintain their viability in times of stress? Can you afford to make them your goals?

Goals should be focused on the important aspects of your business. Great organizations limit the amount of goals to maintain focus on their vision and mission; too many goals muddy the waters, confuse the workforce, and may distract management. Make certain your goals are well synchronized as well. You want goals that complement each other in pursuit of the vision, not contradict or compete against each other.

What are your cybersecurity goals? If you have some, are they feasible, acceptable, suitable and affordable? Do they complement your business function goals? They should. Some of our clients ask us to help them create cybersecurity goals to improve their corporate culture and focus their business on cybersecurity matters. Here are some samples that may help you as you and your senior leadership team contemplates your goals:

- Reduce exposure to cyber risks
- Maintain ability to detect, respond, and recover
- Enable secure information exchange anytime, anywhere by authorized users
- Maintain the security of information and IT infrastructure
- Provide IT systems that are reliable and secure
- Ensure compliance with laws and regulations
- Control all Internet connections
- Maintain positive control over all information
- Ensure all employees are trained in cybersecurity best practices

If goals are the skeleton of your organization, then your objectives are the muscles that make them move. Objectives are the specific, measurable, time-relevant statements of what is going to be achieved and when. They are strongest when they follow the SMART model (i.e., *specific, measurable, achievable, realistic, and timely*).

Let's use our first example goal as an exemplar to show how you can create SMART cybersecurity objectives. We've established a goal to reduce our exposure to cyber risks. We'll assume that you studied Chapters 2.0 and 3.0 well and conducted your cybersecurity risk analysis. Based on your risk analysis and business objectives, the following three objectives are proposed:

Goal: Reduce exposure to cyber risks

    Objective 1: Implement an intrusion detection system by the second quarter of this fiscal year

    Objective 2: Install security patches to software within 24 hours of their release from trusted sources

    Objective 3: Conduct initial cybersecurity training for 100% of new employees within three days of employment and before granting access to information systems

What do you think? Are they SMART? While they are fairly generic examples, they indeed fit the model and are representative of objectives that we've met during our professional careers. Each one can be accompanied by metrics that can show progress toward achieving the objective in support of the goal. Are they representative of objectives you'd like to see in your organization? If so, make sure to establish cybersecurity goals and objectives as you implement your strategy.

Vision, mission statements, core values, goals, and objectives are all essential elements of great modern strategies. But they quickly become worthless unless they are acted upon. You need a plan.

Planning begins with the end state in mind, providing a unifying purpose around which actions and resources are focused.[14] Your strategy, its vision, mission, goals, and objectives set the end state for you to build your plan. If you don't have a plan to rally around, you will fail. We'll discuss cybersecurity planning and implementation in subsequent chapters, but it is important to keep in mind during the building of your strategy that you need to create a plan to translate your vision into reality.

4.3 AVOIDING STRATEGY FAILURE

Many strategies fail, even really good ones. Despite sometimes colossal (and expensive!) strategy building and planning efforts, many companies see their strategies and plans end up on shelves gathering dust or condemned to the waste basket as worth less than the paper they are printed on. Will your new strategy be a winning rally point that leads your organization to new success or will it fail miserably?

We've found there are four main reasons why good strategies fail.

4.3.1 Poor Plans, Poor Execution

Even a good strategy can be torpedoed by good execution of crummy plans or crummy execution of good plans.

Take, for example, the case of Netflix. Netflix was a transformational company that became a "must-have" subscription for many lovers of video products. Originally using a model where subscribers could order any movie they wanted via the web and have a DVD delivered, usually within a day or two, and at relatively low prices, Netflix was quickly pushing "brick-and-mortar" video stores out of business. They were attracting new subscribers daily, and their prospects were terrific—until they rolled out a new strategy.

Netflix looked at the future and determined that the days of providing hard copy DVDs and Blu-ray Disks (BRDs) were numbered. While many subscribers were very comfortable with the current method of distributing digital content via DVDs and BRDs, Netflix recognized that DVDs and BRDs were based on technologies that were rapidly becoming obsolete and that the postal fees to distribute them to subscribers were rising quickly. Their data clearly showed that making their video entertainment available to

---

[14] Joint Chiefs of Staff, Joint Publication 5.0 "Joint Operational Planning," August 11, 2011, http://www.dtic. mil/doctrine/new_pubs/jp5_0.pdf, pg I-1, para 1.a. Accessed on September 17, 2013.

consumers via the Internet required less overhead and made content available to consumers much faster. They also realized there was potential in developing their own content rather than being beholden to the many studios and distributors who controlled the video entertainment products they were offering. Netflix had a bold plan that would retain the hard copy DVD/BRD capabilities for a while yet focus on streaming content including original programming owned and produced by Netflix. The concept made a lot of sense and sounded great in the board room.

But then, they rolled out their plan, which called for the company to set up a subsidiary called "Qwikster" that would offer hard copy DVDs/BRDs, while Netflix would offer just streaming content over the Internet. Increases in subscription fees from US $9.99 per month to US $15.99 per month were proposed to cover the cost of the two services—a 60% jump in price.

Subscribers revolted en masse to the planned changes. "Why fix something that isn't broken?" was repeatedly heard regarding the proposed change to the Netflix business model. While most subscribers were riled up by the proposed increases in fees, many quickly zeroed in on the fact that the plan to separate Qwikster from Netflix was grossly inconvenient for subscribers. For example, subscribers would have to use two different independent web sites to search for content rather than a single site that would permit them to choose how they wanted the content delivered. It was wholly unacceptable to the subscriber base and reports indicated as many as a million subscribers quit the service entirely.[15]

Confidence in the company and its leadership plummeted as well. The stock value plunged precipitously going from a high of US $297.35 on the week of July 8, 2011, to a low of US $52.81 on August 3, 2012.[16]

Netflix went into damage control mode. Their CEO apologized, killed the Qwikster plan, and the company scrambled to assure subscribers, investors that they would right the ship. The recovery took nearly 25 months.

Two years later, Netflix's stock price is trading above US $300 per share and hit a high of US $318.18 in mid-September, 2013. Netflix indeed was able to right the ship, but without a major change to their strategy. Rather, they changed their plan. While the Qwikster proposal is dead and buried next to "The New Coke," Netflix's strategy to retain hard copy DVDs as long as possible, dominate streaming video entertainment over the Internet, and develop original content programming has turned out to be a big winner. In fact, Netflix original programming received 14 nominations at the 2013 Emmy awards.

While Netflix was able to overcome a bad plan and achieve success toward their vision, they dodged a bullet. Most companies don't have that kind of luck and would quickly be swamped. How about your organization? With cybersecurity, the stakes may be very high if you implement a bad plan that results in the exposure of your information. How resilient to a bad plan are you? When do you pull the plug on a stinker and regroup? We submit that if you are SMART about things, you'll be in a better position to detect a sinking ship faster so you don't go down with the ship.

[15] Complex Mag, "NetFlix Lost Almost a Million Subscribers," October 25, 2011, http://www.complex.com/tech/2011/10/netflix-lost-almost-a-million-subscribers. Accessed on September 17, 2013.

[16] http://investing.money.msn.com/investments/charts?symbol=nflx#{"zRange":"7","startDate":"2010-9-20", "endDate":"2013-9-20","chartStyle":"mountain","chartCursor":"1","scaleType":"0","yaxisAlign":"right", "mode":"pan"}. Accessed on September 17, 2013.

4.3.2 Lack of Communication

Lack of communication can defeat a strategy. Recall that strategies assist organizations in establishing priorities. The great ones are simple, well-documented, easy to understand and remember, are based on the current situation with an eye to the future, and are given enough time to bear fruit.

Your employees are essential to carrying out your strategy. Do they understand it? Do they know what is expected of them and how their effort supports the strategy of the organization? Do your employees understand they need the training and education necessary to complete their tasks under the new strategy? Do your midlevel and frontline supervisors understand their role in educating and motivating the employees to ensure your plans are implemented properly?

According to a poll conducted by the Computer Technology Industry, lack of communication is the leading cause of failures in IT projects.[17] Communication up and down the chain of command is necessary to ensure that your implementation plan stays on track. Executives need quality, timely, and complete information to make decisions and must communicate those needs to subordinate levels. Midlevel and frontline supervisors need to understand the information needs of their superiors and be able to translate those needs to subordinates. Moreover, they must communicate information from the frontline employees back up the chain to those who need it.

What is the worst that can happen if you have a great strategy and plan yet don't have great communication with it? Perhaps you will face something like what happened with the 1999 NASA project to send the Mars Climate Orbiter to Mars to gather climatic data. The Orbiter, at a cost of US $125 million, traveled over 400 million miles to get to Mars. Upon arrival, it entered orbit 60 miles too low and was destroyed by the Martian atmosphere it wasn't designed to withstand. In the review that followed the failure, it was discovered the design calculations used to place the spacecraft into orbit were made in imperial measures in terms of pounds force. The software team, however, developed the burn control software using a metric measurement whose units were in terms of newtons. Although the resulting error was less than 0.000015%, it was sufficient to prove fatal to the orbiter and its mission.[18]

While lack of communication can kill the implementation of your plans and derail your strategy, exceptional communication can overcome problems when the plan doesn't go right or hits an unexpected roadblock. Staying with the theme of NASA examples, the Apollo 13 mission serves as an outstanding example of how communicating exceptionally well can have the richest of returns. Confronted with a very limited amount of time to figure out a way to jury-rig the Lunar Module to scrub excess carbon dioxide out of the cabin atmosphere to ensure crew survival, the NASA engineering team developed a duct tape-laden method to make a square peg fit a round hole and simple instructions that the crew were able to use to successfully implement the

[17] Linda Rosencrance, "Survey: Poor Communication Causes Most IT Project Failures," March 9, 2007, http://www.computerworld.com/s/article/9012758/Survey_Poor_communication_causes_most_IT_project_failures. Accessed on September 17, 2013.

[18] Alan Jost, What we have here is … failure to communicate, *Crosstalk: The Journal of Defense Software Engineering*, 19, June 2006.

solution.[19] During a meeting with Astronaut Jim Lovell, the author asked about the incident and how the astronauts knew how to build the solution without ever seeing it. Captain Lovell reported that the crew trusted that their colleagues on earth provided instructions that were complete and simple to follow and hoped that his crew "wouldn't screw it up."[20] As we now know, that trust was well founded and the crew returned home safely.

Do you ensure that communications up and down the chain of command in your organization are complete, simple to understand, and timely? Do your people know your information requirements as a senior executive? Have you told them? Have you checked with them to ensure they indeed understand? Likewise, have your checked with them to ensure you understand their information needs? Successful communications go in both directions seamlessly.

G. Bernard Shaw is oft-quoted as saying, "The greatest problem of communication is the illusion that it has been accomplished."[21] Don't just assume that your strategy has been properly communicated throughout your organization. Encourage the use of every available means to communicate your message throughout the organization. Meet with your people to communicate in person. Send emails. Record a video and post it on your web site. Post your vision prominently on the bottom of your correspondence, briefings, and business cards. Hold regular town hall meetings. Make sure your strategy is well known and understood by everyone in your organization, that everyone understands their role in making it a success, and that your vision is continually positively reinforced. Communicate, communicate, communicate!

4.3.3 Resistance to Change

We have always done it this way!

Those are the seven deadliest words to your strategy implementation.

Your strategy is threatened when it comes into conflict with employees who don't want to implement it. Inevitably, there are some employees who will resist change and either consciously or unconsciously sabotage implementation of your strategy. Know that they are out there in your organization. Expect them to serve as potential roadblocks and act early and decisively to bring them around or show them the door. Include educating and motivating those resistant to change as part of your implementation plans. This is especially important when dealing with ITs and cybersecurity.

The author had an experience involving a curmudgeon who was extremely resistant to change. It was in the early 1980s, and the author was working in an organization responsible

---

[19] Ibid., p. 12.

[20] March 7, 2010. Astronauts Neil Armstrong, Jim Lovell, and Gene Cernan met with deployed Airmen at a military base in Southwest Asia. The author had the honor to meet the astronauts and, like many of the other Airmen, ask questions regarding the astronauts' experiences. Captain Jim Lovell, a distinguished Eagle Scout, remains one of my heroes.

[21] Shaw, G. B., *The Wit and Wisdom of George Bernard Shaw*, Dover Publications, Mineola, NY, 2011, as cited by Fred C. Lunenburg in "*Communication: The Process, Barriers, and Improving Effectiveness*," http://www.nationalforum.com/Electronic%20Journal%20Volumes/Lunenburg,%20Fred%20C,%20Communication%20Schooling%20V1%20N1%202010.pdf. Accessed on September 17, 2013.

for the maintenance of fighter aircraft and electronic systems. I was given the responsibility to automate the maintenance data collection and maintenance analysis functions. We had just received our first desktop computers and developed an architecture that enabled managers to receive desired information directly to their desktop on demand rather than having to travel several miles to the data center to request and pick up predetermined reports. Our initial testing demonstrated huge savings in time, accuracy, and resources. We even saved nearly a semitrailer worth of paper every week as the reports no longer had to be printed out. Implementation was going great! Until we ran right into Edna.

Edna was our senior maintenance data analyst. A former first sergeant in the Women's Army Corps (WAC) during WWII, Edna was nearing 80 years old and had no plans for retirement. "You'll know I am retired when you see my obituary!" was one of her favorite adages. Airmen joked that she was so old that she greeted Lewis and Clark on their arrival to the Pacific Northwest. Edna loved paperwork and her office was a maelstrom of reports and information collected over a long career. When Airmen jokingly used to ask her if she had maintenance information on old retired aircraft like the P-51 fighter (i.e., the WWII "Mustang" fighter), she would dig through her firetrap of paper and proudly produce reports from an era long past. Edna saved everything, refused to archive her files, and proudly pointed out that she had been doing her job since our general "was in diapers." She liked what she did and wasn't going to change what had made her the success she was.

Edna had defeated many of my predecessors. Through her grit and stubbornness, she would wait out Airmen who sought to make improvements, knowing that they eventually would be transferred to other assignments. Armed with a well-maintained and exercised cantankerous and abrasive exterior, she deterred even the most senior officers from demanding she change her ways. I knew getting Edna to change her ways and embrace the proposed automation effort was my personal windmill. Nonetheless, making the change to the new automated system was the right thing to do and essential to our mission. I launched my campaign to make Edna the champion of my mission. "Operation Edna" was underway!

Like any other military officer or engineer, I used an analytical technique to assess my situation. I knew my mission: automate the maintenance functions. I knew the business process inside and out. I also knew the technology inside and out and was convinced that the technology would significantly increase productivity and yield significant cost savings. I had encouraging leadership who shared my enthusiasm for the mission and were committed to its successful implementation. I had developed enthusiasm within the workforce for the effort, garnering valuable support at all levels of command, except for Edna. She was the center of gravity for the maintenance processes and, like a black hole, positioned herself to drag every other maintenance activity into the processes she controlled. Despite her advanced age, as a tenured civil servant, she knew she couldn't be fired, and her health indicated she was going to be there longer than me. I could not wait her out.

People who are resistant to change usually do so because they are frightened.[22] As I analyzed the situation and looked at why Edna would so fiercely resist the change

[22] Exposing Fear of Change, http://www.change-management-coach.com/fear-of-change.html. Accessed on September 17, 2013.

wrought by my automation effort, it became apparent this was not a technical or process issue, this was a human issue. It became evident to me that Edna was afraid of the change and how it would impact her. She was an elderly woman whose job had become her *raison d'etre* and anything that threatened her job was viewed as an existential threat. She would do anything to protect herself, and if torpedoing my project meant that she was protected, so be it.

Understanding the human element is at the heart of addressing resistance to change. I made sure that I spent a lot of time working with Edna, reassuring her that her job was not at risk and that the new automation effort would unleash her to execute her processes and analysis faster and more accurately. I showed her how we had automated her existing process and how she still was an integral part of the maintenance analysis function; her expertise was still needed despite the automation. My team automated every report form she owned in the process, and I personally showed her how she controlled "her data." Several times during the day, I would stop by her office and ask her advice on issues completely out of her arena with the intent of showing respect in the not-too-altruistic hope that it would cultivate some in return. Much to my surprise and delight, her suggestions were insightful and helpful, and I was quick to incorporate them into my plan and gave her full credit in front of our leaders, my peers, and our subordinates. Soon, Edna's cold heart toward the automation started to melt as she recognized that it did not threaten her job, and she became my eager student, demonstrating a seemingly insatiable appetite for knowledge on the computer systems and how she could reinvent her processes using the new tools. To my amazement and that of my superiors and all the maintenance crews, Edna became my biggest champion and declared herself a "born again Geek!"

"Operation Edna" was a success yet represented an appreciable investment in my time and effort that I could have invested elsewhere. As you look at your strategy and the inevitable changes that come with it, be advised that you will have to make an extra investment in addressing those who are resistant to change.

You will be faced with those who say, "We've always done it this way." You'll also find that it is easier for some people to say, "No" rather than "Yes." It is a fact of life. Deal with it.

Understanding that most people face change with a sense of fear is the first step in dealing with resistance to change. Openly address those fears. Communicate with the individual personally. There may be situations where the individual is rightfully concerned that the change will have a negative effect on their employment or conditions of work. Handling such situations with compassion and understanding will go a long way in paving the road to success.

Sometimes, however, you may not be able to overcome their resistance to change and will have to replace the individual to keep your plan on track. Be prepared to implement this as a last resort.

Our recommendation on how to best tackle resistance to change is to treat everyone with dignity and respect, educate and inform, listen to feedback, and clearly communicate the "why" behind your decisions.

We've always done it this way.

4.3.4 Lack of Leadership and Oversight

Ultimately, successful implementation of your strategy relies on leadership and oversight. You and your subordinate leaders are truly the essential ingredient that will determine whether your strategy succeeds or fails.

Ask yourself the following questions: Does your strategy state that you are an organization committed to secure processing of information yet you don't make cyber-security a priority? Do you fail to follow your own policies and guidelines, excluding yourself from mandatory training or procedures? Do you keep cybersecurity on the agenda, ensuring that it is an active topic in every meeting, every presentation, and in every decision? Do you prioritize cybersecurity investments in training, technology, and procedures to ensure you have a cyber-hardened workforce, information environment, and business functions? Are you the evangelist to get the cybersecurity message out to your employees, business partners, and prospective investors? Do you include cybersecurity topics in your memos, reports, and speeches? Do you personally call for and review regular metrics that measure your cybersecurity posture? Do you motivate your colleagues to keep focus on cybersecurity? Do you hire and reward the right people with the right talent to execute your cybersecurity objectives? Are they appropriately aligned and empowered in your organization? Do you address cybersecurity in your Management and Analysis reports to shareholders? If you answered "no" to any of these questions, your leadership is needed to make your cybersecurity program stronger!

Studies clearly indicate how critical leadership "follow-through" is to the successful implementation of strategies. For example, a recent study cited in the *Harvard Business Review* stated "top management has a profound impact on how well employees grasp and support strategy—far greater than any other variable we examined, and far greater than we'd expected."[23] Further, a study by *The Economist's* Intelligence Unit chartered by the Project Management Institute states, "Leadership support is the most important factor in successful strategy execution."[24]

Your strategy cannot be put on autopilot; it requires your constant attention through your personal leadership and the governance activities you enact to provide oversight over your implementation plans. If you "talk the talk" but neglect the "walk the talk," your employees will realize that you are not committed to the strategy and it is not a priority. If it is not your priority, it will not be theirs either.

Providing adequate oversight over implementation of your strategy is essential. Successful leaders set priorities, measure the right things at the right time, and continually communicate up and down the chain of command to ensure that processes

---

[23] Charles Galunic and Immanuel Hermreck, How to help employees 'get' strategy, *Harvard Business Review*, December 2012, http://hbr.org/2012/12/how-to-help-employees-get-strategy/ar/1. Accessed on September 17, 2013.

[24] The Economist Intelligence Unit, "*Why Good Strategies Fail: Lessons for the C-Suite*," http://www.pmi. org/~/media/PDF/Publications/WhyGoodStrategiesFail_Report_EIU_PMI.ashx. Accessed on September 17, 2013.

are effective and efficient. Now, with cybersecurity a priority, processes also must be secure. Does this describe your leadership style? Do you provide the strategy oversight that leads to success? Do your employees know what your priorities are and how you measure success? Are your priorities coherent with your strategy and core values?

Your leadership and oversight is essential to both your strategy and cybersecurity success. Your attitudes, priorities, and interests are reflected in your organization and are reflected in how your people execute your policies and procedures, handle information, and make decisions. You make the decisions on resources, talent assignment and distribution, investments, and priorities. Cybersecurity needs to be a factor in all those decisions. So does your strategy. Be an engaged and focused leader!

4.4 WAYS TO INCORPORATE CYBERSECURITY INTO YOUR STRATEGY

There are countless ways to create your strategy. Numerous techniques are pitched by well-qualified "experts" who specialize in different methodologies that can help your company focus on its future and how to get there. The variety of methods and assistance available is staggering. As an example, a Google search I conducted this morning on "Ways to Build a Strategy" came back with 470 million results. We submit that there is no singular prescriptive method that is best to follow how to create your strategy yet the techniques shared in this chapter lay a good foundation from which you can build.

Incorporating cybersecurity in to your strategy is important in today's Information Age. Information has become a critical component of your business, if not the predominant one. It must be effectively and efficiently managed and properly protected against the risk of unauthorized access, disclosure, tampering, or theft.

Your business may suffer if your information is compromised. Your brand reputation, partnerships, potential investment opportunities, and competitive advantage all rely on the integrity of your information. Therefore, your risk assessment and management program needs to have laser focus on the cyber-based risks to you and your information. Cybersecurity is an essential element of every business today.

Your strategy needs to include cybersecurity as a keystone principle. No matter what your line of work or business sector, cybersecurity touches every activity or business function in one way or another. As we meet with our clients, they readily point out that they are "unique"; one size does not fit all when it comes to cybersecurity. They are correct, but not just because of what they do or how they do it, but because of the risks they face and the risks they are willing to accept. Every organization faces discrete individual decisions regarding cybersecurity.

Nonetheless, we submit there are several cybersecurity considerations that transcend all business sectors and functions that ought to be incorporated in every organization's strategy. No matter what business you are in or what you do, consider the following cybersecurity principles as you build your strategy.

4.4.1 Identify the Information Critical to Your Business

Information has an intrinsic value and is critical to your business, yet not all information has equal value. Do you know what your most important information is? Do you know its value? Like other key assets in your business, information should be subject to asset valuation and focus in your risk management program. Protect what is worth protecting.

4.4.2 Make Cybersecurity Part of Your Culture

The fastest and best responses are those that are done reflexively, such as blinking when something approaches your eye. Cybersecurity can be the same way if you make it part of your culture. Think about it—if everyone on your team is attuned to the need to protect your critical information; understands the threats, vulnerabilities, and risks associated with that information; and acts consciously *and unconsciously* in accordance with your cybersecurity policies and procedures, you likely have a great cybersecurity program that hardens your organization and its information. Policies and procedures must be clearly defined and followed. Continual training and exercises are essential and should be both informative and entertaining to ensure that content is retained and incorporated into daily activities. Everyone must realize that they have a stake in ensuring that the organization and its information are protected. Cybersecurity needs to be "baked in" to everything you do. It needs to be on every agenda, in every process, and in everyone's attention.

4.4.3 Consider Cybersecurity Impacts in Your Decisions

You and your employees make many decisions every day. You decide what to produce and how much, how to align talent, whom to partner with, and whom to avoid. Do you consciously consider how your decisions affect your cybersecurity posture?

Take, for example, a strategic decision whether or not to outsource a function. Traditional business measures will take into account such things whether the function is a core competency, whether it can be accomplished more effectively or efficiently by a specialist, and whether it is an integral part of the business function. What are the cybersecurity impacts of outsourcing a function?

Many people mistakenly think that cybersecurity *only* applies to IT activities, especially during outsourcing deliberations. We submit that cybersecurity considerations apply to *all* outsourcing decisions and in *all* decisions in general. As an example, let's look at outsourcing your accounts receivable function.

Accounts receivable is the method you receive payment for the goods and services you provide to a customer. Many companies outsource their accounts receivable to firms who specialize in billing and collection. But does it make sense for your business? In order to outsource your accounts receivable function effectively, you will have to share a significant amount of information regarding your clients, what products and services you provided, and any other information relevant to your transactions. Is that information potentially sensitive and valued? Do you want that protected? Whom do you want it protected from? These are all representative considerations you should make when determining whether you want to outsource functions.

When you make decisions, consciously ask yourself what the cybersecurity implications are to your decision. Make it part of your decision-making construct. Educate
and train your employees to do the same. Every decision has numerous impacts that
need to be considered. Ensure that you keep cybersecurity in focus for all decisions.

4.4.4 Measure Your Progress

Your strategy will wither and die if you don't measure your progress. When it comes to
measuring cybersecurity attributes in support of your strategy, you are in luck as there
are many to choose from, but choose wisely.

Metrics should measure key processes and functions to enable thoughtful decision-
making. They always result in a cost and should be gathered in support of decisions
rather than for the sake of gathering data. Regrettably, employees around the world spend
countless hours collecting data, building charts and tables, creating briefing slides, and
drafting reports chock full of metrics that are never used or acted upon. That generates
endless waste and drag on their organizations.

You can and should be different. Make sure you are measuring the right things and
make it clear how you will use the information to make decisions. Your superiors and
employees will love you for it!

For example, if your strategy calls for your company to protect its intellectual property and trade secrets from unauthorized disclosure, exposure, tampering, or destruction,
you should tightly control who has both logical and physical access to the information,
including your system administrators and support personnel. As part of your metrics
program, you may measure the following:

- Number of scans against devices containing intellectual property and trade secrets
- Number of attempts to gain access
- Number of unsuccessful attempt to gain access
- Where the unsuccessful attempts are coming from
- Number and type of system vulnerabilities (e.g., vulnerabilities awaiting patches, etc.)

What do you do with your measurements? Often, you may be using them to measure
compliance, but frequently, they drive decisions regarding the apportionment of
resources. For example, if I see an unusual rise in the number of attempts to gain access
to my intellectual property and trade secrets, I would be concerned and want to know
why. I'll start asking a lot of questions. Who is accessing the information? Why are
they doing so? Do they have a need to know? Is the increase in access unusual? Is it
appropriate? I might find through my questions that I have an emergent threat that needs
to be addressed through a new policy, tighter controls, new technology, or better training.
I may even find that I have a situation where unauthorized access occurred, which drives
a whole new line of courses of action and decisions.

You can and should use different measures to evaluate the effectiveness of your
strategy implementation. There are metrics that measure productivity and efficiency,
such as time to create a user account on the network. Others measure outcomes such as
those that compare end results against expected standards, such as measuring whether

network availability achieved its performance standards for the month. A third category consists of those measurements that gauge quality, such as the number of vulnerabilities found in scans (with the fewer the better!). Finally, there are project measures that demonstrate progress of projects over time, such as those that measure the percentage of completion of a project.

Leadership makes the difference when it comes to measuring the right things and making the right decisions. Regardless of what type of organization or business sector you are in, measurement is a key factor in assessing progress toward achieving your strategic goals and objectives. It also is critical in identifying deficiencies that may necessitate midcourse changes or corrective actions.

You can improve inclusion of cybersecurity into your strategy through the wise use of metrics. Measure the right things. Measure those things that are directly linked to your strategy, its vision, goals, and objectives; measure coherency to your core values; measure your progress and your deficiencies. Link your measurements to decisions. Ensure your employees understand why you need the information and what you do with it. When you do, you'll have the right information to make the right decisions at the right time. That is a critical element of your cybersecurity program.

4.5 PLAN FOR SUCCESS

What does a great strategy give you? We submit that it produces what Kaplan and Beinhocker refer to as "a prepared mind."[25]

Your strategy sets the course for your organization to follow. It is the direction you want your employees to follow. It defines your vision of the future, identifies priorities, establishes goals, and sets objectives. It defines the core values that dictate the principles and values that govern how you conduct yourself in your activities. We believe your strategy will prepare you and your employees to move forward toward the successful future you envision. It yields a mind prepared to move to the future.

A great strategy goes nowhere without a great plan. People can get all excited about a great strategy and be consumed by their enthusiasm for its vision, but without the roadmap that a great plan gives, you will never get to where you want to be.

Your plan should define such things as:

1. What will be done
2. Who is responsible for doing it
3. How it will be done
4. What resources are required
   - Information (e.g., critical information, access control requirements and procedures, etc.)
   - Financial (e.g., cost benefit analyses, operating and capital investment options, etc.)

[25] Sarah Kaplan and Eric D. Beinhocker, The real value of strategic planning, *MIT/Sloan Management Review*, Winter 2013, http://sloanreview.mit.edu/article/the-real-value-of-strategic-planning/. Accessed on September 17, 2013.

- Organizational structures (e.g., policies, procedures, processes, management and decision structures, etc.)
- Personnel (e.g., staff, skills, management)
- Environment (e.g., facilities, power, physical security, etc.)
- Technology (e.g., types and capabilities of hardware and software, limitations and controls, etc.)
- Partnerships (e.g., alliances, supply chain, outsourcing, and other third-party relationships, etc.)

5. Risk management
   - How it is measured
   - How it is managed
   - Who makes risk decisions

6. Measuring progress and success
   - What to measure
   - When to measure
   - How to measure
   - How the measure will be used
   - Who is responsible

Planning is done at multiple levels in organizations. We contend there are three distinct levels that ought to be considered as you create the plans to implement your strategy. Each level is important when making decisions and decision-makers ought to know what level of decision they are making as well as the impacts their decision have on other levels.

> The highest level is the **strategic level**. It is characterized by broad scope and scale of the vision, goals, and mission produced at this level. The strategic level has a time horizon measured in years.
>
> The next level is the **operational level**. The operational level is characterized by the execution of a range of operational objectives, frequently defined by functional or geographic boundaries. The operational level has a time horizon that typically is measured in months.
>
> The final level is the **tactical level**. This is the level where most things actually get done. Tactical activities handle your day-to-day tasks that support your goals and objectives. The tactical level's time horizon is in the here and now.

As you build your strategy, we advise that you do not fall victim to the siren's call of those who pull you away from strategy toward discussions of tactics. Once you fall in to that hole, it is extremely difficult to climb out of it to get back on track. There will be plenty of time to argue tactics in the future, and those arguments ought to be fought by the technical experts. Executives need to stay above the fray and focus on the high-level vision that will yield the successful vision you envision.

We'll discuss incorporation of cybersecurity best practices into your planning in later chapters and give some examples that you can include in your plans.

4.6 SUMMARY

In order for an organization to be successful, it needs a thoughtful and effective strategy. Great strategies start with an insightful vision; are simple, well documented, and easy to understand; are based on the current situation with an eye to the future; and are given enough time to bear fruit.

We contend there is only *one* strategy in your business, not an amalgamation of disparate strategies acting in concert or potentially in competition. All corporate activities, including marketing, investing, production, operations, and even cybersecurity, are done in support of the overall strategy. Our intent in this book is not to focus upon overall strategy development. There are many excellent references that deal with that subject. However, we strongly urge that cybersecurity must be incorporated within the overall strategy. Hence, the focus of this chapter is to help understand and guide strategy development from a cybersecurity perspective.

Clearly, to be effective, cybersecurity needs to be an integral part of your corporate strategy. It needs to be a consideration in your corporate plans, policies, and procedures. It needs to be "baked in" to everything you and your company do. Attainment of a cybersecurity vision that melds seamlessly with your strategic plan enhances value, reduces risk, sets the stage for success, and helps to determine how much protection you need.

In this chapter, we focus on asking (and answering) the following key questions relative to folding cybersecurity considerations into the overall corporate strategy:

- Where are we now?
- What do we have to work with?
- Where do we want to be?
- How do we get there?

After addressing these key questions, we discuss how to articulate corporate strategy and your vision of the future (including cybersecurity as a critical component of your vision). Many organizations use a vision statement, a mission statement, and core values as touchstones of their strategy. Vision statements give the high-level "strategic" overview of what you want to achieve. Mission statements tell how you will achieve your vision. Generally, they are short, concise statements that start with the word "To" and capture the essence of the vision. Core values define the principles you will follow as you conduct the activities while carrying out the vision and mission. Each one of these documents can and should reinforce your commitment to cybersecurity as part of your strategy.

Once you've defined your vision, mission, and core values, your next step is to use them as the foundation to create the goals and objectives to achieve your vision.

Goals and objectives are at the heart of planning. There are numerous definitions regarding goals and objectives. Some of these have goals subordinate to objectives and vice versa. For our purposes, we define them as:

- Goal: Statements of what you want to accomplish. For example, "Score a touchdown."
- Objectives: Specific, measurable, time-relevant statements of what is going to be achieved and when. For example, "Get a first down every offensive series."

While there are numerous ways to define goals and objectives, we've found there are also many criteria to gauge their merit. We've found that **goals** are most effective when they are **feasible, acceptable, suitable, and affordable**. Likewise, **objectives** are strongest when they are **specific, measurable, achievable, realistic, and timely (SMART)**. In the chapter, we expand upon these factors.

We identify issues that can contribute to strategy failure, such as poor plans and poor execution, lack of communication, resistance to change, and lack of leadership and oversight. Additionally, we comment on factors that can help to make the strategy succeed: identify information critical to your business; make cybersecurity part of your culture; consider cybersecurity impacts in your decisions; and measure your progress.

A great strategy goes nowhere without a great plan. People can get all excited about a great strategy and be consumed by their enthusiasm for its vision, but without the roadmap that a great plan gives, you will never get to where you want to be.

We discuss items the plan should define, such as:

- What will be done
- Who is responsible for doing it
- How it will be done
- What resources are required
- Risk management
- Measuring progress and success

Planning is done at multiple levels in organizations. We contend there are three distinct levels that ought to be considered as you create the plans to implement your strategy. The highest level is the **strategic level**. The next level is the **operational level**. The final level is the **tactical level**.

An essential component of a good plan is well-articulated policies and procedures, which are addressed in the subject of the next chapter.

5.0

PLAN FOR SUCCESS

Successful generals make plans to fit circumstances,
but do not try to create circumstances to fit plans[1]

General George S. Patton Jr.

5.1 TURNING VISION INTO REALITY

Herb Kelleher, the visionary Chairman Emeritus of Southwest Airlines, is oft-quoted saying, "We have a strategic plan. It's called doing things."[2]

Do you have a plan to "do things"?

You ought to. One of the principal responsibilities of an executive is to plan and execute activities that drive your organization to success. If you don't have a plan or are not vigorously executing your plan, you aren't "doing."

---

[1] *Army Field Manual 5.0*, United States Army, January 2005, pp. 1–1.

[2] Ben Arment, "Strategic Plan: Doing Things,"http://storychicago.com/updates/strategic-plan-doing-things, August 1, 2013. Web. October 2, 2013.

Cybersecurity for Executives: A Practical Guide, First Edition. Gregory J. Touhill and C. Joseph Touhill.
© 2014 The American Institute of Chemical Engineers, Inc. Published 2014 by John Wiley & Sons, Inc.

Having a plan is critical. How many times have you worked with or for people who don't have a plan and fly by the seat of their pants? Were they successful? If they were, could they sustain that success? We submit that those who don't have a plan are destined to "crash and burn" and often take others down with them. To paraphrase the Boy Scouts and Herb Kelleher, "Be prepared"; have a plan and do things!

What's the worst that could happen if you don't have a plan or have one that is failing?

Perhaps we can ask the executives at America Online.

In 2000, America Online purchased Time Warner for US $164 billion in what was the biggest corporate purchase of all time. The strategy behind the merger of the two media giants was sound: America Online would leverage Time Warner's extensive high-speed cable system to bring its product to as many as 130 million new subscribers, and Time Warner would be able to bring its media products such as magazines, books, and movies into the digital world bundled with America Online's content. It seemed like a great strategy—until the time came to implement it.

The newly formed AOL Time Warner was what current Time Warner chief Jeff Bewkes calls "the biggest mistake in corporate history."[3]

What happened?

Ten years later, after years of bitter internal fighting, legal challenges, and plummeting investor confidence, the two companies went their separate ways, licking their wounds from the failed merger. For shareholders, the impact was terrible. After the split, the combined value of the two companies is one-seventh of what it was before the merger.

There are many different views on why the merger failed. Some believe that the merger failed because they had a poor plan and didn't implement it well. Others point to the failure of AOL and Time Warner executives to congeal as a unified team as the principal cause of failure. Still others posit that a lack of communication from management to those implementing the plan guaranteed failure and that leadership was absent in resolving the problems. Our view is that all these things contributed to the failure. Based on the commentary of senior executives involved, we conclude that the strategy to merge America Online and Time Warner was sound, yet the planning and implementation was an utter failure and led to dissolution of the relationship. Steve Case, founder of America Online and the AOL Time Warner Chairman, agrees. He summed up the debate of whether the strategy was bad or the execution of it failed, saying, "It was a good idea, but the execution of it wasn't what it needed to be, and I accept responsibility for that. Everybody involved, I think, needs to accept responsibility for that, but that doesn't take away from the core strategic value of the idea."[4]

You are faced with numerous cybersecurity challenges. You have to protect your critical information to maintain your competitive advantage. What's your plan to protect

[3] Emma Barnett and Amanda Andrews, "AOL merger was 'the biggest mistake in corporate history', believes Time Warner chief Jeff Bewkes," http://www.telegraph.co.uk/finance/newsbysector/mediatechnologyandtelecoms/media/8031227/AOL-merger-was-the-biggest-mistake-in-corporate-history-believes-Time-Warner-chief-Jeff-Bewkes.html. Web. October 2, 2013.

[4] Tim Arango, "How the AOL-Time Warner Merger Went So Wrong," New York Times, http://www.nytimes.com/2010/01/11/business/media/11merger.html?pagewanted=all&_r=0, January 10, 2010. Web. October 2, 2013.

your information? What's your plan to grow your business in a contested cyberspace environment? What's your plan to protect your business?

When it comes to implementing your strategy and ensuring you have the right cyber-security posture, you need to plan and (as Jack Welch advised) "implement like hell!"

5.1.1 Planning for Excellence

Plans are designs or detailed proposals for accomplishing anticipated operations. As an executive, you may not be the one writing the plans, but you are responsible to see that they are created, that they are done right, that they are executed well and, in the event they prove to be heading in the wrong direction, to steer the organization back to the right one. They are the means by which you, the executive, envision the future, lay out creative and innovative ways to achieve it, and communicate to your subordinates your vision, your intent, decisions you've made, and the results you expect to achieve. Your plan answers these questions:

- What will be done?
- Who is responsible?
- How it will be done?
- What resources will be required?

How detailed your plans are depends upon the complexity of the desired operation and the experience of the subordinates who will execute it. Plans can be articulated in a very formal document or through an informal outline. Regardless of what format is used, great plans have common characteristics:

- The intent of senior leadership and the purpose of the plan are made perfectly clear.
- Relevant facts and assumptions are stated up front to provide context.
- Great plans are simple and put into positive terms (i.e., "This is what we will do" rather than "We won't do this.")
- Great plans also are precise and avoid ambiguous language or phrasing that can confuse their readers. They are brief and clear.
- They are complete and contain all information needed to execute the plan.
- They show what will be done, who will do it, and when they will do it.
- They synchronize activities to ensure there are no resource or organizational conflicts.
- They are flexible and allow for adjustments to counter the unexpected and provide measures of progress.
- They define timelines for getting things done.

There are plenty of plan formats and styles that are effective when tailored to the right task and organization. Some are based on the time-phased sequence of activities. Others are based on the fusion of different functions working in concert to achieve a common goal. Use what works for you and your organization! Remember, though, whatever format you choose, make it consistent throughout your organization. Consistency in format yields consistency in understanding, performance, and results.

It is important too to remember that your plan should be evolving continuously to adapt to situations while guiding your subordinates throughout an operation. Hall of Fame football coach Bill Walsh was renowned for his planning and used to "script" the first 25 plays (allowing for adjustments) of the upcoming game.[5] While people focus on the fact that he scripted the first 25 plays, what many fail to recognize is that he planned for adjustments. Walsh knew what he was doing. As the famous German field marshal, Helmuth von Moltke said, "No plan survives first contact with the enemy."[6] Walsh knew that and ensured his plans had the flexibility to adapt as situations changed. As the U.S. Army advises, "Any plan is a framework from which to adapt, not a script to be followed to the letter."[7]

Such planning flexibility is critical concerning cybersecurity. New IT products emerge on the market every week and often their vulnerabilities are revealed shortly thereafter. As they age, venerable and trusted technologies become increasingly susceptible to penetration and exploitation. Your plan needs to include provisions for preplanned improvements to keep you, your policies, your procedures, and (especially) your people hardened against the emerging cyber threats that inevitably will threaten you and your organization. The measure of a great plan is not whether execution of the plan transpires exactly as planned but rather whether the plan produces the desired results in the face of unforeseen circumstances.

5.1.2 A Plan of Action

After writing this chapter, we stepped back and asked ourselves this question: "If we were an executive reading this book and it was the first time we really delved into the subject of cybersecurity, what would our reaction be to the material contained in this chapter?" It might be something like this: "Wow! I'm overwhelmed! These people are trying to make my organization into a computer-driven fortress, and at great expense." There may be a slight grain of truth to that snap opinion, but we see the issue much differently. First, your organization will not survive now or in the future without heavy reliance upon modern IT. Look at your kids. Tell them that they can't have cell phones anymore. You will not enjoy their companionship for very long. In today's world and as far as we can see into the future, the success of your business depends upon the effective management of information.

---

[5] Michael Mink, "*Football Coach Bill Walsh He Prepared, Communicated and Won*," http://news.investors.com/management-leaders-in-success/011801-347555-football-coach-bill-walsh-he-prepared-communicated-and-won.htm, January 18, 2001. Web. October 2, 2013.

[6] Kennedy Hickman, "Franco-Prussian War: Field Marshal Helmuth von Moltke the Elder," http://militaryhistory.about.com/od/1800sarmybiographies/p/vonmoltke.htm. Web. October 2, 2013.

[7] Field Manual 3.0, available at http://www.globalsecurity.org/military/library/policy/army/fm/3-0/ch6.htm. Web. October 2, 2013.

Second, because you have no choice but to make most effective and efficient use of modern information technologies, and because you want to keep these systems safe and secure, you are going to have to formulate a plan to protect your business.

Third, implementation of the plan is going to cost money. The trick is to spend this money wisely. Spend it to prevent problems, not cure them. Moreover, by having risk management systems in place, that is, policies and procedures, your organization can better mitigate cybersecurity risks.

Fourth, in some respects we can understand how a novice executive might be overwhelmed by the formulation of policies and procedures as enumerated in this chapter. However, try to remember that once the policies and procedures are in place, the hard work is done. What remains is vigilance and updates that keep abreast with new developments. View the initial cybersecurity program as a sunk cost. You may even be able to capitalize the cost (this is a question for your CFO and outside accounting firm). You were able to capitalize your computer and network systems, so it may be possible to include the program to protect these systems as an essential part of a capital asset.

Fifth, we understand that you don't want to make information technologies the central focus of your business. You are trying to manufacture chemicals, manage hospitals, run water and wastewater treatment plants, mine coal, produce electricity, and conduct many other jobs that are essential to the health of the economy. But recognize that other businesses that are doing what you do and also have to protect themselves from cyber attacks. They also have to bear the costs of doing so; hence, your objective is to try to do the job better, faster, and cheaper than they can.

Finally, our point is you can't live without computers, you have to keep them safe and secure, and it's going to cost money initially (and also will for your competitors). So pay attention to the material that follows and build a plan that manages risk in a thorough yet cost-effective way.

When it comes to cybersecurity in your business or organization, what things are you going to do? If you've completed your strategy and identified your vision, mission, goals and objectives, you've already laid out a roadmap to follow in building the plans that turn your vision into reality.

Many of our clients ask us for help in building plans to help their business harden themselves against cybersecurity risks. Each one is different and is tailored to the mission of the business, their goals and objectives, the complexity of the task, the talent and resources available, and the culture of the organization. Nonetheless, we think it may be instructive to outline an example plan to show you a methodology to address a common cybersecurity issue.

In this scenario, we return to the BigRX Corporation in Pittsburgh (see Chapter 3.0). As you recall, they have a problem where their web site was found to have a significant vulnerability to SQL injection attacks. Structured Query Language (SQL) injection is one of many web-based attack mechanisms used by hackers to penetrate organizations to steal their vital information and is among the most common attacks seen today. According to the Open Web Application Security Project, a noted worldwide organization focused on improving the security of software, it is the top security flaw in software today.[8] Hackers use SQL injection to take advantage of improper coding of web applications that allow the hackers to inject SQL commands into such things as a log-in form to allow them to gain access to the information contained in your databases.

Here's an outline of the hacker's plan to steal your information using SQL injection: the hacker goes to your web page and finds a log-in page. Normally, when a legitimate user comes to your web page and submits their username and password, a SQL query is generated and submitted to the system's database to be verified. If the credentials are valid, the system allows the user appropriate access. The hacker, on the other hand, rather than inputting a legitimate username and password combination will input a specially crafted SQL command with the intent of by-passing the log-in procedure and granting access to your system. If your system is not properly coded to detect and reject commands injected into forms, the executable code will be sent to the database which doesn't know any better and will do what the command says to do. Voila! The hacker is "in," gains control, and may be able to run rampant through your data, freely reading it (exposing your intellectual property and trade secrets or personally identifiable information) extracting it (to possibly sell to your competitors), deleting it (denying you access to your own information), or tampering with it (causing you to lose trust in your own data.) Regardless what the hacker does, once he or she conducts a successful SQL attack, you have big trouble on your hands.

Suppose you are an executive in charge of the BigRX planning division. The BigRX CEO summons you to his office and tells you by the end of the week he wants a plan to reduce exposure to the risk of SQL injection attacks. What do you do?

As you walk from the C-suite back to your office, you remember the basics of building a plan. What needs to be done already has been spelled out by the CEO: reduce the exposure to risk from SQL injection attacks. Now you have to identify who will do it, how it will be done, and what resources are needed to do it.

You also remember that there are several layers of planning: strategic, operational, and tactical. Each focuses on various levels of specificity in addressing how activities are accomplished. Because this assignment from the CEO (the strategic level) calls for a plan to direct the activities of several cybersecurity functions as part of an enduring process, you will create an operational level plan that concentrates on addressing the activities of the affected functions rather than the tactical level which addresses how those activities are done. Once the CEO approves the operational plan (also known in some organizations as the operational concept), you will work with experts at the tactical level to complete the tactical level plan (frequently expressed as procedures or checklists).

When you get back to your desk, you log in to the corporate network and retrieve the BigRX corporate template for plans (note that this is not just the template for cybersecurity plans, but the template for *all* of BigRX's plans. You are, after all, the head of the planning division):

---

[8] Open Web Application Security Project, "Top 10 for 2013," https://www.owasp.org/index.php/Top_10_2013-Top_10. Web. October 2, 2013.

Plan Title

1. Purpose (What will be done):
2. Situation (Setting the environment/context of why you are preparing a plan): [description of what problems you face, why you need to have the plan, and anticipated impacts if the plan is not followed]
3. Applicability (Who is responsible): [identify all key players responsible for the successful execution of this plan]
4. Concept of execution (How it will be done): "The following shall be accomplished to ensure that all software procured and used by BigRX is free from vulnerabilities to SQL injection.
   a. Process Step One
   b. Process Step Two
   c. Process Step Three
5. Resources required (What resources are required?): [identify all resources required to successfully execute this plan]
6. Governance and oversight (Ensuring it gets done right): [how management will ensure the plan is executed properly, is effective, etc.]

This plan becomes effective upon signature and supersedes any previous direction.

//signed//
Chief Executive Officer

You know what needs to be done. You have the outline to follow. Now's the time to do it!

5.1.3 Doing Things

Less than 30 minutes ago, you were tasked by the CEO to create a plan to reduce exposure to the risk of SQL injection attacks. Although you consider yourself a competent computer user, you recognize that you barely can spell cybersecurity let alone detail a plan to solve a complex technical issue like the one handed to you. Where do you begin?

You start by assembling the right team to support you. Building and maintaining great teams is an essential executive function. You need to find and align the right talent to assess where you are, propose courses of action to get to where you want to be, and help you determine the best course of action to recommend. Choose your team wisely and ensure you get the right mix of technical, business, legal, financial, and other required expertise to address all aspects of the issue you are targeting. Fortunately, you have the full weight of the CEO behind you and are able to quickly gather the right talent from within BigRX and, if you need them, from consultants who can lend specific expertise that BigRX doesn't have on staff.

Well prepared with a great team, the next step is analyzing where you are (just like you did when building your strategy.) Your team quickly determines your web pages and applications are created by contracted programmers who build your web presentations based on the functional specifications BigRX provides. Your current process calls for the code to be received by your IT department, where it is tested for functionality and promoted to the production-level public web site. This process has proven inadequate as it yielded a product vulnerable to SQL injection. That's the problem you have to fix.

Do you have to throw out your existing process and create a new one? Do you have to invest in new technologies? Perhaps not. You may find modifying or updating existing processes is more efficient and effective than starting from scratch. In this case, finding and fixing flaws in the existing process present a SMART model (i.e., *specific, measurable, achievable, realistic, and timely*) potential solution set.

Your investigation indicates the problem is the result of weaknesses in the following:

- Specification: Your counsel points out BigRX didn't specifically tell the vendor you expected them to meet industry best practices and would hold them accountable to produce code free from SQL injection (and other known) vulnerabilities.
- Coding: The CIO indicates the vendor is not coding to standards, thus producing a product that is inferior and exposes your company to unacceptable risk.
- Testing and acceptance: The contracting officers, CIO and CISO, identify that your company did not recognize the code was flawed before it accepted and promoted it to your public web page.
- Internal controls: Your CEO points out that normal monitoring controls detected the problem, but corporate control mechanisms did not react to it until the CEO highlighted it as a major issue.

This helps refine the "what" you need to address to reduce exposure to risk from SQL attacks. Now you have to look at "who" is or should be responsible to address and fix the problem.

The specifications for the BigRX web pages come from a team of individuals who work with your contracting department to identify the requirements for goods and services and issue requests for proposals to prospective vendors. In this particular case, you find that the specifications for this web page and accompanying access to your data were developed by a team of individuals from your business contracting and administrative functions, your marketing team, representatives of your general counsel's office, and the IT department's web shop. You believe the technical group from the IT department should have known to put in the specification something specifically about preventing SQL injection and like vulnerabilities, yet when you talk with the team, they say they **assumed** the vendor would follow industry best practices. You are reminded of Shaw's quote about the illusion of communication being accomplished.[9]

In putting together the plan required by the CEO and publicly noting that he has invested complete authority in you (subject to his approval, of course), you reiterate to the contracting team that they own the contracting process. Thus, you assign responsibility to the contracting officer to ensure that in the future (for this case and all others) contract specifications must include cybersecurity provisions that call for products and services that conform to what your general counsel calls "recognized industry best practices." The CEO issues a strong statement to all employees on this requirement.

So now you, the planning division head, assign your CIO the responsibility to provide the contracting officer with appropriate cybersecurity provisions and technical specificity to ensure that the technical risk of SQL injection is minimized. Further, you assign your general counsel the responsibility to ensure that the contract language addressing the SQL injection protection issue is legally sufficient to hold the vendor accountable.

You address the coding issue similarly. Because you are outsourcing your software development, you rely on the vendor to create the effective, efficient, and secure code that you need. You need to address the need to be effective, efficient, and secure in your specifications. You also need to ensure that your vendor completely understands that they will be held accountable if they fail to meet their agreement to deliver the quality product that meets your needs. In addition to the responsibilities previously assigned, you assign responsibility to the CFO not to pay any vendor for software or services unless they are certified by the CIO and CISO as meeting security specifications.

BigRX should never accept a product or service that hasn't been properly tested to ensure it meets specifications. Regardless of whether it is a medical supply such as a pharmaceutical product or a piece of software, lives could be put at risk if quality standards are compromised. You assign responsibility for testing software for risks and vulnerabilities to the CIO. The CIO is responsible to ensure that software received from vendors is free of SQL injection vulnerabilities. Further, once testing is completed, the CIO will validate the code and notify the contracting officer and the CFO to complete the acceptance and payment process.

Finally, you address your internal controls. Your CEO rightfully wants to know why the SQL injection vulnerability wasn't detected and acted upon. "How did we find out we had an issue?" "When did we know?" "What did we do about it?" All of these questions are ringing in your ears as you look at your internal controls process. Frankly, the technical team did discover the SQL injection vulnerability during a routine vulnerability scan and yet had not acted upon it because, although it was one of a number of cybersecurity vulnerabilities discovered during the monthly scans, it wasn't believed to be a high priority. It is now. To address the internal controls, you assign the CIO responsibility to identify cyber vulnerabilities and potential impacts to the CEO, COO, CRO, and risk committee. Because these vulnerabilities present risk to the organization, you also assign responsibility to prioritize ensuing risk management efforts to the risk committee and the CRO.

You and your team have identified existing processes and key responsibilities. Now it is time to outline how to reduce BigRX's risk exposure to SQL injection attacks and identify what resources will be required.

Reducing BigRX's exposure to SQL injection attacks is a process that starts with developing the requirement to the supplier (the software vendor) to create a program that is free from SQL injection vulnerabilities. When the vendor supplies the code, before BigRX accepts it, it must be tested to ensure that it was done to standard and works as negotiated. Upon successful emergence from BigRX's test and acceptance procedures, the code will be promoted by the BigRX IT staff to the live public page. Once it is live, the IT staff will monitor the code continually using vulnerability scans and other techniques. In the event that a vulnerability is found, the CIO will alert the risk management team, who will decide what next steps to take including mitigating, transferring, accepting, or avoiding the risk.

Seems fairly straightforward, doesn't it?

It can be if you have the means to accomplish the steps in the process. Do you have the right resources in place to execute all the steps properly? Do you have the right talent? You plan on testing the software yet do you have the right equipment and technical staff to do the testing? Do you have the right staff and capabilities to continually monitor the code and other vulnerabilities? As the executive in charge of this planning process, you are responsible to ensure that you have all the required resources identified to make your plan a success.

In this scenario, we assume that the additional testing requirements drive an increase of two full-time equivalent IT technicians with testing experience and certifications from an organization such as the International Software Testing Qualification Board. Your team recommends one of the positions have a Certified Manager of Software Testing (CMST) certification and the other have a minimum Certified Associate in Software Testing (CAST) certification but prefers the candidate have the Certified Software Tester certification (CSTE). This means an additional responsibility to your finance department to fund the two positions and to the human resources (HR) department to recruit qualified candidates. Based on the market estimates of compensation for candidates with these skills, your HR staff estimates it will cost US $100,000 per year per position (including benefits).

The team also determines that BigRX must make its testing capabilities more robust to accommodate the anticipated testing the plan mandates. An additional suite of software, hardware, and network equipment is required to ensure that testing is not done on the customer-facing live system where patient or other sensitive data may inadvertently be exposed during testing procedures. After validating estimates with expert consultants, you identify that creating the required testing capabilities in the BigRX data environment will cost US $750,000. As an alternative, your CIO suggests that you consider subscribing to a cloud-based service that will permit you to create a "dynamic test environment" that will simulate your live production environment using the cloud provider's "virtual machines" in lieu of having to buy your own infrastructure.[10] The estimated cost of such a subscription from a major cloud provider is less than US $200,000 per year. After a careful risk analysis led by the CRO and assisted by the CISO, the team determines that subscribing to a cloud-based virtual testing environment is an acceptable option as it will suitably emulate the actual production environment, a testing database void of actual data will be used, and that encryption

---

[10] "Virtual Machines" refers to a concept where software allows multiple instances of computing to occur on the same computer. In essence, the computer allows several different instances of operating systems to operate simultaneously, with each instance having the "look and feel" of its own independent computing environment yet sharing the same hardware infrastructure. Think of it as a "time share" arrangement where each program uses independent nanosecond slices of time to use the same computing environment, not knowing (or caring) who is using the other time slices.

protecting data at rest and in transit will help protect BigRX and its interests; "the juice is worth the squeeze!" Moreover, the subscription service can be used for testing other software that may be required over the term of the contract period.

Finally, you lead the team in a review of what could happen in the event BigRX is hit with a SQL injection. This is important to consider as one of your potential courses of action is to do nothing and continue on the course you currently are following. This is aligned with "accepting" the risk you already have. Your team walks through a scenario where a hacker uses a SQL injection attack to gain access to your information. You look at what would happen if your information is exposed, tampered with or destroyed. Because you have sensitive patient records covered under both the HIPAA and Privacy Act, you anticipate potential regulatory fines for not properly safeguarding sensitive patient information. Furthermore, as you are a publicly traded company, there will be a potential cost as you disclose the incident in your public filings. You also anticipate litigation in the form of class action or individual law suits from shareholders and patients concerned about the exposure of sensitive information. Your CFO cautions that the cost of your insurance also will likely increase, representing an enduring cost that BigRX will face. Further, you estimate recovery costs including staffing, marketing, and other incidental costs. Therefore, the cost impact of a SQL injection attack on your information system will range from US $5 million up to US $35 million.

Based on the team's findings, you return to the CEO's office on Friday afternoon with the following plan:

 BigRX Plan to Reduce SQL Injection Risk

1. Purpose: To reduce exposure to risk presented by SQL injection attacks.

2. Situation: According to the Open Web Application Security Project, SQL injection is the top security flaw in software today. Hackers and other bad actors exploit this flaw to gain unauthorized access to information. BigRX is the custodian of sensitive patient and corporate information. Compromise of its information through unauthorized exposure, tampering, or destruction will result in regulatory fines, litigation, and or damage repair costs that could range from US $5 to $35 million dollars per event. As a result, the CEO directed creation of a plan to reduce exposure to the risk of a SQL injection attack.

3. Applicability: This plan is chartered by the CEO and is maintained by the planning division. It is applicable to the CFO, COO, CIO, Chief Medical Officer (CMO), CRO, General Counsel, Contracting Officer, and the Human Resources and Marketing staffs, CISO and the IT staff. Each affected organization has responsibility to properly execute the tasks detailed in list item 4 (Concept of Execution) and to report deviations or recommended changes to the plans division.

4. Concept of execution: The following items shall be accomplished to ensure that all software procured and used by BigRX is free from vulnerabilities to SQL injection:

   a. Contracting for software: All contracts for software procurement will be created using the expertise of a product team consisting of qualified representatives of the contracting officer, CFO, CIO, CRO, general counsel, marketing, and

---

sponsoring business unit. The contracting officer or designated representative will lead this team and is responsible to ensure that contract specifications include cybersecurity provisions that ensure that products and services conform to recognized industry best practices. Whenever possible these best practices will be cited specifically in the contracts. The team will identify all software requirements including functional details, performance standards, and security controls. The CIO is responsible to provide the contracting officer with appropriate cybersecurity provisions and technical specificity to ensure that the technical risk of SQL injection is minimized. Each contract also will state that the contract must follow software industry standards for secure software development. Acceptance procedures will be detailed in each contract and contracts will clearly state that the vendor shall not receive payment until the software is validated as having met contract specifications. The general counsel is responsible to review all proposed contracts and will ensure that the contract language addressing the SQL injection protection issue is legally sufficient to hold the vendor fully accountable.

b. Contract modifications: During the software development process it is not uncommon for the vendor to seek clarification on requirements, to demonstrate potential user interface options, or to propose new technologies or techniques for satisfying requirements. The contracting officer or designated representative is the only employee authorized to make changes to existing contracts, and all communication between the vendor and BigRX will be made through the contracting officer. In the event that the vendor seeks additional information from BigRX regarding the proposed software product, the contracting officer will refer the matter to the product team for resolution.

c. Testing and acceptance: Software will comply with identified performance and security standards before it will be accepted by BigRX from the vendor. Software received from the vendor will be tested by the IT staff under the oversight of the CIO to ensure that the software is free of defect or security flaws and meets performance objectives. Testing will be conducted in an off-line testing suite of hardware, software, and network equipment to ensure there is no inadvertent compromise of live production information during the testing and acceptance process. Upon approval of this plan and funding received from the CFO, the CIO will oversee the creation of a test and acceptance environment. Additionally, the HR department will recruit and hire two **certified** software test personnel in accordance with list point 5b of this plan. The certified software testers, who will report to the CIO, will create, coordinate, and execute software testing procedures in support of this plan and other tasks as directed by the CIO. Representatives from the business functions and contracting office will support testing as appropriate to ensure that appropriate functionality meets business objectives and is congruent with specifications issued in the contract. The CISO will oversee security testing procedures and report results to the CIO. In the event that testing is unsuccessful, the CIO will notify the contracting officer, CFO, and sponsoring business unit. Upon successful completion of testing, the CIO will validate the code and notify the contracting officer, CFO, and sponsoring business unit of its acceptance. The code will not be promoted to the live operational environment during this phase.

d. Promotion to live system: All software will be validated and certified by the testing and acceptance process prior to being promoted to the live operational environment and accessible to BigRX's customers and clients. All software changes to the corporate web pages will be scheduled in advance and will be approved by the following corporate officers: CEO, COO, CFO, CIO, CRO, CMO, general counsel, and marketing officer.

e. Recurring performance management: The CIO and sponsoring business unit are responsible to ensure that the operational software continues to meet business objectives. The business unit will continuously monitor the functions of the software to ensure it meets requirements for performance, user interface, and coherency with evolving business requirements. The CIO and subordinate IT staff will continuously monitor the integrity and security of the software and databases through monthly vulnerability scans, regular penetration tests, and strict configuration control procedures. In the event of a detected vulnerability, the CIO will immediately notify the CRO. The CRO is responsible for keeping the corporate risk committee apprised of all vulnerabilities. The corporate risk committee will establish priorities that will govern how vulnerabilities will be addressed.

5. Resources required: The following resources are required to execute this plan successfully:

a. Available resources: The following resources are already available within BigRX and are detailed to execute tasks in support of this plan:

- Finances: The CFO certifies that sufficient funding is available to support this plan.
- Organizational structures: BigRX standard processes and procedures for the creation and management of contracts will support this plan.
- Personnel: The CFO, CIO, CRO, CISO, general counsel, marketing officer, contracting officer, and HR and their staffs have sufficient people training and expertise to execute this plan except as noted in list point 5b.
- Environment: Sufficient office and facility space; heating, air conditioning, and ventilation; and utilities are available to house the software test and acceptance facility in existing spaces allocated to the BigRX IT staff.
- Technology: Public web pages, databases, internal and external network resources, network monitoring tools, boundary protection devices, and other technologies are available to support this plan. Enterprise licensing for network monitoring software and other tools will be applied to the maximum extent possible to ensure consistency across the BigRX enterprise and to manage costs. Additional IT to support this plan are highlighted in list point 5b.
- Partnerships: BigRX has retained access to expert consultants to assist in implementation efforts as required.

b. Additional resources required: The following additional resources are needed to execute this plan:

- Finances: Additional funding is required to create the software test and acceptance capability. Total cost to create the capability is US $400,000 with an annual recurring cost of US $200,000. The cost summary is:
  - Personnel: Total cost estimated to be US $200,000 per year
  - Technology: Rather than create on on-site test and acceptance environment, which was estimated to cost US $750,000 with annual recurring costs of US $25,000 per year and recapitalization costs across a

four-year cycle, BigRX will subscribe to a cloud-based solution offered by MegaMegaData. Using MegaMegaData's virtual machine environment, BigRX will create an on-demand testing capability at a total cost of US $200,000 per year.

- Personnel: BigRX will hire two **certified** software testing personnel to conduct software testing and acceptance procedures. The personnel will be responsible to the CIO to create, coordinate, and execute testing procedures to ensure vendor-supplied software meets contract specifications, vulnerabilities are identified, and the software operates satisfactorily to meet business objectives. Minimum levels of expertise and certification (from the International Software Testing Qualification Board or similarly recognized organization) for the candidates are:
  - Test Director: CMST certification
  - Testing Specialist: CAST certification but preferably the candidate will have the CSTE
- Technology: BigRX will create a software test and acceptance environment which will be an off-line testing suite of hardware, software, and network equipment to ensure there is no inadvertent compromise of live production information during the testing and acceptance process. The following type of equipment is required:
  - Network Infrastructure: For example, racks, power, routers, hubs, cabling, and servers.
  - Software: Specialized software will be procured to simulate high traffic loads and other test conditions.

6. Governance and oversight: The following controls will be implemented to ensure this plan is executed properly and to gauge its effectiveness:

   a. Vulnerability scanning: Monthly vulnerability scans will be accomplished by the IT staff. Results of the scans will be provided through the CISO to the CIO and CRO. The CISO will identify the vulnerability, potential impact, and recommended courses of action in a monthly report. Critical vulnerabilities, that is, those that present an immediate threat, will be reported immediately (not to exceed eight hours from detection) for management review and potential action.

   b. Penetration testing: Upon the approval of the CEO and CIO, the CISO will conduct unannounced penetration testing using organic staff to assess the corporate security posture. These penetration tests will include tactics, techniques, and procedures to assess SQL injection vulnerabilities. Additionally, at least once a year, the CISO, in concert with the CIO, CRO, and CFO will arrange for an independent third-party penetration test of the corporate systems as part of the corporate audit process. Results of penetration testing will be reported to the senior leadership team quarterly or as critical vulnerabilities are discovered.

   c. Management reviews: The following management reviews will be conducted to identify and assess risk in accordance with the corporate risk management process:

- <u>Continuous security oversight</u>: The CISO will maintain procedures to continually monitor the information and IT for vulnerabilities and flaws. Upon discovery of a vulnerability or flaw, the CISO will identify the issue, potential impact, and recommended courses of action to the CIO and CRO.
- <u>Performance monitoring</u>: The CIO will ensure continual monitoring of the information and IT to ensure they are operating in accordance with established procedures and performance standards. In the event of an occurrence that deviates from expected performance standards, the CIO will take appropriate action to restore operations to prescribed standards. The CIO will provide monthly and on-demand IT performance measures to the senior corporate leadership team.
- <u>Risk assessment</u>: The CRO will ensure that identified vulnerabilities are handled in accordance with the corporate risk management process. The CRO will identify potential risks, their potential impacts, and recommended courses of action to the Board's risk committee, who will establish priorities for the CEO and senior management to address the risks present.

This plan becomes effective upon signature and supersedes any previous direction.

//signed//
Chief Executive Officer

On Friday afternoon you brief the CEO, walking him through the proposed plan. His questions are insightful and probing. He wants to know if any of the team members had alternative views, what courses of action were considered that aren't mentioned in the plan, and if the plan is feasible, acceptable, suitable, and affordable. After describing your interactions with the team and the process it followed, you answer his questions and ask for his next steps. He thanks you and your team and directs that you use the BigRX administrative process to send the plan to the senior management team for their formal review and coordination. He requests that you inform them of a briefing you will be making at the Wednesday morning executive staff meeting where it will be a major agenda item. Barring any objections from the senior management team, the CEO will sign the plan after the meeting and your team will implement it.

Congratulations! It looks like you are on the right track. You and your team created a plan of attack that is feasible, acceptable, suitable, and affordable to protect your company from one of the most dangerous cybersecurity threats prevalent in the Internet ecosystem today. After the briefing on Wednesday morning is your work done?

Not quite.

In fact, acceptance of your operational level plan introduces the next steps in planning for success: creating the policies that guide your organization and the procedures that provide the tactical level instructions that make your organization run.

5.2 POLICIES COMPLEMENT PLANS

Policies are the business rules and guidelines of an organization that ensure consistency and compliance with the organization's strategic direction. Policies tell you why you have the policy, its classification, and who is responsible for the execution and enforcement of the policy. They are the "rules of the road" that all employees must follow and are congruent with your strategic vision, your mission, and your core values.

In our BigRX example, did we establish a new policy for BigRX?

Absolutely! The plan to address SQL injection states that BigRX "software will comply with identified performance and security standards before it will be accepted by BigRX from the vendor." Software received from the vendor will be tested by the IT staff under the oversight of the CIO to ensure that the software is free of defect or security flaws and meets performance objectives. Those sound like "rules of the road" that ought to be applied to all software, not just software susceptible to SQL injection. BigRX needs to codify its business rule governing software into a policy that applies across the organization.

Does your organization have policies? Do you have a formal process to document them, educate and train your employees to understand and follow them, and modify them as required? Are they simple to read, remember, and follow? Do your employees know where to find the policies when they need to refer to them for guidance? Are your policies consistent across the organization? Are they viewed as logical constructs that guide the business or roadblocks that impede progress? Do you have policies that address cybersecurity best practices?

Defining policies that govern cybersecurity best practices is essential to the health and well-being of your organization. Due to the complexity of the cyberspace environment and the tight coupling of IT with business processes, there are literally dozens of policies you can and perhaps should have that address best practice "rules of the road" to best protect your business and its information. Regardless of whether you are an executive in a large multinational corporation with a wide variety of operating divisions and product lines or an executive in a small start-up business (or anything in between!), you need to define clearly the policies that direct your employees regarding what needs to be done to secure your business and its information. Rather than detail all the possible permutations of policies available, we provide in Appendix A a list of recommended cybersecurity policies covering a wide range of subjects that you and your leadership team should review to see which ones fit your organization and its mission and subsequently should be defined and implemented.

While there are a plethora of potential cybersecurity policies, which ones are the "must-have" policies that every organization should have? We submit the following 15 policies that are the starting points to a great cybersecurity program and ought to be a part of every company's policy collection.

5.2.1 Great Cybersecurity Policies for Everyone

No matter whether you are running a business, an organization, or even a household, you need to secure your information. The following policies can help protect you and your interests against cyber exploitation.

5.2.1.1 *Acceptable Use Policy.* An acceptable use policy establishes rules that a user must agree to follow in order to be provided with access to a network or to the Internet. These policies have become common practice for many businesses, educational institutions, and government entities and require that all users physically sign an acceptable use policy before being granted network access.

Most acceptable use policies have some common attributes, requiring that the user agrees to adhere to specific guidance on what is acceptable use as well as clearly defined unacceptable use guidance. Sample attributes include the user agreeing to:

- Not use the service to violate any law
- Not use the service to attempt to violate the security of any computer network or user
- That the user acknowledges that the provider retains ownership over the service and the user has no reasonable expectation of privacy as the provider will monitor usage
- Only use the service for official uses specifically granted by the service provider
- To report any attempt to break into their accounts
- To protect their passwords and not grant access to unauthorized users
- Adhere to the network owner's security policies
- Not to use the service to send threatening messages, sexually explicit material, or otherwise unlawful materials or images
- Not to use the service to impersonate another individual
- Not attack the service through malicious or irresponsible activity including port scanning, spamming, and unauthorized network monitoring activities
- Not to introduce unauthorized software into the network or service (which could include malware!)
- Not circumvent any of the provider's security controls
- Not to install or download computer software, programs, or executable files contrary to policy
- Accept that if you violate the policy and unlawful activity is suspected, that your usage information will be disclosed to law enforcement authorities in accordance with legal guidance
- Accept that if you violate the policy that your access will be terminated

You probably have signed numerous acceptable use policies over the years and may not have even noticed. You often see them when you log in to a Wi-Fi connection at your favorite restaurant, coffee shop, hotel, and now even airplanes. Many people don't even read them and just click the "Accept" or "Yes" button to access the service. We don't. We read them, not just for professional curiosity, but because after decades of business and government service, we abhor signing any agreement we haven't read. When you agree to the acceptable use policy, you are entering in to a conscious agreement (read that to mean *a contract*) between you and the vendor. Recall the character "Radar O'Reilly" on the movie and television show M*A*S*H and how he would tell his colonel just to sign the reports without reading them? Be careful and take the time to read your agreements. You never know what some folks will slip into them!

Likewise, carefully review the agreement your organization has with its employees (including you!). Make certain it covers what is important to you and your business. Sit down with your general counsel to ensure it is legally sufficient and clearly communicates your expectations and consequences if the policy is violated. Many businesses include in their acceptable use policy clear notification that violation of the policy will result in disciplinary action depending on the severity of the violation with sanctions up to and including termination of the employee.

Many companies and organizations have found that having a strong acceptable use policy is essential to maintaining good order and discipline in their work place. For example, Dow Chemical fired 24 employees and disciplined another 235 employees for allegedly sending sexually explicit messages and images on company systems.[11] In another example, two city employees of Mentor, Ohio, were fired for improper non-official computer usage when they used the Law Enforcement Automated Data System to check license plates belonging to one of the employee's husband, who was not living with his wife, and used the information to find his address (which just happened to be that of a female friend of the husband). That led to a confrontation between the employee and her estranged husband resulting in an investigation that ended in the employee's termination.[12]

Perhaps one of the most famous examples of the need to have a strong and unambiguous acceptable use policy comes from the case of Michael A. Smyth v. The Pillsbury Company.[13] In this 1996 case, the employee sent his supervisor emails that were considered to contain inappropriate and unprofessional comments resulting in the employee's termination.[14] The employee sued the company, stating that the company had previously and repeatedly stated that its email system was privileged and confidential. He alleged that since his communication was privileged, his termination was unwarranted. The court disagreed, stating, "Once plaintiff communicated the alleged unprofessional comments to a second person (his supervisor) over an email system which was apparently utilized by the entire company, any reasonable expectation of privacy was lost."[15] Smyth lost his case and his job.

In the aftermath of the Smyth case, organizations have improved their acceptable use policies to clearly communicate what use is acceptable and what is unacceptable

---

[11] Stephen Shankland, "Dow Chemical fires 24 in email controversy," http://news.cnet.com/2100-1017-245811.html, November 15, 2000. Web. October 2, 2013.

[12] Joe Guillen, *"Two Mentor police employees fired for computer misuse,"* http://blog.cleveland.com/metro/2007/03/two_mentor_police_employees_fi.html, March 17, 2007. Web. October 2, 2013.

[13] *Smyth v. Pillsbury Co.,* 914F. Supp. 97 – Dist. Court, ED Pennsylvania 1996, available at http://scholar.google.com/scholar_case?case=14078980065370881399&q=%22914+F.+Supp.+97&hl=en&as_sdt=2002, Web. September 30, 2013.

[14] Ibid. The defendant (Pillsbury) alleged in its motion to dismiss the unlawful termination case brought by Smyth that the email communication "concerned sales management and contained threats to "kill the back-stabbing bastards" and referred to the planned Holiday party as the "Jim Jones Koolaid affair."

[15] Ibid.

use. Further, they have unambiguously identified the consequences of unacceptable use, including termination of access, potential legal actions (where criminal activity is detected) and, in the case of employees, sanctions up to and including termination.

We have found the best acceptable use policies are created as a team effort by your business units and marketing professionals, general counsel, risk managers, cybersecurity professionals, and HR department. Your business units and marketing professionals can identify what capabilities and requirements are needed with your automated systems. Your general counsel can identify what behaviors and activities need to be avoided and can help establish guidelines for consequences. Your cybersecurity professionals can identify the threats and vulnerabilities of proposed configurations, services, or capabilities. Your risk managers will identify acceptable levels of risk based on the corporate risk appetite, general counsel guidance, and technical risk factors. Your HR department contributes as well to ensure that any consequences proposed for employees who violate the policy are proportionate to the violation and its impact.

Consider adding another constituency to your acceptable use policy team: a curmudgeon. We find that assigning an individual to play devil's advocate while reviewing policies pays great dividends. They will find the weaknesses in your policies and allow you to fix them before they are published. Grab your greatest skeptic and draft them onto your team. You'll soon find you have a better product.

Finally, while a team may create your business or organization's acceptable use policy, it is management's responsibility to ensure it is correct, complete, coherent with the corporate vision and core values, and can be monitored and controlled. Regrettably, executives often do not even read their organization's acceptable use policy. Policy is a management responsibility. Make sure you read and approve of your organization's acceptable use policy.

5.2.1.2 *Computer Ethics Policy.* Ethics are moral principles that guide an individual's or group's behavior. We believe it is important that you and your organization clearly state your policy regarding the ethical use of computers in your organization.

Some people will argue that you can embed your ethics into your acceptable use policy by identifying prohibited activities. While this is a valid argument, we believe it is shortsighted and dilutes the importance of clearly identifying your ethical posture in support of your core values. After all, ethics prepare you to do the right thing when confronted by questionable circumstances. As an organization with integrity, you want your employees to be best postured to do the right thing, even in ambiguous situations.[16]

Your Computer Ethics Policy should be a reflection of your corporate ethics and core values. They should clearly state what you believe and how your employees should act when using computer resources. The policy should be clear, succinct, and easy to remember.

The Computer Ethics Institute publishes what they call the "Ten Commandments of Computer Ethics." We believe they are an outstanding starting point for you to create your computer ethics policy. We are including them here (in bold print) along with our commentary (in regular typeface) to show you how you can reinforce your company's

---

[16] We are reminded of the following quote by conservationist Aldo Leopold, "Ethical behavior is doing the right thing when no one else is watching-even when doing the wrong thing is legal."

core values with clear statements of what ethical behavior is expected of you and your employees when using computer resources:

The Ten Commandments of Computer Ethics[17]

1. **Thou shalt not use a computer to harm other people**. Computers can hurt people if not used properly. For example, a malicious person could create a "logic bomb" that could destroy your data and information with devastating impact. Stealing, tampering with, or destroying another person's computer, smart phone, mobile device, or information is harmful and should be identified as clearly unacceptable behavior.

2. **Thou shalt not interfere with other people's computer work**. Have you heard the story about the employee who was working on a proposal with a short time-line who got up to go to lunch and left their computers unlocked? If the company won the proposal the employee was certain to be promoted. Unfortunately, another employee who was up for the same promotion saw his competitor (who should have been viewed as a teammate) leave the computer unlocked and went to the work station where he deleted critical files. Fortunately, another employee who practiced proper ethics witnessed the act, the files were recovered, and the unethical perpetrator was dismissed.

3. **Thou shalt not snoop around in other people's computer files**. Do you go to your neighbor's house, open up their mailbox and read their mail? Of course not! Then why would you want to read their emails? Sadly, some unethical people do just that. Locking computers, encrypting data, and setting strong access control procedures can help thwart unethical behavior, yet your policy should be clear to set boundaries on what information people should have access to.

4. **Thou shalt not use a computer to steal**. Your information has value. Using a computer to break into a company's accounts and transferring money or information to an unauthorized account is robbery. Don't tolerate it, and if you discover an instance that you suspect is an example of computer theft, report it to law enforcement officials.

5. **Thou shalt not use a computer to bear false witness**. Sadly, the Internet can be used to besmirch the reputation of individuals or organizations in seconds. Your good name or brand reputation can be ruined by false information. Once false information is published on the Internet, it is exceedingly difficult to correct and eradicate. If you or one of your employees posts false or misleading information on the Internet, you expose yourself or your organization to expensive litigation, probable embarrassment, and ruining of your reputation. Ensure everyone is trained on appropriate communications and consequences. When you find an instance where your ethical standards have been violated, act decisively and quickly to remedy the situation.

[17] *Ten Commandments of Computer Ethics*, The Computer Ethics Institute, Washington, DC, available at http://computerethicsinstitute.org/images/TheTenCommandmentsOfComputerEthics.pdf, Web. October 1, 2013.

6. **Thou shalt not copy or use proprietary software for which you have not paid**. This is important and you need to pay attention to this in your company. With the advent of digital media, copyrighted material is now widespread. Music, pictures, videos, software programs, and digital books are all examples of intellectual property that are protected under the law as proprietary. Your company can (and perhaps should) be sued if you host illegal copies of proprietary software on your network or its storage devices. The penalties can be severe including fines and damages. There are means to scan for some instances of illegal proprietary software, yet your best defense is a well-trained and ethical workforce.

7. **Thou shalt not use other people's computer resources without authorization or proper compensation**. Despite best efforts to secure passwords, some people still write them down and expose them to compromise. The author had an experience where an unethical employee found the user name and password of another employee and used them to access the other employee's account. Once logged in using the other employee's credentials, the unethical employee viewed files he was not authorized to access. Fortunately for the organization, he did not tamper with them. He was discovered when the other employee tried logging in and could not gain access as the network was configured to only allow one access instance at a time. Quick work by the help desk and network administrators found that the unethical employee had logged in from his work station using the credentials of the other employee. A visit by his supervisor confirmed it. In this case, both employees were disciplined. The first for not properly securing their credentials and the unethical employee was dismissed for using the resources without authorization.

8. **Thou shalt not appropriate other people's intellectual output**. This is a lot like the sixth commandment regarding proprietary software. Software piracy is illegal and is theft of intellectual property. You expect your employees to protect your intellectual property and trade secrets. Your ethics program should reciprocate in protecting the rights of others as well.

9. **Thou shalt think about the social consequences of the program you are writing or the system you are designing**. Tim Berners-Lee, creator of the Hyper Text Mark-Up Language that launched the Internet, is quoted as saying, "The power of the Web is in its universality. Access by everyone regardless of disability is an essential aspect."[18] Is your fancy web page usable by people with hearing or visual impairments? Have you even thought about how those who have some form of physical impairment or challenge may be affected by how you display your information? What about the content of your information? Because anyone with Internet access can access publicly exposed information, is the information you expose appropriate for all audiences? Frankly, there is a lot of information, imagery, video, and other items on the Internet the author finds morally reprehensible. Do you want your organization to be viewed as socially responsible in how it interacts on the Internet and internally? We hope so and recommend you include social responsibility in your computer ethics policy.

[18] Worldwide Web Consortium's Web Accessibility Initiative, available at http://www.w3.org/WAI/. Web. October 1, 2013.

10. **Thou shalt always use a computer in ways that ensure consideration and respect for your fellow humans**. This is reminiscent of the Golden Rule: "Do unto others as you would have them do unto you." While it should go without saying that your organization's computer ethics policy includes using the computer in ways that treat others with dignity and respect, it is appropriate and recommended to reinforce this commandment. Training people on proper etiquette in information correspondence is important. For example, many people still do not realize that typing in all capital letters is considered SHOUTING and may be considered offensive. Others do not realize the same courtesies you would place in typed letters ought to be included in emails as well. Interestingly, many people who would never consider barging into the C-suite ironically feel emboldened by the ability to send emails directly to the Chairman or CEO, by-passing their supervisors and various levels of command to share their thoughts on how to run the organization, their favorite recipes, or other information. The bottom line is to reinforce with everyone in your organization that the ethical use of computers includes treating everyone with the courtesy, dignity, and respect.

Your ethics policy can have a powerful motivating effect to your employees. We seek ethical organizations to work for and with and are not alone in doing so. Your employees expect their organization to act in a responsible and ethical manner. So do your shareholders and potential investors as do your business partners and those with whom you have relationships. Nobody wants to work with or for somebody or something that is not ethical.

As an executive, ethics begins and ends with you. Your leadership sets the ethical environment that your employees, peers, partners, prospective investors will scrutinize. If you do not follow ethical behavior when using computer resources, you will be discovered and will be held accountable—even if you are the boss. You must always ensure you maintain your integrity and practice and enforce your organization's standards of ethical behavior.

It is important that your policy clearly state that it is applicable to everyone in the organization: directors, management, and employees. It also should clearly state that your organization has "zero tolerance" for unethical behavior and that any employee found to have violated the policy will be subject to disciplinary action up to and including termination.

It is easy to spell out how not following the policy will be dealt with, but it is important to continually promote ethical behavior when using computers and reward your employees. Many companies now include ethics as a performance measure for their employees. Others have incorporated ethics into non-cash rewards programs, where demonstrated excellence in ethical behavior is rewarded with special recognition before their peers such as earning "rights" to the boss's parking space for a week, a free lunch served by the executive team, a certificate or plaque, or even a paid day off.

The need for corporate ethics is strong. An organization that conducts its business in an ethical manner engenders respect from within the organization and well as from outside. To nobody's surprise, people prefer to work for an organization that promotes ethical behavior. As a result, those organizations enjoy high rates of employee retention.

Likewise, consumers demonstrate brand loyalty to companies that exhibit a strong sense of corporate responsibility and stewardship.

Your computer ethics policy can give you a competitive edge in today's contested marketplace. You'll find that your computer ethics policy establishes your organization as placing a premium on "doing things right" with a clear sense of purpose and social responsibility. It can inspire powerful uses of technology to further your strategic vision while deterring inappropriate and wasteful activities as well. Your computer ethics policy is a great investment.

5.2.1.3 *Password Protection Policy.* Passwords are the keys to your organization's information. They arguably are the keys to your organization's survival, your personal finances, your treasured family records, or (perhaps) even your identity. How good is your password? Are you willing to risk life as you know it on the strength of your password?

You shouldn't.

Twelve years ago, the author discovered that even a cybersecurity professional's password could be compromised given enough time and access. In this case, I lost a case of beer in a password challenge with a notorious ethical hacker who worked for me. The bet was that he couldn't discover my password within the 90-day window our organization had established for password changes.

I was confident. I carefully crafted a complex password that followed the organizational rules designed to make it exceedingly difficult for any potential hacker to compromise a password and gain unauthorized access to our network and its information. I made sure I didn't write my password down, yet made it easy enough that I could remember. Nevertheless, within three months the hacker came back to me proudly displaying my password on a piece of paper. Humbled, I bought him the case of his favorite beer in recognition of his prowess and my seemingly poor choice of password. Before I surrendered the case to him, however, I insisted he walk me through how he discovered my password. What he told me surprised me.

He told me that for the first month of his attack, he learned everything he could about me. He did extensive research to find out where I was born, where I grew up, who my friends were, what social media sites I used, what my favorite sports and sporting teams were, what cars I drove (now and in the past), where I went to school, names of my family members and their birth dates, etc. Some of the information was freely available online, but the vast majority of it was obscure. Some of the information he ended up paying for through specialized search services. Other pieces he gathered by talking with my friends and associates in a deliberate yet seemingly casual conversation. Soon he developed a profile on me that filled in almost all the blanks he needed.

Armed with the information cited above, he started crafting likely passwords based on my personal information, such as where I went to school, favorite sports teams, family names, key dates in my life, etc. He wrote a software program that would create different permutations of that information and regularly attempt log-ins against my account. He had to be careful as our system was configured to lock out the user for excessive failed log-ins and he didn't want me or our security personnel to detect his attempts.

He also attacked our network infrastructure, looking for unprotected locations in our domain controllers where an insider could look to harvest vulnerable password information which he could use to decipher my password and log-in credentials. Because he already was credentialed to be on our network, he had legitimate access to the network infrastructure.

It took him awhile but just a day or two before my 90-day window to change my password arrived, he declared victory and sent me a late-night message from my own email account to prove he did it. He showed me the piece of paper with my password written on it the next day.

How did he do it? As it turned out, my password was indeed complex and long enough that his cracking technique couldn't break through in the 90-day period. Nonetheless, his program indeed had replicated my password and given enough time he would had gotten in through a brute force approach to running all the password permutations that he had for me. The key to his success was gaining access to the encrypted files from the network infrastructure that stored the log-in transactions days before our wager expired. He took his database of my potential passwords, hashed them using the same technique that we used on our network, and compared the retained files from the network against his database to see if any of the hashed information matched what he had retrieved from our network device. He was lucky as one file did match. Soon he had retrieved my password, logged in to my account after midnight, and sent me an email declaring victory. To this day, he is not saying whether he had inside help, but I suspect he did.

What was my password? I am not telling but I'll give you a hint: it contained some of the information identified above during his social engineering research. In the 12 years since the wager, I've never used that password or a similar permutation again. Had I avoided including personal information in my password, he never would have made the match. Remember, you are the target of social engineering at all times. Don't make your password too easy for bad actors to figure out.

We shared some best practices regarding passwords in Section 3.2.2. Your password policy should include them. Here they are again as a refresher:

Password Best Practices

- Try to make your password something you can and will remember.
- Don't store your password on a sticky note by your computer, in your wallet, or in your phone. Keep it as secure as the information it protects.
- Don't make your password easy to figure out (e.g., P@$$W0rd), your spouse's or child's name (e.g., M0mm@of2), or favorite sports team (e.g., $t33LeR$#1). Bad actors run password cracking programs that have thousands of passwords like these already stored in their tables. They also research you and can quickly find the names of your family members and figure out your favorite sports mascots.
- Passwords of 14 characters or more are statistically most secure. Use the maximum strength password that your system will allow.
- Never share your password with anyone.
- Never reuse your username and or password on other accounts.

- Make sure your password has at least two upper case, two lower case, two special characters (e.g., @,#,$,%), and two numbers in it.
- Avoid using typical character substitution (such as @ for 'a', ! or 1 for 'l', and 0 for O) in lieu of letters.
- Change your passwords often. Change your passwords at least every quarter. Now, with automated reminders you can load in your phone, you have no excuse for forgetting to do it.

Passphrases are another form of passwords that many people use to create complex and lengthy passwords that are easier to remember than scrambled and difficult to remember passwords. For example, we prefer to create our password based on a song title, affirmation, or other phrase. We can remember the following phrase, "In 1979 Pittsburgh was the 'City of Champions' because the Steelers won the Super Bowl and the Pirates won the World Series." We are able to use that phrase to give us the following password:

I1979Pwt"CoC"btSwtSBatPwtWS

This password is statistically complex with a whopping 27 characters and meets the best practice objectives of a minimum two upper and lower case letters, two numbers, and two special characters. Still think you can't remember a long password? We bet you can when you use this technique.

When it comes to your password policy, *in addition to the best practices we already identified*, there are several other best practices you should incorporate into your policy:

Password Policy "Must-Haves"

- <u>Account lock-outs</u>: Use a "three strikes and you are out" policy to lock accounts after successive unsuccessful log-in attempts. While an attacker can create a denial of service by deliberately creating three failed log-ins, the risk of a hacker cracking your password by repeatedly attempting all possible password combinations is not worth leaving your system unprotected. Make sure your procedure to unlock is secure and as convenient as possible.
- <u>Separate administrative and user passwords</u>: System and network administrators are among the most powerful people in your organization. They have access to your organization's most valuable information and treasured resources. Make it your policy that they must use separate passwords for their system and network administration duties than they use for their standard user functions (such as email and office duties.) There have been several occurrences where bad actors launched spearphishing attacks directed at system administrators in an effort to expose or compromise their passwords. Once the bad actor gained control of the system administrator's system they gained root access and had complete control over everything the administrator controlled. In contrast, if the bad actor compromises the administrator's standard user account, he has standard user access, which should minimize the potential for damage. When you separate the administrative and user accounts, you reduce your threat exposure.

- Force password expiration: Executives hate having to change their passwords every 90 days. Everybody does. Nonetheless, you need to do it as it is the right thing to do to protect your information. Ensure your policy mandates password changes and enforce it. You'll find many senior managers will attempt to get waivers and keep their passwords constant. Hackers love that! Since they seek to compromise the "Big Fish" presenting a static target like a password that doesn't change often just makes the hacker's job easier. Leaders ensure the policies apply to everyone, especially them, and enforce the policy across the organization.

- Don't recycle passwords: Let's see, it is October and we are in the fall quarter so we are going to reset our passwords to my usual autumnal passwords. Good idea, right? Think again. Hackers like to bank passwords and one of the first things they do when trying to access your account is use passwords you've previously used to access your accounts. Many organizations have adopted the best practice of not allowing previous passwords to be used again. Consider making it your policy not to accept any password that has been used in the past 10 passwords.

- Avoid transmitting passwords via email: This should be obvious but isn't as many organizations send their passwords out via nonsecure email systems. If you have to send a password to someone by email, make it your policy that the next log-in forces a password reset by the individual.

Don't make things easy for hackers by allowing weak passwords. Your password policy should be one of the strongest and most enforced cybersecurity policies you have. It also can be one of the most difficult to gain support for across the organization. Your leadership is essential. Make sure your password policy follows best practices and everyone, including senior leadership, follows it throughout your organization. Finally, make sure you follow these same password best practices at home as well as in the office. Strong at work and strong at home keeps you strong all the time.

5.2.1.4 Clean Desk Policy. Could the following situation happen in your organization?

In 2012, a New Orleans hospital janitor and his girlfriend plead guilty in federal court involving the theft of information from the hospital where he worked. According to the FBI, the janitor stole computer printouts containing confidential patient information such as names, social security numbers, dates of birth, phone numbers, home addresses, and other personal information that was intended to be shredded. The hospital is covered by the Health Insurance Portability and Accountability Act (HIPAA) which protects patient information collected by a health care provider. The janitor took the information to his girlfriend, who used the information to create online accounts with companies using the names of the hospital patients contained on the printouts. Once the girlfriend had created the accounts, she ordered merchandise that she had shipped to her residence for her use and for others.[19] The girlfriend subsequently was sentenced to 27 months in prison while the janitor received three

---

[19] Federal Bureau of Investigations Press Release, http://www.fbi.gov/neworleans/press-releases/2012/pair-pleads-guilty-to-stealing-patient-information-to-be-used-for-personal-gain, January 5, 2012. Web. October 2, 2013.

years probation with a special condition of six months community confinement followed by six months home incarceration.[20]

Could a janitor or other unauthorized individual steal hard copy records off of a desk or trash can in your organization? Could they use that information to potentially harm you, your business, or your clients? What type of litigation would you face from those claiming damages due to the exposure of their personal information and how much would it cost you? What would happen to your brand reputation? How do you thwart such potential bad actors? You need a clean desk policy!

You and your organization need a clean desk policy that specifies that during periods when the desk is unattended, such as after work hours or during extended lunch breaks, all work papers, including sticky notes, note pads, and digital media (e.g., diskettes, thumb drives, SD cards, etc.) need to be cleared from the desktop and secured in locked drawers.

You may be wondering why a book on cybersecurity says you need a clean desk policy.

It is because cybersecurity is about risk management and the papers on your desk contain valuable information that you don't want to put at risk of theft, exposure, unauthorized access, tampering, or damage.

Clean desk policies help organizations comply with important information security regulations such as the ISO 27001/17999 standards, and legislation such as the Privacy Act and HIPAA. In addition to presenting a positive and professional impression of the workplace, it also fosters and encourages better organization of information as employees deliberately have to manage all of their information. This can pay off for you and your organization as employees are likely to be more efficient in retrieving paper documentation, will be more likely to use digital documentation rather than more expensive paper-based documents, and be less frustrated in searching for information. Besides, auditors love it too.

Your clean desk policy should include your computer monitors. Your policy should include logging off of the network and turning off monitors. Many organizations push computer patches to workstations after normal work hours; so turning off the computers themselves may not be practical, but there is no reason why your computer monitors should not be blank and turned off to save precious power.

Clean desk policies should be short and unambiguous. Your policy should include such items as follows:

- Always clear your desk before leaving your workspace for meetings, meals, and at the conclusion of your work day.
- Always lock your computer using a password-protected screensaver when away from your desk during the work day.
- Allocate time in your calendar to secure your paperwork properly.
- If in doubt, throw it out. Because of the increasing number of incidents where valued information is harvested from dumpsters and recycling bins, your policy should dictate that all discarded paper must be shredded.

[20] WVUE Fox8 New Orleans, Former Ochsner janitor, girlfriend sentenced in scheme to use stolen patient information, http://www.fox8live.com/story/17401136/former-ochsner-worker-and-girlfriend-sentenced-for-stealing-patient-information?clienttype=printable, April 26, 2012. Web. October 2, 2013.

- Consider scanning paper items and filing them electronically in corporate electronic files in accordance with your corporate information management plan, which may include "cloud storage."
- Lock your computer, desk, and filing cabinets at the end of the day and when you are away from your desk.
- Log off your computer at the end of the day and turn off your monitor.
- Lock away portable computing devices such as laptops, tablets, smart phones, or other mobile devices.
- Treat mass storage devices such as CDROM, DVD, or USB drives as sensitive and secure them in a locked drawer.

Enforcement of the policy needs to be clear and unambiguous too. It doesn't matter if you've written the best policy document in the world if you don't enforce it. Walk through the work spaces of your employees to do spot checks. When you see instances of noncompliance, use your chain of command to ensure that it is fixed and follow-up randomly and often. Follow the policy yourself. Be clear that violation of the policy will result in disciplinary actions up to potential termination.

5.2.1.5 Use of the Internet Policy. Do you have employees who violate copyright laws? Perhaps you do and don't even know it. Perhaps they don't know it themselves. Let's say you have an employee who is building a briefing about innovation and wants to include the clip from the movie *Apollo 13* where the engineering team creates a workaround solution to the carbon dioxide air scrubber problem cited in Chapter 4.0. They find a site hosted in Eastern Europe that has the clip, they download it, and embed the file in their PowerPoint briefing. The briefing looks great and presents your organization's message on innovation extremely well. Success! Right?

Wrong. The employee's actions present two problems. First, movies, music, images, and other intellectual property are valuable commodities and you must obtain the rights from the actual owner to use them. Without appropriate rights and permissions to the digital content, your employee just broke the law and put you and your company at risk. What's the worst that can happen? Under federal law, if company computers were used in the commission of a crime, law enforcement officials can seize the computers as evidence. How long can you and your organization survive without its IT infrastructure? Second, because the employee used your corporate computers to acquire copyrighted material, your company is liable and potentially exposed to a lawsuit from the party that legitimately owns the rights to the material. Thirdly, many (if not most) sites that host bootleg or otherwise illegal media are known to have poisoned the files with embedded malware that surreptitiously inserts malware such as backdoors into your system. Not only may you be fighting criminal action or intellectual property lawsuits, but also you may be fighting with bad actors over control of your own network.

Downloading files from the Internet may not be your only problem. What happens to your organization's productivity if employees misuse Internet resources? The author was the CIO of a large organization and noticed that network performance was dropping. Personnel were complaining about slow email delivery times and having to wait excessive amounts of time for web pages to load. The problem was creating serious productivity losses. Analysis of metrics found that over 75% of web searches were landing at web sites that specialized in sports. Further research indicated that over 80% of our available bandwidth was being used by bandwidth-intensive streaming video. It was apparent that our employees were using their Internet connections to stream their favorite games (and potentially movies) over the Internet to their desktops. When the data was presented to the head of the organization and his directors, they ordered a policy to limit web-surfing and to block streaming video. Not surprisingly, productivity soared, network performance improved dramatically, and the boss and directors were delighted.

The Internet is a great tool that gives you and your employees access to the world's information. You want to use it to enhance your operations, not hinder them. Your employees should be using it to their advantage and that of your business. You need a policy that addresses the appropriate use of the Internet in your organization.

What should your Internet use policy say? While every organization is different, there are several common items we believe are important to include in your policy:

- Purpose: Tell your employees why you have an Internet use policy. Most companies remind their employees that access to the Internet is a privilege required for business, not a right. Do any of your employees need that kind of reminder?

- Applicability: It applies to everybody.

- Threats: It is important to remind your employees that the Internet can be a risky place, and sometimes is a haven for bad people with malevolent intent. Remind them not only about the risks presented by malicious code, but also risks presented by copyright violations and lack of productivity if access is misused. Some employees may be surprised by the obvious, so clearly state the threats using explicit examples of policy violations.

- What's allowed: Tell your employees what Internet use is permitted. Examples include email, web access, and electronic data exchange (e.g., file transfer protocol use to exchange large files). Clearly state that use of the Internet is only for official business use. If you allow use for personal reasons (such as to receive emails from your children's school), clearly spell out under what conditions you will allow your corporate resources to be used.

- What's not allowed: This is a critical part of your policy. You need to define not only what is not allowed, but also why it is not allowed. Common examples of prohibited Internet use include:

  ○ Illegal activity: Clearly state you will not tolerate illegal activity such as copyright infringement or child pornography. Such actions will result in severe disciplinary action including termination and notification to law enforcement officials.

  ○ Immoral activity: Clearly state that you will not tolerate immoral activity such as using the Internet to view, acquire, or disseminate pornography or material which negatively represents race, creed, sexual orientation, or genders. You own the network and its liability.

- ○ <u>Unauthorized access</u>: Make sure your employees know that they are to focus on their work and the information they need to accomplish it. Using corporate resources to access unauthorized information not needed for the performance of their duties will subject the employee to disciplinary action.
- ○ <u>Information misuse</u>: Your information has great value. You need and want it treated right. Not using it for its designed purpose, disclosing it without authorization, or tampering with it will not be tolerated.
- ○ <u>Information exposure</u>: Make it clear that you will not tolerate the disclosure, exposure, or transmission of your organization's intellectual property and trade secrets, including information considered confidential, proprietary or otherwise sensitive, without proper authorities. Identify what those authorities and controls are.
- ○ <u>Content management</u>: Clearly state that you will not tolerate the creation, posting, transmission, or voluntary receipt of any information that is considered threatening, harassing, offensive, hateful, libelous, or otherwise unlawful. You own all the information on your network. Make sure it is the type of information you want, need, and is relevant to your business.
- ○ <u>Information partnerships</u>: Does your company endorse your competitors? Does it endorse organizations whose values conflict with your core values and interests? If you permit employees to embed hyperlinks to *non-approved* organizations or sites in their email correspondence or on the web pages of your corporate network, you may just be seen as endorsing that site and embracing all it stands for. Similarly, because nearly all web sites collect "cookies" that identify users, even a visit by one of your employees to a web site that contains controversial or offensive material may be considered a tacit endorsement of that site and its content. You need to control your partnerships. Be careful when writing public-facing copy and tightly control with whom you link using the Internet.
- ○ <u>Gambling</u>: You could place this under the immoral activity, but we believe it deserves a special notation on its own in your policy. Any form of gambling needs to be specifically prohibited in your policy. There are numerous statistics that demonstrate that on-line gambling adversely affects business productivity and often leads employees to theft, embezzlement, and fraud as they seek additional resources to fritter away. Don't be a statistic. Clearly prohibit gambling using your corporate resources.

We believe that responsible use of the Internet can help make your business soar. When used properly, it increases the velocity and precision of your business, opens new and profitable markets, and develops and nurtures treasured relationships. You need the Internet, you want the Internet, and you have to have it to succeed in today's marketplace. Nonetheless, improper use of the Internet can have a hugely negative impact on you and your business, exposing you to risks that include civil litigation, criminal investigations, and malicious exploitation by hackers and other bad actors. Clearly establishing your policy on Internet use is vitally important. It prepares your work force to use the Internet properly to gain and maintain the competitive advantage you seek.

Take the time to review your existing Internet use policy. After reading this section, if you find some things missing in your policy that you believe need to be in it, feel free to incorporate the concepts from this section into your policy. If you don't have a policy, now is the right time to create one and tell your employees what your expectations are regarding proper Internet use.

5.2.1.6 Employee Internet Use Monitoring and Filtering Policy. How do you ensure that your employees follow your Internet use policy?

Many employees are under the impression that they have the right to privacy while at work. They believe their emails are private communication protected under law. They believe they should have free and unfettered access to Internet resources. Similarly, they believe that if they want to visit their Facebook, Twitter, or other social media site, they should be able to do so as long as it doesn't interfere with their work duties. Moreover, these same employees often believe that they have a reasonable expectation of privacy when they use their work computer to do their personal email using a Gmail, Yahoo, Hotmail, or similar commercial web-based email service.

Are they right?

They could be if you don't give them adequate formal notice that you monitor all Internet traffic entering or exiting the organization, that the employee has no reasonable expectation of privacy, and that you filter Internet traffic to block sites and protocols that you deem inappropriate and/or noncongruent with your business and its objectives. You need a policy that clearly defines the following:

- Any information that is created or resides on your corporate network or its devices becomes the property of the organization.
- The organization **will** monitor **all** communications, including emails, Internet web browsing, mobile devices, fax machines, and telephony, to ensure that employees use the services in a safe and responsible manner.
- The employee does not have a reasonable expectation of privacy when using corporate resources.

Such a policy can be controversial and the legal implications vary from state to state, country to country. Your general counsel should be part of the team that creates and scrutinizes your proposed policy before you make it official. The general counsel should be able to identify any weaknesses in your proposed policy's language as well as identify areas you may have missed in your draft. We've had great success when we include our best legal minds early in the process.

Revealing that you monitor and filter Internet use is important. Some employees have challenged corporate policies on monitoring and filtering citing the Fourth Amendment's prohibition against unreasonable searches and seizures. The courts have long held that employers indeed can monitor their employees' use of the Internet and email provided that the employees are notified in advance that such monitoring is conducted. Once you've made the appropriate notification, such as in your

Employee Internet Use and Monitoring Policy, the courts have ruled that the employee no longer can claim a reasonable expectation of privacy when using their employers' computer systems.

Does monitoring your employee's Internet use present any ground-breaking initiative? Hardly. In fact, a 2007 survey by the American Management Association and the ePolicy Institute found that two-thirds of employers monitor their employees' web site visits in order to prevent inappropriate surfing. The survey also found 65% use software to block connections to web sites deemed off limits for employees.[21] Similarly, a study by the Society for Human Resource Development found that 74% of reporting companies monitor use of the Internet.[22] So, if you are not already monitoring your employee's Internet use, you are in the minority.

Your policy should include the following:

- Purpose: Your policy establishes how you will monitor Internet use in your organization. The policy is intended to ensure employees use the Internet in a safe and responsible manner, and that employee web use can be appropriately monitored or reviewed.

- Applicability: As usual, it is applicable to everyone who uses your corporate network. This is crucial on legal grounds. Nobody who uses your corporate resources can be exempt from this policy. That includes directors, officers, management, employees, contractors, vendors, or even visitors. Everyone is subject to monitoring. Put it on your log-in screen to remind everyone of that fact.

- Who will conduct monitoring: Most companies use their IT department and automated tools to conduct their monitoring, although some out-source it to a third-party vendor. Regardless of who conducts your monitoring, clearly identify who will conduct the monitoring, what their charter is, and what they do with their findings. Actions based on the findings should be reserved for management and not delegated to the IT staff or proxies.

- Reports: This is important as it specifies types of reporting, to whom, and when reporting will be accomplished. For example, when the author was the CIO of a large organization, we had a policy that stated we would provide continual Internet monitoring. My staff used automated tools and could quickly prepare usage reports upon demand, although I found that quarterly reports were a good fit for our business rhythm. If the staff detected aberrations during the quarter, the policy called for them to bring it to my attention immediately for a management review. On more than one occasion, I took the findings to our board with a recommendation to change our policy due to emerging threats.

- Record keeping: This is important too. You want your employees to know that you are keeping records that clearly identify who did what, when they did it, and

[21] Fact Sheet 7: Workplace Privacy and Employee Monitoring, Privacy Right Clearinghouse, https://www.privacyrights.org/fs/fs7-work.htm. Web. October 2, 2013.

[22] Mary Nestor-Harper, Demand Media, "Is It Legal for Employers to Monitor Employees at Work?," The Houston Chronicle, available at http://smallbusiness.chron.com/legal-employers-monitor-employees-work-16563.html. Web. October 2, 2013.

how they did it. This provides a powerful deterrent for those tempted to violate your acceptable use and Internet use policies. Your policy also should stipulate how long you retain your monitoring records. The length of record retention depends on the type of organization you have. While many organizations retain monitoring records for up to 180 days, the author was in an organization that retained records for over seven years, which coincided with the period of a large and highly competitive contract. Select the record keeping duration that best fits your needs. Make sure it is a corporate decision involving your general counsel, not one delegated to the IT staff.

- What will be filtered: A security best practice is to "Deny All, Permit by Exception." Using this technique, all Internet sites and protocols will be blocked except those specifically allowed by the organization. This is initially a painful policy to implement, yet extremely effective in reducing your threat exposure. Educating your work force about it in advance is crucial and is key to its success. Your policy must include the means for employees to request access to certain web sites, services, and protocols and receive quick action (which can be measured via your metrics program). The following are typical things that are filtered by rule in most organizations:
  - Adult/sexually explicit material
  - Gambling web sites
  - Video gaming sites
  - Hacking web sites
  - Pop-up advertisements
  - Chat and instant messaging
  - Anything dealing with illegal drugs
  - Intimate apparel and swimwear
  - Peer to peer file sharing
  - Personals and dating services
  - Social network services
  - Spam, phishing and fraud, and spyware
  - Tasteless and offensive content including violence, intolerance and hate (aka dirty word search filtering)
  - Web-based email
- How to change a filter rule: This is an important item you need to address in your planning with procedures that provide quick and accurate results. For example, you may find that your business is considering acquiring a company in Africa that has access rights to strategic materials vital to a new manufacturing process. The web site for that company resides in an Internet address range that is currently blocked by your business because it previously was outside your market and is the home of some unsavory hackers. As a result of this emergent business relationship, you want the specific IP address for the company to be open for use as well as several other addresses known to be the company's suppliers. You do not

need nor want the whole range of addresses opened, just the targeted addresses. Your policy should specify who has authority to request a filter change, who is the approval authority, who has responsibility to make the changes, and when the changes will be made. Your procedures will spell out the tactical level mechanics of the policy and are not necessary to be included in the policy, but you may find value in including the procedures if your corporate culture demands it. As a point of interest, in organizations where I served as the CIO, our standard for unblocking an approved web site was faster than you could get a pizza ordered and delivered (i.e., 30 minutes or less).

- Enforcement: As with all your other policies, clearly state the consequences of not complying with this policy. Typically, noncompliance with this policy will be met with sanctions up to and including termination.

Does your organization monitor its Internet usage? If so, when was the last time you reviewed the data found? Here's a tip that may lead to revealing information you may never see from your IT staff: ask to see how many attempts were made to reach sites or services you are blocking based on the rules above. You may be surprised by the results.

You will be confronted by a situation where you have to discipline an employee for improper Internet use on corporate systems. Make sure your actions can hold up in court. Have a strong and unambiguous use monitoring policy. Clearly communicate it to your employees and have them acknowledge the policy in writing, whenever possible.

5.2.1.7 Technology Disposal Policy. What is your policy to deal with your old computers after you no longer need them?

Some organizations merely do a simple disk wipe, if that, and then try to sell them. Be careful. That could be a recipe for disaster for you and or your business.

There have been numerous reports of people who found sensitive information on used computer hard drives. One of the most notorious was when British researchers who bought a used hard drive on eBay found it contained highly sensitive details of U.S. military air defense systems. They found the test launch procedures for the Terminal High Altitude Area Defense (THAAD) ground-to-air missile defense system. The disk also contained security policies, blueprints of facilities, and personal information on employees including their social security numbers. Based on the information contained on the drive, the researchers traced the drive to Lockheed-Martin, who designed and built the THAAD system.[23] Think of the fun Lockheed-Martin executives had dealing with the notifications of personally identifiable information (PII) disclosure, impacts on brand reputation, and potential litigation in the aftermath of this incident.

Is this an isolated occurrence? Sadly, no. According to a six-month study conducted by Kessler International, a New York computer forensic firm, over 40% of the computer hard drives they bought on eBay were found to contain "personal, private and sensitive information—everything from corporate financial data to the Web-surfing history and downloads of a man with a foot fetish." Of the information retrieved by Kessler

---

[23] Pete Warren, "*Anti-missile defence details found on secondhand computer*", http://www.theguardian.com/technology/2009/may/06/data-loss-lockheed-missile-defence, May 7, 2007. Web. October 2, 2013.

International, researchers found personal and confidential documents, including financial information, emails, photos, corporate documents, Web browsing histories, DNS server information, and other miscellaneous data.[24]

What happened to your last home computer? Who's using it now and what personal information of yours might they have access to? What about your last work computer? Is there anything valuable on that drive?

Can you wipe hard drives sufficiently to erase information permanently so that it cannot be retrieved by the next owner of your computer (who could be overseas!)? Some people believe that using industrial strength tools that delete your files and overwrite them at least seven times are sufficient. These tools often are freely available on the Internet. Frankly, most of these tools are very effective in wiping drives and making information retrieval increasingly difficult to achieve, but not impossible. That's why all of the tools are distributed without a warranty. The only *guaranteed* way to prevent someone from retrieving your information from your old hard drives is to physically destroy it or degauss it (magnets). While your risk appetite and the strength of the tool choices available likely will govern your decisions, how you articulate your policy on technology disposal will give your employees the guidance needed to take the appropriate actions as you retire obsolete systems.

We believe that the disposal of your technology needs to be accomplished in the context of your risk management program. We believe the key factors for your consideration are what type of information is contained on the hard drive, what its value is, and what is the risk of exposure. Here's where the CIO, CFO, CRO, and General Counsel need to work together to provide cogent guidance to the work force on how to responsibly dispose of equipment. Your decisions need to make good business sense and your policy needs to reflect that same good sense with clear direction that your employees can easily understand and follow.

What do we recommend you include in your technology disposal policy? While we remind the reader that every organization is unique and should tailor its policies to meet its organizational objectives, we have found that the following is a productive construct (model) to follow when developing your technology disposal policy:

1. <u>Determine how you value information</u>: This is critically important and governs your next steps. Consider placing your information into three categories:
   - <u>Category 1</u>: This is information that is most sensitive and cannot fall into unauthorized hands under any circumstances. Loss or exposure of such information may result in an existential event for you or your business. Examples of this type of information may include your critical intellectual property and trade secrets; key financial information including account numbers and credentials; security information such as account names and stored passwords. Many people also associate a certain monetary threshold to delineate what information falls into this category. For example, a small

---

[24] Lucas Mearian, "Survey: 40% of hard drives bought on eBay hold personal, corporate data," Computerworld, http://www.computerworld.com/s/article/9127717/Survey_40_of_hard_drives_bought_on_eBay_hold_personal_corporate_data?taxonomyId=19&pageNumber=1, February 10, 2009. Web. October 2, 2013.

business may determine that the loss or exposure of information valued in excess of US $250,000 may make it a category 1 event.

- Category 2: This is information that is very sensitive and disclosure of which will cause significant harm to you or your business. Examples of this type of information may include business plans, architectures, designs, and confidential information. PII frequently falls into this category such as social security numbers. Using the monetary threshold example, the small business may determine that a potential information loss ranging from US $100,000 to US $250,000 would make this a category 2 event.

- Category 3: This is information that is valuable yet its disclosure will not cause appreciable harm to you or your business. Examples of this type of information would include routine correspondence, uncorrelated data, most photos and images, and replaceable or depreciated information. From a monetary threshold perspective, it is valued by our exemplar small business as having a value less than US $100,000.

2. Determine how to dispose of technology by category: There are several ways to dispose of your old technology. Your policy should help your staff identify the methodology consistent with your corporate risk strategy. Continuing our example, consider the following construct (model) when determining how to dispose of your old technology:

- Category 1: The IT staff will remove the hard drive from category 1 assets and destroy the hard drive through degaussing (magnets) or by physical destruction. Other media, such as thumb drives, containing category 1 material will be handled the same way. Two-person control (i.e., someone to destroy the drive and someone to witness it) is required along with documentation certifying the drive's destruction. All other components of the system may be salvaged for resale in accordance with the corporate disposal policy.

- Category 2: The IT staff will use an "industrial strength" disk wiping program that meets the National Industrial Security Program Operating Manual (NISPOM) and DOD 5220.22-M standards. The IT staff will execute the program on three separate occasions on the drive to ensure that all information is reasonably expected to be erased and irretrievable. Upon completion of the disk wiping and its certification, all components of the system may be salvaged for resale in accordance with the corporate disposal policy.

- Category 3: The IT staff will use an "industrial strength" disk wiping program that meets the National Industrial Security Program Operating Manual (NISPOM) and DOD 5220.22-M standards. The IT staff will execute the program on the drive to ensure that all information is reasonably expected to be erased and irretrievable. Upon completion of the disk wiping and its certification, all components of the system may be salvaged for resale in accordance with the corporate disposal policy.

3. Printers and copiers: Modern printers and copiers all have storage devices on them that retain information on items you've copied or printed. They must be

sanitized prior to disposal. In the event that your staff or a bonded consultant is not able to sanitize the device satisfactorily in accordance with the directions above, the device should be destroyed in accordance with this policy.

4. Determine who will dispose of the technology: IT departments are not very good when it comes to getting a good return on your dollar in selling your excess technology. They are usually very busy just keeping up with the technology on hand let alone of disposing the older stuff. That's why it is important that you have someone else in your organization responsible to dispose of it. Whether it is someone in your financial department or your logistics department (whom we prefer), make sure they are equipped with the requisite training and equipment to quickly check the out-going devices to ensure that all your information is sanitized from all digital media (i.e., hard drives, thumb drives, even CDs still left in their drives!)

5. Enforcement: As with all your other policies, clearly state the consequences of not complying with this policy. Typically, noncompliance with this policy will be met with sanctions up to and including termination.

Your technology has value even when you no longer have a need for it. Monitors, computers, servers, network devices, peripherals, and printers all have value. It may be worth your effort to sell these items to someone who has a need for them but ensure that your organization has the right policy and controls in place to prevent your valued information from heading out the door with your obsolete technology. Another method for transferring usable equipment is to give **sanitized** equipment to employees in recognition for exemplary efforts, or as holiday or bonus gifts.

5.2.1.8 Physical Security Policy. You may have the best boundary protection in the world for your information, but if you don't have the right physical security controls, you may open yourself to actually make it fairly easy for bad actors to gain and exploit your information to their advantage. Ensuring your information is protected from **physical** attack is an important part of your cybersecurity risk management program.

Picture this scenario: A maintenance worker arrives at your facility. This is not unusual as you often hire third-party vendors to perform a variety of tasks including janitorial services, facility and equipment maintenance, and even reload your snack bars. The maintenance worker has what appears to be printout of an email from one of your senior IT managers ordering that all computers be inspected for potentially faulty fans and random performance monitoring devices be installed on some machines. Although that senior IT manager is on vacation, it appears to be legitimate. After all, your IT shop is very proactive to ensure that your IT systems always are available and he appears to have a legitimate work order. The maintenance worker is uniformed with his company golf shirt and khaki pants, is extremely polite and professional, and even shows you several web sites that indicate problems with the fans that cool the processors in the computers. Several of the sites show how failed fans caused processors to overheat and computers to fail. You don't need that headache. He explains that by inspecting the fans he can tell if the computer is at risk of catastrophic failure due to overheating.

The good news, he says, is that his company is a certified third-party vendor for the fan company and can replace them at no charge to your company. What do you do? What does your physical security policy say you should do?

Your policy should address visitor and contractor access and should guide you to deny entry and not give the maintenance worker access to your computer without you personally verifying through official channels that the maintenance worker is authorized to access your facility and your computer.

Think the scenario is far-fetched? Regrettably, it isn't. In fact, there are numerous incidents where bad actors brazenly have entered facilities with the intent to steal information. In many instances, rather than attempt to break into your systems by hacking into your computer, they find it easier and more effective to just gain access to your home or office and steal your computer, return to their lair with the purloined equipment, and harvest the information from it at their leisure. If you don't have your computer appropriately protected with a strong password or other user authentication technique and have your data on your hard drive encrypted, your information is now in the crook's hands.

A bank in England recently dodged a bullet when confronted by a scenario similar to the one detailed above. According to press reports, a man posing as a third-party maintenance worker entered Santander Bank's branch in the Surrey Quays Shopping Centre and attempted to fit a monitoring device on the back of a computer in the bank.[25] The device was a small box that plugged into one of the USB ports on the back of the computer, much like you use to plug in a mouse or keyboard. The box was equipped with a key board video monitoring device that would record what displayed on the monitor and transmit it to the bad actor's control center, which could be in a car outside the facility or potentially in another country well beyond the reach of your law enforcement officials. The alleged perpetrators were apprehended and reportedly no Santander information was exposed, but the threat of a physical attack to your information is acute.

The author knows this first hand. When serving as a military officer in charge of the networks supporting the nation's fleet of tanker and cargo aircraft, his unit was subjected to a multidisciplinary threat assessment that tested the entire base and its ability to properly deal with threats. Entering the inspection, we were very confident in our ability to protect the information stored in our network. We had the best boundary team in the business. We had previously undergone several penetration tests that had failed to breach our defenses. We had the technical countermeasures in place to detect attempts to access our information, alert our crews who were on duty around the clock to protect our information, and to deny access to unauthorized users. We thought we had all the bases covered.

Until three o'clock one Wednesday morning.

It turned out that the inspectors[26] went around looking for unlocked and unattended facilities in the middle of the night. They checked every building and every window and

---

[25] Ted Thornhill, "Hacker Gang used tiny screen-grabbing 'bug' installed by bogus repairman in audacious cyber-heist plot to steal millions from Santander bank,"http://www.dailymail.co.uk/news/article-2419805/Police-thwart-plot-hackers-syphon-money-Surrey-Quays-Santander.html#ixzz2gemnPh9B, September 13, 2013. Web. October 2, 2013.

[26] The military conducts inspections to ensure compliance with policies and regulations very similar to auditing in the commercial sector. In this example "inspectors" are acting as very engaged "auditors."

found one door unlocked in one of the squadrons on base. They entered the squadron facility and took some potentially sensitive papers that were left out on a desk which they would provide to the base commander a couple of days later just to see if anyone noticed they were missing and sounded an alarm (which regrettably didn't happen). They also climbed under one of the desks and replaced an uninterruptible power supply with what appeared to be a new one. It was a wolf in sheep's clothing!

In fact, the seemingly inert uninterruptible power supply was a monitoring device cleverly designed to look like a useful appliance. The inspectors had augmented the device to fit their nefarious plans. They took the cover off the unit and added a network interface card, a small processor, a wireless cell phone card, and a USB port to the device. This gave them the ability to plug directly into a USB port on the desktop computer stored beneath one of the squadron desks and beam out any retrieved information to a monitoring station off-base. It was devious. Nobody would look under a desk and notice that the uninterruptible power supply also was plugged into the computer. After all, there was a tangle of cords under the desk. What's another cord, right?

Fortunately, my crews detected the device when it popped up on the network. We had activated a feature called "port security" that enabled us to prevent any new devices from being granted permission to enter our network without the expressed permission of our network managers. It was locked down so tight that we knew the building and room that the request for access came from. Our crews even knew which plug in the wall the request came from. Because it was so late at night, the lieutenant on duty sent his biggest, baddest sergeants to investigate. They went to the building and found the lights on, a door propped open, and one of the inspectors sitting under a desk talking with his buddy on a cell phone as he was cleaning up his work. When he looked up to see two very unhappy, very tall, and very large sergeants standing over him, he surrendered and the device was removed.

We were lucky. We had strong logical controls in place that offset the weak physical security that the squadron exposed the entire base to. The wing commander made certain that his subordinate commanders and all base personnel were more attentive to physical security after that incident. However, we recognized that you can never let your guard down. The base established new policies to conduct better end-of-day security checks, stepped up security police checks to ascertain that facilities were properly secured at night, and implemented random security checks throughout the day.

I was proud of my team. We were the first that had actually caught the inspectors in the act and prevented them from gaining access via this means. Did we get an atta-boy in the inspection report? Actually, no. The inspectors cited us for failing to follow our own policy to call the security police to apprehend the perpetrators rather than sending our technicians. As it turned out, while the biggest, baddest sergeants were effective, they weren't the best response. We changed the policy to include sending a technical team to survey potential damage **after** the cops apprehended the villains. The wing commander also ordered changes to policies that included checking under desks and in other spaces for new and unusual equipment that didn't belong as well as training for all personnel on physical security controls.

Could this happen to you? Absolutely. We found from the inspectors that they didn't invent the technique to install a bogus uninterruptible power supply. They used a technique that criminals and spies use to gain surreptitious access to networks and information. Other techniques they use include gaining access to unlocked communications and server rooms where they can tamper with existing devices or add their own to penetrate your network and harvest your information. They also routinely harvest sensitive papers, thumb drives, CDs, and even unplug and remove hard drive storage units from unattended desks, just as criminals often do. They gauge the unit's security posture based on the opportunities presented to them as well as the response of the personnel. Nothing piques the interest of an auditor or an inspector more than *having nobody challenge them* as they attempt to enter a closet containing your sensitive network equipment, sit down at a work station, or walk out the door with your vital information.

For several years, one of us worked at a nuclear weapons facility that required virtually all personnel to have "secret" clearance from the Atomic Energy Commission; some even had "top secret weapons" clearance. We were required to remove all classified documents when leaving our offices and store them in a hardened, fire-proof combination lock safe. On the evening shift, plant security went through each office and routinely surveyed for unsecured classified documents, tried to open the safe, and before they left the office, they spun the combination lock dial. Violations were very few, but when they did occur, there was absolute hell to pay. One poor soul who was "caught" by his carelessness, told me that the course that he was obliged to take in "Security Reorientation" reminded him of a Chinese water torture movie that he saw. Neither he nor his department mates ever had another violation.

How do you prevent criminals, auditors, or perhaps even your own employees from just walking up and stealing your information? We suggest that the prescription starts with a comprehensive physical security policy that addresses such things as facility controls, visitor and contractor access, employee credentialing, equipment removal, and emergency procedures including evacuation.

5.2.1.8.1 FACILITY CONTROLS. If you are like most people, you lock up your valued possessions when they are unattended. Most people lock their houses when they leave for the day and many do at night when they are sleeping. You lock up "your stuff" to keep it out of the hands of those who don't have your permission to use it.

Nowadays, many people put considerable thought and investment into protecting their assets. They install sophisticated sensors and alarms around their house to deter criminals and alert authorities when breaches occur. The normal front door lock largely is a thing of the past with augmentation from deadbolts providing an additional trusted layer of physical protection against intruders. You want to and need to feel secure in your home and not only are these measures prudent investments to make but also they may be essential.

Once inside the sanctity of the protected space you call the office or home, what other physical security controls do you have? What is your policy? What rules have you established to control your domain and the information in it?

We have found there are several easy-to-implement rules you ought to include in your physical security policy that can better secure your home and office. While some may not apply to everyone, they are pretty good rules to follow and include in your policies:

- Don't put your valuables in plain sight for everyone to see: Temptation is a mighty bad thing. Would you place a Rodin statue worth US $10 million in your living room window and leave for vacation? Would you leave confidential information on your desk and leave for lunch? Your information has value. Protect it and limit its exposure.

- Lock up things not in use: Your mom taught you to put your things away. Do as she said and secure your valuables when they aren't in use. That goes for your valuable information too! Thumb drives, CDs, storage devices, and important papers all contain information that has value. My mom told me not to leave money lying around as my brothers would pick it up and spend it. She was right. Lock up your information when it is not in use, and when you have an asset of great value, invest in a big, steel, fire-proof safe.

- Visually inspect: Be vigilant and train your employees to be vigilant too. Things that are out of place ought to be investigated. For example, if you find a window, door, or other entry point unlocked that normally should be locked, notify authorities. If you see a desk is unattended and important papers are exposed or a computer is left unlocked and turned on, do something about it. Your policy should spell out what to do when you see something that is out of place or unusual.

- Double check: Whenever we travel, my wife likes to check the room where we stayed before we leave to ensure we haven't left anything behind. Then she insists I do the same in case she missed something. Having two people check important "can't fail" items always is a good policy, especially when protecting your vital information.

- Control who comes into your facility: Do you freely permit strangers into your home? We don't recommend it nor do we recommend you permit strangers into your work place either. Control who enters your facilities and keep them under appropriate surveillance and control until they leave.

- Check everything coming in and everything going out: You may consider it important to check everything coming in to your facility and everything going out. For example, if you are at the Centers for Disease Control, you want to make sure that all the samples coming in have the proper safety controls to prevent contamination and exposure and you definitely want proper safety controls on the way out as well! Admittedly, it is not practical to check everyone in every facility, but for those facilities having very high-value information and operating in a high-risk profile environment, this type of policy is appropriate. Do you think that the U.S. Army regrets not implementing these types of controls at the facility where Private Bradley Manning worked?

- Clutter is bad: Not only does a cluttered workspace portray an unprofessional image, it makes it extremely difficult to manage and retrieve information. Despite the plaintive cries of those who revel in the joys of building nests of paper around them, allowing clutter to accumulate presents risk of theft, information loss, and decreased productivity. There are even some who would argue it presents a safety risk. Just like your language in church, you have to keep things clean. Articulate a clean desk the policy throughout your organization.

- Train everyone to know and follow the rules: Ignorance is not bliss. In fact, ignorance is the leading cause of inadvertent information disclosure as well-intentioned employees allow important information to escape control. As an executive, you need to ensure that you have the right policies in place to support your strategy and plans and align the right talent to execute them. Make sure your employees know the rules and follow them!

5.2.1.8.2 VISITOR AND CONTRACTOR ACCESS CONTROLS. Your employees have varying levels of access to information in your organization, hopefully based on their roles and need to know. But how about visitors and contractors? How do you control their access to information while they are in your facilities?

Having a visitor and contractor access control policy is essential to protect your information. Your employees should know what your rules are for handling visitors from the moment they arrive on your property until the moment they leave. Similarly, while contractors may be important contributors to your team, they remain employees of other firms and require special handling and consideration.

Every organization is different and has varying levels of security controls depending on the assets of the company. Banks possessing lots of cash are more likely to have 100% escort of uncleared personnel into areas where that cash is exposed than a retailer who wants the customer to roam through the store in search of a purchase. Nonetheless, both the bank and the retailer have controls over what areas the visitor can visit, what levels of escort are required (e.g., customers visiting the retailer are unlikely to be permitted to visit the back offices without an escort), and what information is exposed to the visitor.

What is your policy for visitors and contractors? Do you allow them to roam freely throughout your organization? Do you insist your visitors be escorted? How can you tell the difference between employees, visitors, and contractors? If someone picks up a piece of paper on a desk or sits down to a computer terminal, how do you know whether they are on your team or potentially working against you? What's your policy?

We've worked in a variety of organizations ranging from areas that handled highly classified material all the way down to nonprofit activities. Despite the wide variance in security controls that we've encountered, here are several best practices that you include should include in your visitor and contractor access control policy:

- Designated parking for visitors: Having designated parking spaces for your visitors not only is good form but also it makes good sense from a security standpoint. Providing parking in a controlled location permits your security personnel and greeting party to observe the activities of your visitors as they arrive. Include in your physical security policy provisions that will protect your facilities and personnel by placing bollards or barricades between your parking areas and facilities to provide protection from potential physical threats posed by vehicles and their cargo.

- Reception: Greeting a visitor with a courteous reception should be part of every organization's policy. If you want to "wow" a visitor, make it your policy to provide a professional and friendly reception, but don't ignore the importance of security. Make it your policy to have your visitor sign-in and your receptionist verify the visitor's identity through hands-on inspection of a government-issued photo identification card such as a driver's license. The author fondly recalls one highly skilled receptionist that made it a pleasure to hand over your driver's license to her. She would inform the visitor that she was instructed to see their driver's license as part of the sign-in procedure. She would look it over to verify that the identification card and visitor matched and then she would complement the individual for how good their photo was, ask how they pronounced their name (that's pronounced Two-Hill, right?), or would say that she couldn't believe someone so young had a driver's license. She was disarming and made the visitor feel that she really cared about them—because she did! Also, make it your policy that your receptionist checks that the visitor is expected (e.g., has an appointment) or is welcome (e.g., the person they came to see or designated proxy is available and willing to see the individual). Have a procedure to deal with unexpected visitors that you or your team is not ready to receive. Getting their contact information and assigning someone to follow-up with them is always best. Finally, never let a visitor roam through your organization to meet with one of your employees. Always ensure that your employees meet the visitor in the reception area and escort them to the designated meeting location.

- Visitor agreement: Many organizations include in their policy a requirement to have visitors sign a visitor agreement when they check-in to the facility. These agreements often require the visitor to agree to security provisions such as that they will stay with their escort at all times, that they will display their visitor badge at all times, that they agree not to record or photograph in the facility, etc. Have your general counsel review any and all agreements, including your visitor agreement, before they are presented to ensure that they are suitable, appropriate, and complete.

- Badging and identification: Many organizations recognize that it is difficult to tell the difference between visitors, contractors, and employees. That's why many make it their policy to issue visitors special name tags or other devices to show that they are visiting the organization. Many organizations use color-coding to make it easier to distinguish who's who. Typically, visitors requiring an escort are coded "red," those who can have access unescorted to **select** areas are coded "yellow" (be cautious), and those who are fully cleared are coded "green." Furthermore, often a second color (usually on the bottom half of the badge) identifies the person's affiliation, for example, visitor—black; contractor—blue; and employee—white.

A best practice is to issue the special visitor badge in exchange for the visitor's photo identification card, which will be returned upon check-out and the turn-in of the visitor badge. In some organizations where highly sensitive information is handled, many organizations make it policy to have the escort announce that there are visitors in the area and to secure sensitive material. Other organizations add onto that policy by illuminating flashing lights or other visible signals that indicate visitors are in the area. The intent of these actions is not to embarrass the visitor but, rather, focus the work force on their responsibilities to safeguard valued information. Nearly all visitors appreciate the disciplined approach to security and some even revel in the special attention they receive. Regarding contractors, providing special badging (see the color-coding discussion in the previous paragraph) to identify contractors is very appropriate and recommended. We've often visited organizations for meetings where we couldn't tell which attendee was a contractor and which was an employee. For clarity's sake we had to pointedly ask what roles the individuals were fulfilling. Some people believe that contractors are selected to join the team and therefore should be afforded the same privileges as full employees. We disagree. While contractors almost always are highly valued teammates, they remain employees of other organizations and likely are not authorized to access the same level of information as your employees. Make your policy simple and provide special badging (as above) for visitors, contractors, **and** employees so that everyone can tell the difference and posture accordingly to protect your information.

- Visitor electronics: The sensitivity and value of your information will dictate how you handle electronic devices in your work place. In areas where highly sensitive or valuable information is handled, it is good policy to prohibit electronic devices such as cell phones, smart phones, music players, iPods, cameras, tablet computers, thumb drives, and other similar media—**for everyone**. Be very careful to define your policy regarding what electronic devices you will allow visitors to bring into your work spaces.[27] Many people forget that most phones have cameras built into them which can quickly and easily photograph unprotected information. Likewise, smart phones can rest in one's pocket unnoticed recording or transmitting your conversations without your knowledge or permission. Regardless of the type of information in your organization, it is good policy to provide a secure locker in your reception area for your guests to leave their cell phones and other electronic devices while you are meeting with them. It also is good policy that you should be polite and have your own devices silenced or turned off and appropriately stored during the meeting as well.

- Emergency procedures including evacuation: It should be your policy that in the event of an emergency, all sponsoring employees are responsible for the safe evacuation and accountability of their visitors. Your policy should designate a location for visitors and their escorts to meet during evacuations and emergencies. It also should assign responsibility to a designated individual to account for all visitors using the visitor log maintained at the reception location. Make it clear in your policy and the visitor agreement that visitors will not leave the premises without properly checking out in accordance with your policy, even in times of emergency.

- Visitor check-outs: We believe that bidding your visitor farewell is as important as your greeting them to your facility. Make sure your policy includes not only the exchange of credentials (i.e., the visitor returns their visitor badge in return for their identification card) but that they are asked how their visit was. Ensure too that the visitor is not leaving the facility with any unauthorized

---

[27] Similar policies have existed in chemical and petrochemical plants for many years related to flammable material. Many companies, for example DuPont and Dow, explicitly prohibit cigarette lighters, matches, and other such material from entering their manufacturing facilities.

material, such as papers, thumb drives, or CDs. If a departure search is anticipated or is deemed necessary, then the visitor agreement should provide for
this eventuality.

- <u>Network or System Access</u>: Contractors often are granted access to the networks
of their host organization. In many cases, contractors actually operate the corporate networks serving in important functions such as the system administrators,
network administrators, and help desk. Your policy should call for each contractor
to adhere to all cybersecurity policies as do your employees. From the acceptable
use policy to the network management policy, contractors authorized to access and
use your network in the performance of their contracted duties should follow your
policies and be held accountable. Your policy also should identify clearly that
whoever in your organization sponsors the contractor (i.e., established the requirement for the contract) is responsible for monitoring and control of the contractor.
This is critical. Failure to provide adequate positive control of contractors can
result in information mishandling, breaches, disclosures, or worse. Ensure that
your policy clearly identifies your rules for network and system access and what
permissions are authorized for contractors and visitors. Regarding visitors, a best
practice is to establish a separate wireless network solely for visitors to access the
Internet. Separate from your corporate network, this password-protected network
provides your visitor the ability to access the Internet, yet insulates your critical
information from unauthorized exposure.

- <u>Tours</u>: Many organizations receive requests for tours or host them for clients. Your
policy should identify who in your organization is responsible for tours, how they
are to be managed and the security controls that will be implemented to ensure that
your vital information is secured and protected. Best practices for tour management
focus on having a plan for each and every tour that includes the following:

  - Purpose: Every tour has a purpose. Spell out what your objectives are for this tour.

  - Assignment of Responsibility: Be clear who will do what, when they will do it,
  and how success is measured.

  - Notification to Employees: Ensure all employees know that there will be a tour,
  who will be visiting, what the purpose of the visit is, where it will be conducted, what areas it will visit, and what times the visitors will be there.

  - Security Instructions to Employees: Clearly communicate to employees what
  their responsibilities are to ensure that safety protocols are taken to secure your
  information. For example, if visitors are in a particular area, you may instruct
  employees to remove all sensitive material and information from view prior to
  their arrival.

- <u>Enforcement</u>: As with the other must-have policies we've identified, you need to
spell out in your policy that failure to comply with this policy will result in disciplinary actions up to and including termination. For contractors and visitors, they
should understand that violation of your security policies could subject them to
legal action and criminal charges. Your policy ought to include a caveat that holds
the sponsoring employee responsible for the conduct of their visitor as well and
clearly state that in the event that the visitor violates the policy the employee will

be subject to disciplinary actions up to and including termination. Organizations with that caveat tend to have better control of their information, pay better attention to their visitors, and their sponsoring employees do a much better job in escorting their visitors.

5.2.1.8.3 EMPLOYEE CREDENTIALING. Does your organization use badges or other means of identifying employees? Do you have uniforms (e.g., everyone in Target knows that the person wearing a red polo shirt and khakis slacks likely is an employee), name tags, or other identification? In today's business environment, many companies include employee credentialing as part of the physical security posture and have specific policies governing employee credentialing.

Many companies use systems that combine an employee identification card with security controls to grant employees access only to the areas they have a need to enter. As example, one of our clients has a manufacturing arm. They limit access to the manufacturing facility only to those employees who have a need to be there. Administrative personnel and others who do not have a specific need to be in the manufacturing facility are denied access by the facility automated security card system, which is a standalone system not connected to the Internet. This is a good system and increasingly is becoming the norm for medium to large businesses.

Some of our friends in small business debate whether they need to invest in employee credentialing. In many instances the answer is no. Depending on the size and type of your company, your business practices, and your security requirements, you may find there is no need to invest in credentialing. For those who do find they have a requirement to credential employees, here are several best practices to consider including in your security policy:

- Issuance: Make all employees sign an agreement approved by your general counsel detailing their responsibilities for their corporate credentials. They should be fully aware that credentialing identifies them as a representative of your organization, and that any misconduct by them will reflect unfavorably on the company. Tell them in no uncertain terms that you will react severely if they bring disgrace upon your organization.

- Use: Employees should display their credentials when they are in the workplace and carefully secure them when outside of the workplace. Wearing your employee credential in the parking lot or about town is an invitation for trouble. In fact, it is well known in the intelligence community that foreign intelligence sources look for individuals leaving sensitive facilities who continue to wear their employee credentials. These foreign agents would photograph the individual and craft false credentials using the photograph as a template. Industrial spies are no different.

- Tail-gating: Tail-gating on the highway is trouble and it is in sensitive facilities as well. If you have facilities that require badge access for everyone in the facility, make it your policy that everyone has to use their badge to enter the facility. Penetration testers routinely attempt to gain unauthorized access to facilities by entering right behind a legitimate employee while displaying (or not) falsified credentials. This tactic also is sometimes referred to as "drafting."

- Termination procedures: Ensure your policy has provisions for employee termination. There are numerous examples of woes companies had when they failed to disable employee access to computer networks and facilities after the employee left the organization. Regardless of whether an employee resigns, retires, or is fired, you need to have a policy that immediately removes their access to corporate resources and disables their credentials upon termination. Your policy should assign specific responsibility to ensure that the credentials are disabled and physically collected (usually assigned to the HR department).

5.2.1.8.4 EQUIPMENT REMOVAL. What is your policy for removing equipment from your facility? We believe Wisconsin's Random Lake School District wishes they had an enforced policy regarding removal of valuable computer equipment from schools grounds.

In 2013, IT Specialist Mark Utsby allegedly began to take iPads and tablet computers from the school district. According to press reports, he took them home and later sold them on Craigslist, reaping nearly a quarter of a million dollars. After Utsby resigned from his position in the school district, the superintendent visited his office and discovered that the equipment was missing. He called Utsby who claimed the equipment was there when he left yet a custodian had seen Utsby taking boxes from his office out to his car. Only after a visit from the local police department did Utsby confess to taking the gear.[28] His case is still pending as of this writing.

Do you have a policy that governs removing equipment from your facilities?

During the course of his professional career, the author learned firsthand the importance of having an equipment removal policy and proper enforcement.

While serving as the CIO of a large depot and maintenance facility in California during the late 1990s, the author received a call from law enforcement officials who reported one of their sources indicated that an employee at the facility was trying to sell computers taken from the facility. That employee worked in my organization!

The employee operated the computer warehouse. As the depot and maintenance facility was slated for closure, he was responsible to receive all excess computers, wipe their hard drives with the approved software that would sanitize them of sensitive information, and processing them for reallocation, salvage, or resale. Working with our general counsel and law enforcement officials, we permitted police investigators to establish video surveillance in our warehouse where the employee was observed putting our computers in his car trunk. Police in surveillance vehicles then followed him home where they filmed him transferring the computers into his garage. Later, an undercover agent was able to purchase one of the computers leading to the employee's arrest.

We pressed charges and, in accordance with our collective bargaining agreement, suspended him pending conviction. He pled guilty and received a suspended sentence and probation. That didn't save his job, however, as we immediately terminated his employment.

---

[28] Bret Lemoine, "*Former Random Lake School District employee faces theft charges,*" http://fox6now. com/2013/09/05/former-random-lake-school-district-employee-facing-theft-charges/, September 5, 2013. Web. October 2, 2013.

After that incident, at my recommendation the CEO instituted a policy of random car checks to thwart any other criminal activity. These checks proved extremely effective and discovered that the work force was stealing tools, office supplies, and other valuable items. With the closure of the facility imminent and those not willing to transfer to other locations losing their jobs, many thought they were entitled to take what they considered excess materials. They were wrong.

Ever since then, I've made sure to publish an unambiguous equipment removal policy. Simply stated, the policy is that no equipment leaves the corporate facility without written authorization of senior management. The policy applies to all employees and they receive training on it upon initial employment and annually. The policy calls upon all employees to participate in the loss prevention program and to question anyone who is observed attempting to take equipment from the facility. Random checks of employees by security personnel are part of our enforcement mechanism and employees know that theft will result in their termination and/or criminal prosecution.

Our policy has been extremely effective over the years and yielded several "saves" by eagle-eyed employees who alerted security personnel to thwart potential thefts. Importantly, this procedural control helped secure the information that was contained on that equipment. Computer hard drives, storage media, and even storage on printers and copiers contain your valued information. Implement a policy that controls equipment leaving your facility.

5.2.1.8.5 Emergency Procedures Including Evacuation. People are more valuable than your information. In times of crisis and emergency, your policy should safeguard people as your first priority. Nonetheless, you should establish policies and procedures that ensure that your vital information is secured in emergencies.

Regrettably, theft during emergencies and evacuations is not unusual. Criminals have long sought to loot unattended properties during evacuation. Typical theft targets have usually centered on tangible property easy to "fence" or resell. Now, in our digital marketplace, information has taken its place next to cash, jewelry, and electronics on the thief's wish list of items to steal.

While many companies recommend you pack up your sensitive information and computers during periods of evacuation and crisis,[29] your policy should clearly identify your priorities in guiding employees to make the right decisions when confronted by crises. Remember that information can always be replaced. People cannot. Therefore, the safety of employees should always come first.

As a general rule, it remains our policy that during evacuations, whenever possible, employees lock their computers; as time permits, put their sensitive papers into drawers and lock them; and then quickly make their way to their designated evacuation point.

5.2.1.9 Electronic Mail Policy. We presented some important information in the Acceptable Use (5.2.1.1), Internet Use (5.2.1.5), and Employee Use Monitoring and Filtering (5.2.1.6) policies that highlight recommended policies that govern appropriate use of Internet-based resources including electronic mail (aka email). Unfortunately,

---

[29] Krissy Schwab, *"Do You Have to Evacuate Your Home? Don't Leave These Items Behind!,"* http://www. quickenloans.com/blog/evacuate-home-dont-leave-items, October 29, 2012. Web. October 2, 2013.

some people believe that use of the Internet only applies to using their browser for web navigation and that email is a separate function distinct from "Internet use." Because there are a significant number of people who fall into this category, we've found it important to reinforce the policies cited with a specific policy regarding email.

The purpose of your email policy is to preserve your organization's professional image and brand reputation; moreover, it applies to every employee, including you!

Your email policy also helps to protect against threats. By making your staff aware of your rules regarding proper handling of emails, you can reinforce your defenses against spear-phishing attacks, information breaches and disclosures, and other potentially dangerous threats. This will help to improve your compliance posture and reduce your liability. For example, if an employee engages in misconduct involving your email system that violates your policy, the fact that you had taken steps to prevent inappropriate use may help to avoid legal liability and allow focus on the misconduct itself.

In your policy, clearly reinforce that the company owns any communication sent via email or that is stored on company equipment. You should state that corporate computer systems, including email system, are provided for official use and that any other use must be approved in writing by management. Even though it is stated in other policies, reiterate that management and authorized staff have the right to and will monitor email and other information on corporate resources and that the employee does not have a reasonable claim to privacy when using corporate resources. This is critically important in maintaining positive control over your corporate resources and is a legal best practice.

In addition to defining approved and prohibited use of email, your policy should include guidance on personal use of the email system. You should be alert to the fact that personal use of your systems competes for corporate resources and clearly define what is acceptable and what is not. Following best practice, reminding personnel that the email system is for official business and their emails will be monitored generally motivates employees to avoid use of business email for personal use. Another best practice to include in this section of your policy is that any personal correspondence must be stored in a separate folder from work-related correspondence. This makes e-discovery easier and reinforces the distinction between official and personal use. Finally, because any electronic mail originating from your company will have the "look and feel" of official correspondence sanctioned by your company (because it is!), ensure that your employees know and understand that any "personal" email MUST comply with corporate guidelines for email content and composition.

We have found an important addition to email policies is a section on email etiquette. Your policy should define the style you expect your employees to follow when sending email. For example, you may prefer formal correspondence and expect your employees to address clients as Mister, Miss, Mrs. or by title (e.g., Doctor or General). If this is the case, then your policy should clearly define what rules of etiquette you expect them to follow. Another important etiquette rule to define is what your expectations are regarding responding to emails. Like the famous "answer the phone within three rings policy" that many companies follow, it is important to establish rules that spell out the expected time to respond to electronic mails. For example, some organizations expect that their employees will respond to senior management and customers within 24 hours. While the sender may not have all the requested information, a personal and professional reply within 24 hours indicates that employee is engaged, is polite, and is paying attention to the request. This improves customer confidence and management trust, enhancing business objectives.[30]

There are several other rules that your email policy should address that will improve your organization's business functions. These are best practices that should be well-defined to support your operations:

- Account creation and removal: Define the rules about establishing accounts. Most organizations do not create accounts and permissions until the employee has completed all requisite training and in-processing through the HR department. Similarly, once an employee has retired, transferred, changed positions, or has been terminated, HR should be responsible to notify the IT department immediately to terminate access to corporate network assets including email.

- Directories and personas: Your policy should define the style of your screen name and what information will be shared in directories. For example, your policy may dictate that everyone in the company will use a firstname.lastname naming convention such as Abraham.Lincoln@ExecutiveMansion.Gov. Your policy also may define how your persona is displayed when shown in the receiver's electronic inbox such as *Abraham Lincoln, President*. The format of the email signature box also must be defined and should be consistent for all employees. Further, you may state that your organization will include names, titles, office symbols, desktop and mobile telephone and fax numbers, and electronic mail address in its directory. Your policy needs to be consistent across the organization as access to this information increases the velocity and precision of your operations. An important safety tip: consider avoiding publishing the desk top and mobile phone numbers of senior executives to all employees. As an alternative, provide the secretary's number as the primary contact number. If the top brass want their confidants to have access to their private numbers, they will control such access themselves (within the guidelines of corporate security).

- Electronic mail forwarding: Email forwarding is a leading cause of spam in organizations. It also is a leading cause of unauthorized information leakage. Use your policy to define what information can be forwarded. Deny by rule automatic forwarding of emails to accounts outside of your corporate domain. This means that any forwarding of electronic mails must be accomplished by the conscious decision of an employee rather than batch forwarding to accounts outside your control. This procedural "firebreak" better controls your information and protects it from inadvertent disclosure.

[30] One fly in the ointment regarding prompt replies to email correspondence is the fact the many employees are not very discriminating when sending emails. Frankly some people simply send emails to cover their butts, and consequently send carbon copies unnecessarily to everybody. This clogs the system and it takes time to clear the detritus. When the culprits of clogging are identified, they should be "reeducated" and given a stern warning. If that doesn't work, they should be allowed to practice their email skills elsewhere.

• <u>Electronic mail retention</u>: The author had a boss who used to say, "the 'E' in 'Email' stands for 'evidence'." Perhaps he was right. If you fail to retain electronic correspondence properly, you may be putting you and your company at risk. Consult with your general counsel to determine appropriate requirements for the retention and storage of information, including electronic mail. Clearly identify those requirements in your electronic mail policy.

5.2.1.10 Removable Media Policy. Do you want your employees to infect your network with viruses and other malicious code? Of course not. But nonetheless, many organizations continue to allow uncontrolled access to network devices by removable media such as thumb drives. Recall from our previous discussions on the Stuxnet case that the attack vector supposedly was through an infected thumb drive.

Removable media not only is a concern regarding infections but also of exfiltration of information. Recall the cases of Private Bradley Manning and Edward Snowden, who used removable media to steal sensitive corporate information to the great detriment of their employers. Can you afford such an information breach?

To control the threat of infection and information breaches, many companies create policies that dictate how removable media is used in their businesses. Given the threat environment, if you do not already have such a policy, publish one, educate your work force, and implement it as soon as possible.

As with most policies that we mention in this section, there are two factors that drive the relative strength of your policy. The first is the value of the information you wish to protect. That should be gauged by the highest valued information resident on your network. In general, the higher the value of the information to be protected, the higher the level of control. The second factor is your risk appetite. If you determine that your business process controls; employee training, loyalty, and discipline; and network segmentation provide adequate mitigation of unauthorized information exfiltration, you may decide to accept the risk of permitting removable media on your networks.

Many organizations have determined that the risk is too high and disable all Universal Serial Bus (USB) ports on their network. For example, the US military famously took such an action in November 2008 in the aftermath of a significant virus infection traced to an infected thumb drive.[31] This is not practical for many businesses but it is effective. Network administrators can disable USB ports by policy across all devices on the network and only open them up under carefully controlled and monitored circumstances. This is a very effective security technique, unless your network administrator is like Edward Snowden and violates the policy.

As a practical alternative to disabling all USB ports, many organizations have come to adopt a removable media policy that features the following attributes that have become recognized as best practices. Consider adopting the following rules as part of your removable media policy:

[31] Michael Barkoviak, "*Pentagon Bans USB Drives After Virus Hits Computers*," http://www.dailytech.com/Pentagon+Bans+USB+Drives+After+Virus+Hits+Computers/article13427.htm, November 24, 2008. Web. October 2, 2013.

- The organization will provide employees with removable media: Controlling what media is used imparts effectiveness, efficiency, and security. By limiting the variety of removable media used in the organization, your network defenders can focus their efforts. Your security professionals can procure the media, wipe it of any potential dangerous code, and configure it to meet your security specifications before labeling it and issuing it to employees. Protect your information by maintaining positive control over the media it is contained on, including that which can be removed and migrated.

- Do not plug any media not provided by the organization into USB ports: You shouldn't trust removable media that your security personnel haven't checked out. Nonapproved media should not be allowed on your network. With the cost to clean up viruses and other infections continuing to rise, this simple rule makes good business sense.

- Disable auto-play: Some devices can be configured to automatically execute their programs as soon as they are plugged in to the USB port. This is very bad if the program is a RAT kit, worm, zombie file, or other malicious code. To thwart this threat, most networks disable the ability to automatically play executable files. Common network management tools allow your network administrators to disable this capability across every device in your network. Include this rule in your policy.

- Automatically scan anything connected to USB ports: Wouldn't it be great if every time someone plugged a removable media device in a USB port that it would be scanned by the network for compliance before allowing its connection? Fortunately, there are products on the market that allow your network administrators to implement such a rule set. Make it policy and invest in this capability.

- Provide a removable media screening capability: How many times have you been to a conference or meeting where you received a thumb drive or disk containing information? If you are like us, you have lost count. How do you know that the media is clean and free of malicious code? You don't. In fact, you should follow the adage famously linked to Ronald Reagan: "trust but verify."[32] Create an offline capability where employees can bring media received by outside sources to one of your security professionals who can perform a deep scan to ensure it is safe of malicious code without putting your information at risk. Upon clearance from your security team, you may consider allowing the media to connect to your network. This rule only works if you have the resources in place to provide this service quickly, so ensure that you address this through the lens of your risk management program.

---

[32] Suzanne Massie, a writer on Russia, met with President Reagan as part of his preparations for meetings with Soviet President Gorbachev. She claims to have taught him the Russian proverb, "Доверяй, но проверяй" (trust, but verify) advising him that "The Russians like to talk in proverbs. It would be nice of you to know a few. You are an actor—you can learn them very quickly." President Reagan subsequently used the term so many times that many people believe he created it! For more information, see James Mann's "The Lady Who Warmed Up the Cold War" at http://www.thedailybeast.com/articles/2009/03/10/the-lady-who-warmed-up-the-cold-war.html. Web. October 2, 2013.

- <u>Scan removable media that has been connected to non-organization sources</u>: Often your work force, particularly sales and marketing personnel, will take their removable media to other locations and connect it to another computer. Many others may take removable media home and connect to their home computers to work at home. To paraphrase virtually everyone's mom, "you don't know where that computer has been!" It could be infected with a nasty virus or comparable malicious code that could infect your removable media. When your employee plugs the newly infected media into your resources, that virus or malicious code now infects your network. Protect yourself, your business, and your valued information. Scan everything before allowing it on your network.
- <u>Train your work force</u>: Your work force is your first line of defense when it comes to cybersecurity. Inculcating a culture of cybersecurity pays rich dividends. When your work force recognizes and appreciates cybersecurity threats, vulnerabilities, and impacts they are more inclined to adhere to policies, enforce them, and not tolerate violations. Invest in educating and training your work force as your investment will pay off in countless positive ways.

5.2.1.11 Remote Access Policy. For years "road warriors" have traveled on business. After long days with clients, they return to their hotel rooms, connect to the hotel network, and remotely access your corporate network to deal with their electronic mail and access corporate information in preparation for the next day's events. Likewise, many employees would head home after a long day in the office, have supper with their families and, after getting the kids to bed, remotely access your corporate network to catch up on their electronic mail and get some additional work done in preparation for the next day. Remote access to your network has become a fact of life for many employees, including you.

Remote access policies have evolved over time and are very organization-specific. Typical services addressed include electronic mail and file access. As you've been introduced to numerous cybersecurity principles in the context of risk management, now's a good time to review your corporate remote access policy. Is it easy to understand? Is it written in a style directed at the remote user or toward the technician enabling the capability?

Your remote access policy should be written in a style that is easy to understand and is applicable to both the remote user as well as the technical team that will implement and maintain the technology that enables the capability. Your policy does not need to detail the technical procedures that underpin the implementation. Those procedures are best documented in operating instructions maintained by your technical staff. Rather, your policy should focus on the broad rules needed to provide a useful capability to maximize the productivity of your work force. Include the following best practice rules in your policy:

- <u>Tightly control who has remote access</u>: Not everyone needs remote access. Only grant remote access to your corporate resources to those who have a legitimate and vetted need. Make it clear that remote access to your corporate resources is solely for official business and is limited *only* to those who are specifically authorized to use the services.

- Train those who have remote access: "Road Warriors" and those who use remote access to your systems are more likely to expose your corporate resources to risks. As such, they need to have heightened awareness and understanding of risks and countermeasures. Make sure they are equipped to recognize risks, use the right tools and procedures to mitigate them in accordance with your policies, and perform at the levels you expect.

- Properly provision services: While some companies only grant remote access through corporate-provided devices, most now allow access through corporate devices, home computers, or any Internet-connected device. This drives several security concerns. How do you ensure the person attempting access is a legitimate user and not a bad actor or imposter? How do you know the remote device accessing your network isn't infected with malicious code and will spread that infection when it connects to your network? Address through your policy what services you will provide, what security mechanisms will be employed, user responsibilities, and what rules you have regarding devices and procedures.

- Use authentication: Use two-factor authentication access procedures whenever possible. As an example, mentioned earlier, the author's bank allows me remote access to my on-line banking, but requires me to provide two forms of identification. First, I have my password, which I protect (something I know). Secondly, I am provided a tool that generates a code specific to me which changes every 30 seconds (something I have). When I log-in, I provide both the something I have and the something I know to verify my identity. This technique has become a best practice to ensure that only authorized users access systems.

- Use anti-virus and anti-malware software: Include in your policy a statement that all devices remotely connected to your corporate resources must be configured in accordance with your policies using approved antivirus and antimalware software. This includes home computers and mobile devices, which are an ever-increasing preferred method of remote access for today's dynamic work force.

5.2.1.12 Mobile Device Policy.   Most of our clients rightfully are very concerned about the security risks posed by mobile devices used by their employees. The risks presented by ubiquitous tablet computers (such as Nexus 7s, Surfaces, and iPads) and smart phones (such as Android, Windows, and iPhones) are plentiful and cause many executives to pause when deciding how to invest and incorporate them into their business process.

These executives are wise to consider threats posed by mobile devices. There are many publicly available hacking procedures available on the Internet that can show "wanna-be" hackers how to intrude into unprotected mobile devices. News reports of criminals exploiting mobile devices heighten awareness of the threats and anxiety over them. What should you do?

Make mobile devices a key part of your business strategy and figure out a way to use them effectively, efficiently, and securely.

Mobile devices greatly enable and improve the productivity of your business and work force. They improve the productivity by enabling greater connectivity and information sharing, nearly anyplace and nearly any time. They provide access to the world's information and resources in a way that was unfathomable 20 years ago. They are redefining the business environment and are increasing the velocity and precision of business. You need to leverage the power of mobile devices while preserving the effectiveness, efficiency, and security of your business processes.

You likely are hearing a lot about BYOD. BYOD stands for "Bring Your Own Device" and refers being able to use whatever mobile device you have to do your business.

Some companies anguish over mobile device policies and immerse themselves in technical gibberish that distracts from what the policy should be doing: that is, defining the rules that make your business more effective, efficient, and secure. Some erect horribly restrictive and technically complex policies that make using mobile devices a chore not a mission enhancement. Others argue that small businesses don't need a mobile device policy because they don't have a lot of devices and likely don't have a lot of infrastructure. Nonsense! Everyone in your organization who uses mobile devices in the execution of their duties needs to know what your business rules are for using these devices. Don't make things too complicated. Stay away from focusing on the technology and focus on the business impacts and risk when building your policy.

Regardless of the size or focus of your business, your mobile device policy should address common themes that govern how the devices will be used, how your information should be protected, and how your risk is managed. Here are some best practice areas to consider in creating your policy:

- Devices: What mobile devices will be supported? Only certain devices or whatever the employee wants? We've found that organizations with traditional IT departments seem to default to a position where they only will support a designated set of devices. These organizations often can't demonstrate the agility that their constituents desire, and when the newest device hits the market, it causes a collision between business functions and the IT staff that you inevitably are called in to referee. This friction is a distraction you and your business don't need. The organization needs to expect that technology will change and new products will emerge. Your policy should allow for accommodation of new technologies and be device agnostic.

- Information and risk: Your risk appetite will drive most of your decisions regarding your mobile device policy. Understanding your information and the risks to its potential exposure, disclosure, tampering, or destruction is essential. Questions to answer include: What is the sensitivity of the information being handled by the devices? Is sensitive information stored on the device? Should you allow storage of information on the device? What is the impact if the device and its information fall under the control of unauthorized persons? What is the risk and how much does it cost if this information falls into the wrong hands? What is the likelihood that this will happen?

- Regulatory compliance: Depending on the information on the device, you may find there are regulatory controls that govern how you must protect that information. Those regulations may drive you to implement specific controls. For example, the HIPAA requires native encryption on any device that holds data subject to the act. That means you are required to use only devices that have capability to encrypt files containing information covered under HIPAA. Your policy should address regulatory compliance requirements.

- Acceptable use: Your policy should clearly spell out your rules regarding acceptable use of mobile devices. Further, you should be equally clear in identifying prohibited activities. Your brand reputation may be compromised by those who misuse mobile devices. Ensure your policy spells out acceptable use and consequences for not following the policy.

- Data plans: Who pays for mobile device service? Will the organization pay for the data plan at all? Will you issue a monthly stipend or will you require the employee to submit expense reports? Who pays for these devices? The answer to these questions depends on your corporate culture, its available resources and priorities, and even the duties of the employees. Some companies offer stipends to authorize mobile workers yet provide devices fully covered by corporate plans to those employees designated as "must-have" users. Your policy should define what your organization will pay for, how, and when.

- Legal considerations: Does your acceptable use policy apply to business conducted on mobile devices owned by your employees? What does your policy say regarding your rules for doing business on these devices? Can you monitor employee activity? If so, what are you monitoring and why? What privacy can the employee expect? What data is collected from the devices? Are your rules enforceable? Your policy should address the legal requirements you consider relevant to your mobile device users.

- Services: What corporate services and information can mobile device users access? Some businesses only offer electronic mail while others offer richer abilities, including access to office files and full user privileges. Other organizations are able to monitor the mobile devices remotely for troubleshooting and security purposes and push patches to keep them current. Some even have the capability to remotely wipe or disable lost or stolen devices. The size, complexity, and resources available in your organization will drive what you can and cannot provide. Nonetheless, your policy should clearly state what services are provided.

- Security: There are a plethora of security questions your policy should answer. What are your policy's security measures? Do you require all devices have password protection enabled? Do they need to have current antivirus software? If so, who is responsible to keep them updated? Do you require the device to automatically lock itself if it hasn't been used in five minutes? Do you require that any data on the device be encrypted? If so, what encryption software do you use, who manages the key, and how is it kept up to date? How do you back-up the data to ensure that it doesn't get lost if the device is lost, stolen, or damaged? Is it your policy to configure the device to automatically wipe itself after ten failed log-in attempts? If the device is lost or stolen, who is authorized to use tools to locate the device and or remotely wipe it? How do you know who is using your devices and what they are doing? Do you care? How do you enforce your security requirements? Do you allow connection to commercial Wi-Fi? There also are several security best practices you should consider such as directing that employees turning off Bluetooth and Wi-Fi when not in use as these transmission capabilities expose the devices to potential threats. Another best practice is to prohibit use of public computers when conducting corporate business as some public computers are known to host keylogging software and other agents that can be used to compromise your defenses. The security section of your mobile device policy should be comprehensive yet easy-to-understand and easy-to-follow.

- Maintenance: Your policy should address the care and maintenance of mobile devices. What happens if your device malfunctions or a mobile user has a problem? What happens if an employee forgets their password and is locked out of their device? Who does your employee call when they need help (especially in a hurry)? In an era where employees may provide their own mobile device, this presents special challenges for your work force as well as your IT staff. Your policy should make it clear who is responsible for what capabilities.

- Business processes: Your policy should include clear rules regarding business processes used to support mobile device usage. As an example, who is authorized to have the organization provide for or subsidize mobile devices? What is the process to request such support and who approves it? Who budgets for mobile devices and pays the bills? Who is responsible to validate the bill before it gets paid? If an employee is terminated, how and when are their mobile services terminated as well? How is your inventory managed and accounted for? How and when do you audit to ensure that your policies are properly followed? Your policy should detail the key business processes that make your mobile device usage a true business enhancement.

- Applications: There are literally thousands of applications available that can be installed on mobile devices. What is your policy for installing applications on the devices? What applications are permitted and which are forbidden? In a BYOD environment where your organization subsidizes employees, the lines of authority and ownership are blurred, so not only do you need to be clear about what your rules are but also you must be within your rights. Best practices for mobile devices owned by the organization call for managers to only install applications that are essential to conduct their business on the device.

- Back-up and recovery: In a BYOD mobile device environment, back-up and recovery increasingly is an employee responsibility. Be sure to spell out roles and responsibilities on how important corporate information is managed on mobile devices including its back-up and recovery. Many organizations use cloud-based services to host information generated on mobile devices and through a combination of automated and manual processes ensure that such information is pushed from the mobile devices to the secured storage locations. Ensure that your policy addresses how information is backed up and remains recoverable.

Mobile devices are great tools that can help your business regardless of its size and composition. Whether you have corporate-issued mobile devices, bring your own, or have a combination of the two, it is important that your organization identify its mobile device rules to ensure you remain effective, efficient, and secure.

5.2.1.13 Software Policy. Do you know that September 19th is International Talk Like a Pirate Day?

Did you also know that software piracy costs the software industry about US $59 billion per year?[33] According to the Software and Information Industry Association (SIIA), the unauthorized copying of personal computer software for use in the office or at home or sharing of software among friends is the most pervasive form of piracy encountered abroad and in the United States.[34]

How much can illegally copying software cost you?

Consider the case of End Corp. as reported by SIIA. "John" was the head of a new division of End Corp., a small company with about 45 PCs. John was hired to reduce expenses for the company, so he decided to cut corners on his software licenses. John would only authorize the purchase of one copy of each software program. His rationale was, "we bought it, and we can do what we want to do with it." John's plan seemed to work until the day that one of his employees called the software publisher for technical support for the pirated software. The publisher knew they were not licensed for multiple users so they called SIIA. End Corp., facing the possibility of a copyright infringement lawsuit, agreed to pay a fine of US $270,000 for the illegal software. In addition, End Corp. was required to destroy all illegal software and re-purchase what it needed to be legal. The total cost to End Corp. for failing to comply with the copyright law was in excess of US $500,000.[35]

Software piracy is not confined to the office. James Baxter of Wichita Falls, Texas, recently was sentenced to 57 months in prison and ordered to pay restitution in excess of US $400,000 as the result of his pirating and resale of software from his home.[36]

Could what happened with End Corp. and Mr. Baxter happen to you? Let's hope not!

Effective software management is an essential business function. Software represents a significant line item in most organization's budgets as businesses spend a significant amount of resources to acquire, operate, and maintain software. Your organization needs a strong policy with teeth to effectively manage the software that fuels your business. You and your company need to clearly define your software policy.

Your software policy should be succinct. Its purpose is to define the rules for the effective and efficient management of one of your most valuable assets (software) while protecting you and your business from the illegal or inappropriate use of software.

---

[33] Fahmida Y. Rashid, "*Software Piracy Costs $59B in Lost Revenue, May Be Even Higher: Survey*," http://www.eweek.com/c/a/Security/Software-Piracy-Costs-59-Bn-in-Lost-Revenue-May-Be-Even-Higher-Survey-272553/, May 12, 2012. Web. October 2, 2013.

[34] Software and Information Industry Association, "*Real Life Examples of Software Piracy*," http://www.siia.net/index.php?option=com_content&view=article&id=338&Itemid=351. Web. October 2, 2013.

[35] Ibid.

[36] TimesRecordNews, "*Wichitan sentenced in software piracy case*," http://www.timesrecordnews.com/news/2012/feb/28/wichitan-sentenced-software-piracy-case/, February 28, 2012. Web. October 2, 2013.

Your software policy should include the following attributes:

- Applicability: Your policy is applicable to all employees, contractors, and anyone else who has access to your network and its devices.
- Budget: Assign responsibility for budgeting for all software. Normally, this is a function assigned to the CIO. CIOs will look for opportunities to reduce software costs through the purchase of enterprise licensing agreements, which generally are less expensive on a per user basis than list costs. Centralized budgeting also provides better visibility of software costs, something every executive team appreciates.
- Acquisition: Centrally manage your software buys instead of distributing them across your organization. In addition to economies of scale and better visibility into software costs, channeling all software buys through the CIO has proven to yield improvements in business processes as the CIO optimizes the flow of information across the organization.
- Licensing and registration: Your policy needs to state explicitly that all software will be appropriately licensed and registered. Be clear that your organization respects copyrighted laws and will not tolerate any illegal or inappropriate instances of software on your network or its devices. Define standards for auditing compliance with licensing and registration and assign responsibilities to ensure that your policy is carried out. This is important as a recent study by the International Data Corporation found that the vast majority of pirated software contains hidden malicious code that opens your computers and networks to attack and exploitation.[37]
- Software installation: Issue-specific guidance on how software will be installed on your network. Empower your CIO by assigning specific responsibility for the validation and approval of software to the CIO; no software is installed on your network or devices without permission of the CIO.
- Storage and documentation: Master copies of your software licensing and registration materials as well as software documentation need to be maintained. Your policy should assign responsibilities to execute these tasks to a software licensing manager, who usually reports to the CIO.
- Inventories: Assign responsibility for the maintenance of the master software inventory to the CIO and provide the necessary resources that enable the CIO to execute these duties.
- Auditing: Normal procedure in today's network environment is to conduct auditing by periodically scanning the network and devices and comparing the official inventory against the fielded software instances. If you have software fielded that you do not have sufficient licenses (or no licenses) for, you have trouble.
- Upgrades: You can only upgrade what you own rights to. Include in your policy a statement that your organization will upgrade software through a

[37] International Data Corporation, *"The Dangerous World of Counterfeit and Pirated Software, How Pirated Software Can Compromise the Cybersecurity of Consumers, Enterprises, and Nations … and the Resultant Costs in Time and Money,"* March 2013, p. 3.

deliberate process managed by the CIO but controlled through the corporate decision-making process. This is important as software upgrades often can be disruptive to business functions. It is important to choreograph upgrades to maximize positive effects while minimizing negative ones. Strictly prohibit any upgrading of software packages outside of the official organizational process. This will provide better business continuity and reduce your exposure to software piracy.

• Copying and distribution: Don't mess around with this item as failure to control copying and distribution of software is one of the leading causes of "inadvertent software piracy". If your organization needs to copy or distribute software, it should only be done under the oversight of the software licensing manager and the CIO. Include that in your policy.

• Shareware and freeware: You may have employees who come to you and say they've found a great new software tool they want to install to enhance their ability to do their job. You may even have a teenager at home who says the same thing about a program he found that can help with his homework. They may even tell you the software is "free." Be skeptical. While there indeed are numerous programs available, they all come with a cost. Shareware is intellectual property that is copyrighted. Most owners of Shareware offer you a test-drive of their software and, if you like it, you pay them a fee. Freeware, on the other hand, is indeed free but often has no documentation and no path for any maintenance. Cybersecurity professionals and CIOs are very cautious at the thought of installing Shareware and Freeware on their networks and devices as the quality and security of the software is rarely guaranteed and usually not as reliable as licensed and registered software. If you are using Shareware or Freeware on your systems, be careful. If you are in a corporate environment, make sure both your CIO and general counsel have reviewed and approved of the software before you install and use it.

• Using company software on home systems: Your policy should strictly prohibit the copying and use of software licensed to the organization on home systems without the expressed written consent of senior executive management and the software license manager. While you have a strict policy, that doesn't mean that you shouldn't provide for employees to receive legally procured software sponsored by the organization. For example, many companies recognize that employees often will perform work-related activities on their home computers. To mitigate the risk of infection from unprotected home systems, many organizations will provide their employees licensed and registered antivirus and antimalware software purchased under enterprise agreements. Some companies go further by providing business applications that allow the employee to use their home computer much like they would at the office. These are good policies and pay off in enhancing productivity while minimizing risk to the business. While you may not want to invest in home productivity packages for all employees, considering licensing for home computers is increasingly become a great investment for companies that can afford it.

An easy-to-follow software policy can keep you out of big trouble. Make sure you effectively and correctly manage your software both at home and in the office.

A final suggestion on your software policy: review it annually to ensure it is up-to-date. We suggest every September 19th.

5.2.1.14 Access Control Policy. Do you lock your doors at night? Most people do and practice physical security access controls to prevent unauthorized access to their homes and facilities.

What about your information? Do you have an access control policy that addresses who can have access to your information? Do you define who can see it, who can edit it, and who can delete (aka destroy) it? You need an access control policy.

Access control for information is usually implemented in three ways:

- Role-Base Access Control: In Role-Based Access Control, access decisions are based on an individual's roles and responsibilities within the organization or user base.

- Discretionary Access Control: Discretionary Access Control is a means of restricting access to information based on the identity of users and/or membership in certain groups.

- Mandatory Access Control: Mandatory Access Control secures information by assigning sensitivity labels to information and comparing this to the level of sensitivity at which a user is operating. It ensures the enforcement of organizational security policy without having to rely on voluntary web application user compliance. This frequently is used in systems such as in government where you have mandatory segregation of information such as Top Secret, Secret, Confidential, and unclassified information.

We recommend that as you create information you determine who can use it, view it, modify it, or delete it. If those privileges are assigned to people performing certain roles, such as your internal auditors, implement a role-based access construct. If your scheme calls for all personnel in a group to have access, such as the HR department having access to personnel records, then implement a discretionary access control system. Finally, if you require tight controls over information, invest in a mandatory access control construct where the threat of human error is minimized.

5.2.1.15 Network Management Policy. The final "must-have" policy governs how you manage your network. Many organizations have come to the startling realization that their network is the circulatory system of their business and their fortunes rise and fall with the efficiency, effectiveness, and availability of the network and the information access it provides. For nearly all businesses, denial of service means denial of income. You need a strong policy that ensures your network is professionally managed to deliver the capabilities and results your organization needs.

Some organizations publish network management policies that only apply to the IT staff. We believe this is a mistake. Your network is used by everyone on your team.

Everyone who has network access has a stake in the effective management of your network. Everyone needs to understand "the rules of the road" for your network. Therefore, we believe it is essential that your network management policy clearly states that it is applicable to everyone in your organization.

There are numerous best practices for network management that enhance business productivity, maintain network integrity, and preserve information security. We highly recommend your policy should include the following best practice policy principles:

- Deny all, permit by exception: When you buy network devices such as firewalls, they may arrive out of the box configured to let everything through. Hackers know this and one of the first things they check is to scan your system to see what ports and protocols (think of these as the gates or doors into your system) are open so they can gain access. Your policy should only allow what you need to enter or leave your network. Make it your rule that you will deny all traffic except that which you specifically give permission. Have procedures with management oversight that allow employees to request opening ports and protocols or visit web sites to conduct official business.

- Least privilege: This principle is commonly applied in many organizations and is synonymous with "need to know." Least privilege means that you only grant privileges at the minimal level required to do the job assigned. You may ask why implementing the least privilege principle as part of your policy is a big deal. Let's look at problems associated with administrator privileges on computers. Many employees demand administrator privileges on their client computers so they can configure their environment to fit their style, troubleshoot their own systems, or install their own software. Granting administrator rights to noncertified personnel is a dangerous practice and is not recommended. A significant risk vector from malicious software comes from giving users administrative rights on their client computers. When a user or administrator logs on with administrative rights, any programs that they run, such as browsers, email clients, and instant messaging programs, also have administrative rights. If these programs activate malicious software, that malicious software can install itself, manipulate services such as antivirus programs, and even hide from the operating system. It can run through your entire network in milliseconds. Users can run malicious software unintentionally and unknowingly, for example, by visiting a compromised web site or by clicking a link in an email message. Only grant privileges based on the legitimate need to perform the duties you've assigned. Direct the principle of least privilege as part of your network management policy.

- Secure operating systems: Using standard, security-focused guides to configure your operating systems is a best practice that can enhance your security and ensure your network is operating at optimum performance levels. We recommend that your policy call for secure operating system configurations that only install what is needed and turn off all unnecessary services. This ensures your system is best configured to withstand attack, reduces your attack surface, and reduces what needs to be maintained.

- Application security: Whether you consider yourself one, you are a computer operator. Your applications are what you and your employees operate every day. Your policy should include rules regarding applications and their security. Implement least privilege to reduce the effectiveness of attacks that execute with the privilege of the current user. Ensure you have a means of performing input validation to ensure only the right information is input into your applications. This reduces your risk of attacks (e.g., SQL injection) from malformed data input. Test applications in a segregated test environment before putting them on the live network. Install only what you need to reduce your attack surface. Ensure use of secure protocols and block everything you don't need. Use application "whitelisting" which means that you will only allow applications you have approved to operate to run on your network. Encrypt all data at rest to secure your information. Make sure your applications and data entry are as secure as possible.

- Vulnerability management: Ensure your policy includes rules directing the continual auditing and self-inspection of your network for vulnerabilities. Vigilance is the watchword when monitoring your network and its devices for vulnerabilities. Your policy should call for vulnerability assessments on a regular basis, especially when new systems or applications are deployed or change configuration. This ensures system vulnerabilities are detected and that systems are not placed into service with deficiencies that should be corrected. Don't let your network vulnerabilities hide in your IT organization! Your policy should direct a comprehensive vulnerability tracking and review process that is integrated into your corporate risk management process. Your policy should also call for the automated patching of software across your organization. This is a best practice that not only decreases your exposure time to threats but also significantly reduces the cost of patching. If you are not already doing automated patching and verification, we suggest you do a business case analysis to see if it is the best fit for your organization (it usually is). If you aren't scanning your network for vulnerabilities from the inside and outside, you are missing something that someone else will find and exploit. What they find may just put you out of business!

- Malware filtering: Why would you deliberately let malware enter your network and poison your information if you could stop it? The good news is that there are many network procedures and tools that can filter code that bears the tell-tale signatures of malicious code and stop it from entering (or exiting) your network. Using such devices as proxy servers, you can filter mobile code such as ActiveX and Java scripting to provide a control mechanism to strip potentially malicious executable mobile content from entering your network. Your policy ought to include spam filtering as well to strip unwanted and potentially dangerous emails. Many commercially available filters are increasingly sophisticated and can complement your efforts to thwart spearphishing by detecting and containing emails containing spearphishing markers. Finally, make sure your policy calls for the use of antivirus software protection. Ensure procedures for the installation, use, monitoring, and updating of antivirus software, and threat signatures are a core component of your policy.

- Sensor architecture: Does your IT staff monitor what traffic is on your network? Do you have an intrusion detection system in your facility? How about on your network? The best run networks make an investment in intrusion detection and protection systems. Many organizations find that monitoring network traffic through well-placed sensors (including the network devices themselves) can help them identify problems as they are occurring so that they can be appropriately addressed. They also can detect malicious activity. For example, the ability to deploy threat-specific detection signatures that trigger immediate alarms when they detect traffic of interest is a key component of most intrusion detection systems. We recommend that your policy address how you will sense when "something's not right" and what you will do about it.

- Centralized logging: Your network devices generate a lot of valuable information you may not even know that exists. Nearly all devices create files that record what the device did so that administrators can review these "logs" as part of their maintenance procedures. These are treasure troves of information that can be critical in performing threat and attack assessments. In fact, they have proven to be so valuable that hackers deliberately target them to erase any evidence they were in your system. These "log files" are valuable and your policy should call for the transfer and storage of critical system logs to a centralized secure location with adequate back-up. It is important that you preserve these logs as official records as they often are requested by auditors and as part of legal discovery processes. Failure to produce the log files may be viewed as a sign of "network malpractice," deliberate malfeasance, or incompetence. Include centralized logging and positive control over log files as part of your network management policy.

- Threat/incident analysis: Regularly conducting threat and incident analysis should be a keystone of your network management policy and should complement your overall risk management plan. While your policy should call for continual monitoring by trained technical personnel, it should also call for the retrospective analysis of threats and incidents involving management to increase the organization's effectiveness in responding to new and evolving threats. We recommend your policy call for quarterly management level reviews of threats and incidents as well as minimum annual board level reviews of network threats and incidents.

5.2.2 Be Clear about Your Policies and Who Owns Them

Your policies govern your business and how it is run. Creation and enforcement of policies is an essential management function. Your cybersecurity policies are no different than any other policy in your organization. Do not fall victim to the trap that because many cybersecurity issues involve highly complex technical concepts that they fall into the realm of the IT staff. If you believe this, you and your organization will fail.

Your organization's cybersecurity policies are not owned by your IT staff. They belong to management and should enhance business while accepting appropriate levels of risk approved by senior levels of management using the established corporate risk management processes.

Users in organizations that defer all cybersecurity policies to their IT staff often report frustration with what they view to be an overly cautious and restrictive network environment that stifles the introduction of new and potentially highly productive capabilities, denies access to desired products and services, and presents a "Just Say No" attitude. Meanwhile, in these same organizations, the beleaguered IT staff is frustrated as well. Charged with defending the network and its information "against all enemies, foreign, and domestic" they are measured by management by how well they defend the network and its information, not necessarily by how effectively their network enables the organization to thrive, grow, and be profitable. This is a management failure. Don't punt your management responsibilities to the IT staff!

We highly recommend that you make a point that management owns all policies including cybersecurity-related policies such as those we've outlined in Section 5.2.1. Senior management such as the CEO or Chief Operating Officer should sign the policies into effect, not the CIO or CISO. This reinforces that the policies are applicable to all employees and are in support of the corporate strategy.

Your policies need to be well documented and coordinated through your general counsel. They should be easy to understand and complete. They may be the best policies the world has ever seen, but if your employees don't read and follow them, they are worthless. Therefore, we recommend that you insist that your employees read your policies and sign that they acknowledge and understand them. The fact that it is so important that they must acknowledge receipt and sign an agreement that they understand the policy is an effective measure that protects the organization against certain liabilities and reinforces to the employee the need to pay attention to the policy.

A final discussion on policies regards your partners, prospective mergers, and possibly clients. Many of us have partnerships and other relationships where we share information to enhance our business posture. When it comes to cybersecurity, the policies of your partners and those you with whom you share information are very important and warrant your focused attention to ensure your information is well protected.

Make certain your partners and those with whom you share information have the right policies and procedures in place to adequately safeguard your information. Before you make any commitments, ensure you clearly define your information management and security requirements. Perform the due diligence and exercise due care to ensure that your information is adequately managed and protected, even when it is in the care of your prospective partners. Involve your general counsel throughout to ensure your surveys are complete and appropriate. Review your prospective partner's policies and procedures to make certain they provide the adequate controls necessary to meet your organization's standards (you may find they exceed your standards or present a better way of doing things!) In the event they do not meet your standards, ensure your management knows this and understands the implications so they may determine next steps.

Policies complement your strategy and its plans. They are the business rules and guidelines of an organization that define consistency and compliance with the organization's strategic direction. Policies address what the policy is and its classification, specify who is responsible for execution and enforcement of the policy, and articulate why the policy is required. They are the "rules of the road" that all employees must follow and are congruent with your strategic vision, your mission, and your core values. With the right plans and policies in place, you and your organization are well postured to implement your plans with the tactical level procedures that convert your vision into reality.

5.3 PROCEDURES IMPLEMENT PLANS

Here's another instance where we read the chapter again from the perspective of an executive struggling to determine how to construct an effective and efficient cybersecurity program at a cost that matches his organization's risk profile. The previous major section in this chapter, Section 5.2.1, was a gold mine of information on the types of policies that organizations could and should adopt in constructing a great cybersecurity program. The details that we included will permit readers to use our material almost verbatim, in many cases, to put together their program. Hence, you may be surprised that we took the opposite approach for this section, the procedures part of the chapter. Like you, we've found the keys to leadership are to give your people a well-defined mission; precisely articulated goals and objectives; clear, unambiguous policies; and sufficient time and money to do the job. Then you permit them to get on with the job using the procedures that they devise to accomplish it. You will monitor their progress, approve procedures they develop, and by and large get out of their way, except to give them help when they need it. That's why we do not provide the same level of detail for procedures that we did for policies.

Procedures define the specific instructions necessary to perform a task or part of a process. They are tactical level instructions that can take the form of a work instruction, a desk top procedure, a quick reference guide, a checklist, or a more detailed procedure. They detail who performs the procedure, what steps are performed, when the steps are performed, and how the procedure is performed.

Procedures to implement your cybersecurity plans and policies are critically important. They should be precise, clear, and reliably and consistently produce the desired results. Procedures must be consistent with your policies and directly support your plans and objectives. As a manager, you are responsible to guarantee that your organization has the proper procedures to execute the tasks assigned by your plans. You are responsible to ensure that they effectively, efficiently, and securely produce the results your organization needs to succeed.

Because of the tactical nature of procedures, we will not delve deeply into them. However, there are numerous cybersecurity procedures you should be aware of and practice daily, both in the home and the office. Some of the more common include:

- How to turn your computer on and off
- Account creation and termination
- Password creation and protection
- Log-in/Log-off procedures
- Application use instructions
- Use of the READ process when reviewing emails (*Relevant*, *Expected*, *Authenticated*, *Digitally Signed*)
- How to file electronic records (e.g., emails and electronic documents)

- How to back-up and recover files
- Procedures to secure your workstation and office space during absences

Procedures such as these fall into a category that many people refer to as "basic cyber hygiene." They become so ingrained in our psyche and behavioral patterns that they become second nature and seemingly obvious. Following them almost becomes instinctive. At home you'd consider brushing your teeth, bathing, combing your hair, and putting on clean clothing before you leave home part of your daily hygiene ritual. They are something everyone expects you to do and when they are not followed people notice—and not in a good way. Practicing basic cyber hygiene is something managers everywhere should practice and enforce throughout their organizations.

Do you follow your organization's cybersecurity procedures? Do you enforce adherence to procedures? If you don't, you are exposing you and your organization to risk and that may be a risk your shareholders don't find acceptable. Do things right and follow procedures.

5.4 EXERCISE YOUR PLANS

The author recently met with a group of corporate directors during a conference on cybersecurity. All were distinguished executives who had long and distinguished careers in business. As we sat around the lunch table many reflected on how athletics in their youth had helped shape their leadership skills and gave them the stamina to excel in their careers. Each one of them said they regretted they had not maintained their level of fitness. They said that had if they kept up their exercise, they would be stronger, able to accomplish more, and perform at higher levels.

Does this describe you too?

The famous Notre Dame head football coach Knute Rockne supposedly said, "Practice makes perfect" yet Hall of Fame coach Vince Lombardi added, "Practice doesn't make perfect. Perfect practice makes perfect." Whether you are maintaining your personal fitness level or your cybersecurity posture, you have to practice to achieve the level of perfection your organization and your shareholders expect.

Plans that sit on shelves just gather dust and are worthless. Regularly test them and your people to check proficiency and compliance!

An example is with disaster recovery and business continuity plans. Organizations that have plans that have never been tested usually flounder when confronted by crises. This was readily apparent in the aftermath of Hurricane Katrina, which slammed the Gulf Coast communities of Mississippi, Louisiana, and Alabama in September 2005.

The author assumed command of Keesler Air Force Base in Biloxi, Mississippi, two years after the hurricane had ripped its way through the community. The base alone suffered over US $1 billion in damage, and two years later was still rebuilding and recovering. As the largest single employer in the region, Keesler Air Force Base was an important part of the community and, as its commander, the author was a member of many local chambers of commerce.

Over the course of my first couple of months in command, numerous business leaders in southern Mississippi recounted their hurricane experiences with me. I deliberately picked their brains to find their "lessons learned" in the hope that we could improve our plans and procedures at the base, so we'd be better prepared when the next hurricane struck.

More than one business leader shared that they found their disaster recovery and business continuity plans were insufficient to allow rapid recovery in a disaster the magnitude of Katrina. Several said they wished they had provisioned back-up and recovery of their data to locations much further away, stating that many had contracted for data storage in New Orleans, which less than 60 miles away, also was struck by Katrina. Others said they thought recovery from the storm would just entail restoration of power. "We thought we'd just have to turn the computers back on, reload from disk, and we'd be good-to-go" said one executive. He never thought that his facility and all the computers and disks in it would be swept away and out to sea in the storm. "We never thought to practice that everything would be wiped clean. We thought we were too far inland. We thought wrong."

In the aftermath of Hurricane Katrina, the Mississippi Gulf Coast showed its strength and resiliency. Community and business leaders partnered with government leaders to create plans that would better preserve lives and properties in the event of another major disaster. Together we rehearsed those plans on a regular basis through a series of table-top exercises as well as a series of tactical on-scene exercises that ensured our plans were well synchronized; presented feasible, acceptable, suitable, and affordable solutions; and that we had sufficient abilities to command, control, and coordinate our efforts.

Our efforts paid off on September 2, 2008, the Mississippi Gulf Coast region was hit by Hurricane George. While not as powerful as Katrina, George packed a wallop, causing millions of dollars of damage. We safely evacuated both nonessential military and civilian personnel from the threatened region, rode out the storm in our shelters, and were able to recover in days. Our plans worked as practiced and nobody got hurt—success!

Exercises and testing are critical to maintaining the strength and health of your organization. They will give you a true and accurate picture of what you currently can do and set the stage for decisions on next steps to achieve your goals and objectives.

Test your plans regularly. Plan for the worst and for the most likely and exercise those plans to gauge their effectiveness and the proficiency of your staff. Don't tolerate noncompliance! There will be those who do not take exercises and testing seriously and fail to follow plans and procedures. Be clear about accountability and enforce discipline in support of your plans and procedures.

Also, be prepared to pull the trigger on "cowboys." "Cowboys" refers to those who do not follow plans and procedures, believing they have a better way of doing things. Unfortunately, "cowboys" don't like to be bothered by such things as the corporate process for making changes to plans and procedures. "That's just worthless bureaucracy!" said one employee my organization terminated for failure to comply with safety procedures. When you execute a plan, you know that it is a framework for action toward a desired result, not a checklist. Nonetheless, you need to know that you can count on people who can work within that framework to achieve success. Practice until you are perfect and keep an eye out for "cowboys." You may be better off without them.

5.5 LEGAL COMPLIANCE CONCERNS[38]

We always have our general counsel review our plans, policies, and procedures. Not only do lawyers have trained eyes for details but also they are keen to find weaknesses in how messages are conveyed, and can provide valuable assistance and advice on how to make your plans, policies, and procedures better. If you don't have your general counsel involved in developing your plans, policies, and procedures, you will not have the best products possible.

When creating your plans, policies, and procedures you should assign responsibility to your general counsel to ensure that you are compliant with all legal and regulatory requirements. In addition to national laws and regulations, many states and municipalities have specific laws, regulations, and ordinances that may affect how you do business there. Your general counsel should help you to navigate through the wide variety of legal issues to keep you compliant and competitive.

In addition to the disclosure requirements identified in the Security and Exchange Commission's *Corporate Finance Disclosure Guidance 2 (Cybersecurity)*, there are several pieces of legislation that you ought to be aware of that affect your plans and procedures. They include:

- The Sarbanes-Oxley (SOX) Act of 2002[39]: The SOX Act was created in the aftermath of several notorious corporate accounting and finance scandals and is intended to provide greater accounting and governance controls over publicly traded companies. While SOX drives many IT compliance and security initiatives, its cybersecurity requirements are vague at best. Nonetheless, to pass a SOX audit, your company must implement security best practices for any system that touches anything related to your financial reporting and accounting systems. Many general counsels will tell you that may include your entire network infrastructure, including your network log files. The impact is that you cannot cut corners with your cybersecurity posture. Because SOX calls for executives and management to be held accountable, you should invest in best practices to protect your information, your business, and yourself.
- HIPAA of 1996[40]: HIPAA was created to achieve three objectives: protect health insurance for individuals when they change or lose their jobs, protect the health care privacy for youths 12–18 (even from their parents), and provide for the security and privacy of health care records. From a cybersecurity standpoint, the last objective is the most groundbreaking, as the Act requires a host of security

---

[38] The authors are not lawyers nor do we play them on the stage or television. (We also are not accountants.) The information conveyed in this section is a reflection of our interpretations of the law and practical experiences in the business environment. Do not take our recommendations as official legal guidance or advice. Always consult a bar-certified lawyer for legal advice.

[39] http://www.gpo.gov/fdsys/pkg/PLAW-107publ204/html/PLAW-107publ204.htm. Web. October 2, 2013.

[40] http://www.cms.gov/Regulations-and-Guidance/HIPAA-Administrative-Simplification/HIPAAGenInfo/downloads/hipaalaw.pdf. Web. October 2, 2013.

requirements that drive significant investments to achieve compliance with the Act's provisions. For example, the Act specifies that all systems that possess Personal Health Information (PHI) must have intrusion protection systems. All PHI must be encrypted and the integrity of the data must be ensured. When PHI data is exchanged between medical providers, two-way authentication is required to ensure information is exchanged only with trusted and authorized partners. There are significant documentation requirements under the act that have a cyber-security impact. For example, all system documentation must be available for audits and include all configurations and system setting information (in writing!) Also, you must document all risk analysis and risk management programs that may be audited by regulators. HIPAA cybersecurity provisions are not inconse-quential. If you have or even think you may have PHI data on your systems (and your HR department may and not even know it), then you are well advised to have your general counsel and internal auditors perform a comprehensive review to determine your liability under this law. You may be surprised by the results and have to adjust your plans accordingly.

- The Gramm–Leach–Bliley Act (GLB), also known as the Financial Services Modernization Act of 1999[41]: The GBLA set new laws regarding financial services. Like HIPAA, it established new rules and regulations regarding the privacy of financial information that have a significant cybersecurity impact. For example, it establishes a privacy policy agreement between financial institutions and their individual clients to protect the consumer's personal nonpublic information. The Act calls for systems containing nonpublic personal information (such as your name, account number, and balance) to have an information security plan, a thor-ough risk analysis, and demonstration of the ability to monitor and test the plan to ensure its effectiveness. The intent is to protect the clients and their privacy. Is your organization the custodian of information protected under the GLB Act? If so, do you have sufficient cybersecurity controls in place to achieve compliance?

- The Privacy Act of 1974[42]: The Privacy Act of 1974 defines what information is personally identifiable and governs the collection, maintenance, use, and dis-semination of PII in federal information systems. It is a groundbreaking piece of legislation that many states have adopted as well with some passing laws that direct the protection of PII on information systems operated and maintained by public and private organizations as well. You likely have PII information either in your own HR department's records, your pay system, or perhaps in your client records. Ask your general counsel to research what your responsibilities are regarding PII. Where you do business and with whom (including state and federal governments) may drive cybersecurity costs above and beyond what you originally anticipated. Ensure that you have all the bases covered and perform your due diligence and due care regarding privacy information.

[41] http://www.ftc.gov/privacy/glbact/glbsub1.htm. Web. October 2, 2013.
[42] http://www.justice.gov/opcl/privstat.htm. Web. October 2, 2013.

5.6 AUDITING

Don't just take your CIO's word that everything is under control; audit your organization.

In addition to traditional auditors, who check your compliance with rules, regulations, and your policies and procedures, there are other cybersecurity-specific auditing capabilities you ought to add to your methods to ensure you have an accurate and unbiased view of your current cybersecurity posture.

The first is to include Certified Information System Auditors (CISAs) to your staff. Individuals with this certification have completed a comprehensive examination, have demonstrated over five years experience of professional information systems auditing, control or security experience, follow a code of ethics for information system auditors, maintain their proficiency through continuing professional education (minimum 20 hours per year and 120 hours every three years), and adhere to the Information Systems Auditing Standards maintained by ISACA.[43] Adding CISAs to your internal auditing team can present an in-house capability to provide a thorough analysis of your information systems and their ability to comply with plans, policies, procedures, and regulatory guidance.

A second capability is to hire independent Penetration Testers (aka Pen-testers) to attempt to penetrate your networks or specific information systems. Many organizations retain Pen-testers to deliberately test new capabilities and configurations by attempting to penetrate them. Look for Pen-testers who maintain the Certified Ethical Hacker certification as they too have undergone a recognized disciplined process to achieve their skills and operate under an international code of ethics. Pen-testers also ought to enter into a specific agreement with your organization that they will do no harm to your system or its information. Your general counsel should be part of every negotiation and contract involving Pen-testers to ensure that your organization's best interests are preserved. We recommend you run a penetration test annually or every time you have a major system upgrade or new configuration.

A third capability is to hire multi-disciplinary "red teams." These teams often include Pen-testers yet supplement them with other skilled professionals who evaluate other aspects of your security posture, including your physical security, corporate culture, communications security, administrative procedures, and contracting. They are sneaky and devious (on purpose) and often are very successful in finding problems

---

[43] ISACA, http://www.isaca.org/Certification/CISA-Certified-Information-Systems-Auditor/How-to-Become-Certified/Pages/default.aspx. Web. October 2, 2013. ISACA, formerly known as the Information Systems Auditing and Control Association, is a professional organization that establishes and maintains cybersecurity-related professional certifications. Another noted organization that certifies cybersecurity professionals is the International Information Systems Security Certification Consortium (ISC2), which maintains similar credentialing programs. Look for certification from one of these organizations when interviewing candidates for your cybersecurity positions.

you did not even know existed. We recommend you consider hiring a multi-disciplinary red team every couple of years or whenever you have a major change in personnel, policies, products, or information systems. When you do hire them, we recommend they report directly to senior management such as the CRO, Chief Security Officer, Chief Operating Officer, or CEO.

When to audit is a decision involving the board of directors and senior executive management. We recommend you audit your organization at least annually, when you make major system or configuration changes, when you introduce new products or capabilities, and after major adverse events. Remember Sun Tzu's adage to "Know Yourself." Use auditors whenever possible to better "know yourself" and improve your cybersecurity posture.

5.7 SUMMARY

- One of the principal responsibilities of an executive is to plan and execute the activities that drive your organization to success. You need a **plan** to execute your strategy.
- Plans are designs or detailed proposals for accomplishing an anticipated operation. As an executive, you may not be the one writing the plans, but you are responsible to see that they are created, that they are done right, that they are executed well and, in the event they prove to be heading in the wrong direction, to steer the organization back to the right direction.
- How detailed your plans are depends on the complexity of the desired operation and the experience of the subordinates who will execute it.
- It also is important to remember that your plan should evolve continuously to adapt to situations, while guiding your subordinates throughout an operation. Such planning flexibility is critical concerning cybersecurity.
- The measure of a great plan is not whether its execution transpires exactly as planned, but rather whether the plan produces the desired results in the face of unforeseen circumstances.
- Policies complement plans. They are the business rules and guidelines of an organization that ensure consistency and compliance with the organization's strategic direction.
- Policies tell you why you have the policy, its classification, and who is responsible for the execution and enforcement of the policy. They are the "rules of the road" that all employees must follow and are congruent with your strategic vision, your mission, and your core values.
- Defining policies that govern cybersecurity best practices is essential to the health and well-being of your organization.
- While there are a plethora of potential cybersecurity policies, we've found that there are 15, which are "must-have" policies for every organization:

Great Cybersecurity Policies for Everyone

1. Acceptable Use Policy	6. Employee Internet Use Monitoring and Filtering Policy	11. Remote Access Policy
2. Computer Ethics Policy	7. Technology Disposal Policy	12. Mobile Device Policy
3. Password Protection Policy	8. Physical Security Policy	13. Software Policy
4. Clean Desk Policy	9. Electronic Mail Policy	14. Access Control Policy
5. Use of the Internet Policy	10. Removable Media Policy	15. Network Management Policy

- Your organization's cybersecurity policies are not owned by your IT staff. They belong to management and should enhance business while accepting levels of risk approved by senior levels of management using the established corporate risk management processes (and where appropriate, oversight and concurrence by the board of directors).
- Your policies need to be well documented and coordinated through your general counsel.
- Insist that your employees read your policies and sign that they acknowledge and understand them. Having your employees sign an agreement that they understand the policy is an effective measure that protects the organization against certain liabilities and reinforces to the employee the need to pay attention to the policy.
- Make sure your partners and those with whom you share information have the right policies and procedures in place to adequately safeguard your information.
- Plans that sit on shelves just gather dust and are worthless to your organization. Regularly test them and your people to check proficiency and compliance! Exercises and testing are critical to maintaining the strength and health of your organization. They will give you a true and accurate picture of what you currently can do and set the stage for decisions on next steps to achieve your goals and objectives.
- When creating your plans, policies, and procedures you should assign responsibility to your general counsel to ensure that you are compliant with all legal and regulatory requirements.
- In addition to the disclosure requirements identified in the Security and Exchange Commission's *Corporate Finance Disclosure Guidance 2 (Cybersecurity)*, there are several pieces of legislation that you ought to be aware of that affect your plans and procedures. They include:
  - The Sarbanes-Oxley (SOX) Act of 2002
  - The Health Insurance Portability and Accountability Act (HIPAA) of 1996
  - The Gramm–Leach–Bliley Act (GLB), also known as the Financial Services Modernization Act of 1999
  - The Privacy Act of 1974

- Don't just take your CIO's word everything is under control; audit your organization!
- In addition to traditional auditors, who check your compliance with rules, regulations and your policies and procedures, there are other cybersecurity-specific auditing capabilities you ought to add to your methods to ensure you have an accurate and unbiased view of your current cybersecurity posture.
  - Include certified information security auditors (CISAs) on your staff.
  - Hire independent Penetration Testers (aka Pen-testers) to attempt to penetrate your networks or specific information systems. Run a penetration test annually or every time you have a major system upgrade or new configuration.
  - Consider hiring a multi-disciplinary red team every couple of years or whenever you have a major change in personnel, policies, products, or information systems. When you do hire them, we recommend they report directly to senior management such as the CRO, Chief Security Officer, Chief Operating Officer, or CEO.
- Audit your organization at least annually, when you make major system or configuration changes, when you introduce new products or capabilities, and after major adverse events.

6.0

CHANGE MANAGEMENT

Through 2015, 80% of outages impacting mission-critical services will be caused by people and process issues, and more than 50% of those outages will be caused by change/configuration/release integration and hand-off issues.[1]
Ronni J. Colville and George Spafford

6.1 WHY MANAGING CHANGE IS IMPORTANT

The only one who likes change is a baby with a dirty diaper. So says a popular adage often cited by technology wonks. Nonetheless, change is a very good thing when managed properly. In your business, change brings you new capabilities, better efficiencies, and creative new ways of doing things. Change erases poor processes rife with wasteful steps, eliminates toxic leadership, and retires substandard products. Thus, change can be a very good thing.

[1] Ronni J. Colville and George Spafford, "Configuration Management for Virtual and Cloud Infrastructures, The top seven things to consider," Gartner Consulting, http://datasecuritycompliance.blogspot.com/2013/05/configuration-management-for-virtual.htm, cited at http://www.netstandard.com/the-human-side-of-network-downtime/. Accessed on February 28, 2014.

Cybersecurity for Executives: A Practical Guide, First Edition. Gregory J. Touhill and C. Joseph Touhill.
© 2014 The American Institute of Chemical Engineers, Inc. Published 2014 by John Wiley & Sons, Inc.

Change also can introduce significant risk to you and your organization. In fact, periods of change are where most risk is introduced. Changes in personnel, process, and products represent great risk to you and your business. You need to be keenly aware of change and be prepared to manage it as part of your risk management process.

Your risk environment is complicated. You and your organization face an ever-evolving threat landscape replete with increasingly sophisticated cyber threats and malicious bad actors. Attacks can range from cyber "weapons of mass disruption" such as distributed denial of service (DDoS) attacks and zombie infestations up to and including finely tuned, exquisitely researched, and implemented focused attacks specifically targeted against you and your information. There are many capable bad guys out there who can ruin your day (and your business) just by hacking into your computer systems and gaining access to your information.

But what if somehow these bad actors were aided inadvertently by someone in your organization? What if your own organizational processes were a mechanism that enabled a bad actor to gain access to your vital information? Sadly, this happens all too often. The Gartner Group estimates 65% of all cyber attacks exploit misconfigured systems.[2] We actually think that figure is too low. We submit that the likelihood that someone specifically targets you or your business is fairly low, yet with hackers and the curious using tools like Nmap and Nessus to scan the Internet continuously for vulnerabilities and Metasploit to exploit discovered vulnerabilities, if you have a misconfigured system, chances are *very* good that someone will find it and potentially exploit it.

Cyber attacks can wreak havoc on your business and drive huge losses, cause potential litigation, and lead to loss of precious momentum. So can system downtime caused by your own people. Gartner research estimates that the average cost of downtime for a small- or midsized business is approximately US $42,000 an hour, but for larger companies or e-commerce models, this number can easily reach six figures.[3] Most businesses, including yours, cannot afford to absorb the damage that downtime produces.

Regrettably, most downtime is found to be a self-inflicted wound. According to the Yankee Group, an IT research organization, over 62% of all network downtime is caused by configuration errors.[4] Downtime is a denial-of-service attack against you and your business that costs you precious time and money. For your IT staff, it is a catastrophe and professional embarrassment. It also is a time when even more configuration errors that can expose you and your business to additional risks inadvertently may be introduced as the staff scrambles to restore service as fast as possible. We advise that the best thing executives can do during periods of downtime is to remain calm, ensure the IT staff has the right resources (i.e., time, people, and tools) to properly restore services, and, after restoration, order a thorough vulnerability scan to ensure that the "fix action" did not introduce a new vulnerability.

---

[2] Carolyn Duffy Marsan, "Hidden Threat on Corporate Nets: Misconfigured Gear, June 8, 2009, http://www.networkworld.com/news/2009/060809-corporate-networks-threats.html. Accessed on October 13, 2013.

[3] NetStandard, "The Revenue Drain of Downtime, March 26, 2012, http://www.netstandard.com/the-revenue-drain-of-downtime/. Accessed on October 13, 2013.

[4] Marsan, ibid.

With IT systems and their associated code being so complex, one can see how easy it is to mistype a digit or miss a step in a procedure. Nonetheless, with the consequences of downtime or exploitation presenting such a high risk to you and your business, you cannot afford such missteps. You need to ensure that any changes to your security baseline are closely managed and risks controlled.

Many businesses are required to have change management controls as directed by such laws and regulations as the Sarbanes–Oxley Act and the HIPAA. If you are part of an organization operating or supplying critical national infrastructure, change management controls are mandatory to ensure safety and security. Shareholder and regulators both look to see whether management is effective in making information secure in accordance with mandatory controls and industry best practices. Failure to comply and deliver satisfactory results clearly (and appropriately) is viewed as management failure.

Billy Crystal says, "Change is such hard work." It indeed is hard work, yet change is a fact of life that occurs in both your home and business environments, in available technologies, and in personnel and processes. In order to manage your cybersecurity risk, you must tightly control your change management process.

6.2 WHEN TO CHANGE?

Many people know they have to make changes, but don't know when to pull the trigger. They look at the risks associated with a proposed change and determine they don't have the appetite to accept the risks. Others are comfortable with the way things are and don't see the need to change. After all, they have benefited from their current processes and products and see no need to change. A third group recognizes the need to change and is continually looking for new and better ways to posture their business to produce high-quality products valued by an ever-increasing market, but haven't yet decided on their next steps. Do any of these describe you and your company?

Obsolescence: Obsolescence is a significant driver in spurring change. For example, chances are very good that if you ask one of your employees under the age of 25 to use a typewriter and carbon paper to type a memo, they will look at you as though you have three eyes and are speaking to them in Klingon.[5] How many businesses still have typewriters? When was the last time you used carbon paper to make duplicate copies of your documentation? Have you even tried *buying* carbon paper lately? While it is still made (in much reduced quantities), it is increasingly difficult to find in stores. Good luck too in finding typewriters because computers and printers have virtually rendered typewriters obsolete. Even if you have a secret love affair with your trusty IBM Selectric III and want to continue to use typewriters and carbon paper (if you can find them), the obsolescence of the technology does not make it a good fit for today's competitive business environment. You need to change.

The same type of obsolescence drives change in IT leading to profound cybersecurity effects. For example, like us, you still may have dozens of 5.25 floppy disks and

---

[5] For us older guys, if you really want to have fun, ask the youngsters if they know what a Thermofax is? We'll bet you a nickel they will Google it!

3.5 diskettes containing your documents from the 1980s and 1990s. Unfortunately, you may no longer have a computer that has the capability to read those disks as such media has become obsolete. What if those are important company records that suddenly and unexpectedly become subject to an e-discovery search as part of a lawsuit? What if someone in your business (or at home) decides to discard those disks without properly destroying their contents and they fall into the wrong hands? For the motivated "dumpster diver," your discarded disks may prove to be a treasure trove of information. Remember, even if the information is old, it may contain information such as account numbers, routing identifiers, and personally identifiable information that a bad actor can use to their advantage and your detriment.

Obsolescence doesn't only apply to storage media and hardware. In fact, it happens with software at a much greater and faster rate and often drives significant investments in time, staff, and other resources to ensure that your software remains current and protected against threats. This applies to both your home and your office.

Let's look at a home example first. You may have a Dell computer loaded with the Windows XP operating system. You may use the computer to manage your finances using your trusty Quicken software, play solitaire and other fun games that you've mastered over the years, and type letters to friends using your WordPerfect word processing software and printed on your ancient Epson printer (it is a good thing you stocked up on its ink cartridges as they are getting more expensive and difficult to find!). What do you do if the computer dies? Can you migrate your software over to a new computer? Probably not as the software was written for a specific operating system that is no longer supported by Microsoft and most likely wouldn't work on your new computer's chipset. Even if it did, you probably wouldn't want it as the security of older code is suspect and the newer products are orders of magnitude better in terms of security and performance. In fact, if you do have that trusty Windows XP computer, we recommend you back up all your information and make certain you do not connect it to the Internet as there are legions of bad actors ready to exploit old systems out there looking for it!

Examples in the office aren't this easy and straightforward. Office environments typically are more complex and involve significant numbers and types of computers, network devices, and software packages that all have to work together to support your business effectively, efficiently, and securely. Each of these pieces of hardware and software experience has different and independent cycles of required upgrades and patching that maintain your cybersecurity posture as well as market currency. Choreographing the upgrades so your network continues to work well in support of your business while remaining secure from the vulnerabilities that expose you to risk is a daunting challenge and a principal concern of your IT staff, your CIO, and risk managers. It should be yours too!

Let's say, for example, that your organization not only has its corporate intranet to handle internal business but also it has a robust web presence with web pages that advertise your products and services. Because you operate in a competitive global market, your web page is critical as one of your core services is to provide timely and accurate information to your current and prospective customers. Unfortunately, several of your firewalls are misconfigured; you are running out-of-date software; and you have a backlog of uninstalled security patches that have yet to be installed, including the latest version of the Microsoft Server operating system used by your servers. Would you be surprised if hackers discovered these deficiencies during their persistent scanning of the Internet? We wouldn't. In fact, failure to keep your systems properly configured and hardened with the latest security patches is like wearing a virtual "Kick Me" or "Hack Me First" sign on the Internet. Sadly, this was the situation that NASDAQ found itself in when they were hacked in 2010.[6] While you can certainly run your systems with old software and misconfigured systems, you run with significant risks. What is your risk appetite? Can you afford to get hacked?

Best value: Best value is the second cost driver that influences change. As an executive, you make decisions every day that involve best value. You decide how to allocate resources for the best effect, what processes to use, and how to invest toward the future.

The author has a personal example that illustrates how best value determinations affect change. Several years ago, the author was recently assigned to a new position as the IT director for the Air Force's human resources organization. The organization was still reeling from a major failure in the rollout of the new military personnel system under the previous leadership regime and was gun-shy of any new changes in technology. When the author arrived, he found a mishmash of servers, software, and network equipment from nearly every major vendor in the market. It was as if the organization had decided to say "yes" to every salesperson who ever visited. The equipment filled the server room to capacity and strained the facility's HVAC and electrical system. The host of contract and direct employees needed to operate and maintain the systems continually jockeyed for precious desk space, heightening an already peaked stress level.

I knew we could not continue to operate this way and ordered a junior officer to lead an assessment of the operating environment. I trained the officer how to determine its total cost of ownership and set him loose to assess each system. What he found made our next steps clear. We were extremely inefficient. The servers in the fleet were beyond their manufacturer-recommended service lives, were expensive to operate and maintain, were operating below capacity, and were spread across racks that were not filled to capacity. Given the officer's report, I commissioned two independent studies to evaluate consolidation of servers using a technique called virtualization. One would be conducted by in-house personnel, while the other would be conducted by a contract consultant who specialized in server consolidation and virtualization. The reports came back quickly with nearly identical recommendations: consolidation and virtualization would save us over US $1.5 million in direct costs, would allow us to reduce staff by 25%, and would reduce our facility footprint by over 65%, helping us avoid costly upgrades to our HVAC and power systems. It clearly was indeed the best course of action. The fact that their recommendation was the same as an "expert" gave leadership and my staff the confidence to take the next steps toward the change we elected to make. After gaining

---

[6]Jonathan Spicer and Basil Katz, "Nasdaq's Poor Computer Security Led to Cyber Attack, Probe Says, November 18, 2011, http://www.huffingtonpost.com/2011/11/18/nasdaqs-poor-computer-security-cyber-attack_n_1101142.html. Also see the FBI's announcement on July 25, 2013, of their indictment of the individuals who may have been responsible for the attack, http://www.fbi.gov/newyork/press-releases/2013/manhattan-u.s.-attorney-and-fbi-assistant-director-in-charge-announce-charges-against-russian-national-for-hacking-nasdaq-servers. Accessed on October 13, 2013.

approval from our agency head and the other managing directors, we moved forward with the change (managed exceptionally well by that junior officer) and executed the plan without any downtime to business operations.

We made the change because it clearly was a best value proposition. We were able to operate at an equal-to-or-better level of service, we significantly reduced operating and labor costs, and we avoided costly facility investments if we stayed the course with the existing systems. We changed for the best and our leadership and key stakeholders appreciated the results! Do you look at best value when making change decisions? Do you look beyond the initial price tag to see if there are any second- or third-order effects (such as what we found with power and HVAC)? Are your hardware and software solutions the best value for you and your business? What changes are you looking to make?

Competitive advantage: The third motivation in making change is to maintain competitive advantage. A great example is found with baseball and "Big Data." "Big Data" refers to the huge amount of unstructured and structured data available through automated systems and the analytical tools that are used to convert raw data into actionable information. As popularized in the book and subsequent movie "Moneyball,"[7] "Big Data" in baseball saw its start in 2000 with Oakland Athletics general manager Billy Beane's adoption of analytical information to guide personnel decisions. One of the principal responsibilities of general managers is to form the teams; they acquire, trade, and assign the players. Beane knew from his owners that he had a limited budget to form his team, and he wanted to find the best value players to remain competitive. He hired a team of analysts led by his assistant, Paul DePodesta, who used computers to analyze the mounds of baseball statistics to determine the best candidates to fill the team given the resources available. Not only did they find the best players to meet their budget; they also used statistics to help guide game strategy, such as providing actionable intelligence on batter's vulnerabilities to certain pitches and strike zone locations, determining whether bunting was an effective tool, and positioning of players in the field in certain situations. Using "Big Data," Oakland was able to field a team that fit within their limited budget and became highly competitive against teams with much higher payrolls.

Soon, other baseball teams took notice of the value of "Big Data," most notably the Boston Red Sox. Known for the "Curse of the Bambino," an 86-year championship drought some say was caused by then-owner Harry Frazee selling the popular pitcher and emerging slugger Babe Ruth to the New York Yankees, the Red Sox, who had won five of the first 15 World Series, found themselves relegated to also-rans much of the following 80+ years. That changed with the hiring of Theo Epstein in 2002 and sabermetrics expert Bill James in 2003. Using information derived from "Big Data" analytics,

---

[7] Michael Lewis, *Moneyball: The Art of Winning an Unfair Game*, W. W. Norton & Company, New York, NY, 2003, and *Moneyball* (the movie), Culver City, CA, 2011. We loved the book and highly recommend it. The movie was terrific too although the part of Peter Brand (portrayed by Jonah Hill) was a concatenation of many actual personalities. Some say that Paul DePodesta, Beane's assistant, was the man who used "sabermetrics" (an acronym of the "Society of American Baseball Research" metrics) to launch the "Big Data" effort in major league baseball. Perhaps that is true but we believe Bill James and Earnshaw Cook deserve special mention as the pioneers of baseball "Big Data."

the Red Sox formed a team of legitimate stars such as Pedro Martinez, Manny Ramirez, and Curt Schilling and overperforming cast-offs from other teams such as David Ortiz, Mark Bellhorn, and Kevin Millar to compete against the powerful New York Yankees. Many people believe that analytics had a hand in the trading of the team's popular short-stop, Nomar Garciaparra, at the trading deadline in July. Garciaparra, still considered a legend in Boston, was hitting .321 with 5 home runs and 21 runs batted in during a mere 38 games. Those are great statistics for anyone, yet Garciaparra had been hurt and missed numerous games during the season, and the data indicated he was likely to con-tinue to miss more games as he aged. Epstein traded Garciaparra and received shortstop Orlando Cabrera and first baseman Doug Mientkiewicz in return. While Boston fans mourned the loss of Nomar, the Red Sox thrived. They won 22 of their next 25 games, earned a play-off spot, destroyed the Yankees in an epic four-game sweep after being one pitch away from elimination, and swept the mighty St. Louis Cardinals to win their first World Series title since 1918 when Babe Ruth pitched them to victory in games 1 and 4 against the Chicago Cubs (who now own the longest streak of futility in baseball and hired Epstein away from the Red Sox in 2011).

Are you thinking about making a change in your organization to gain or maintain your competitive advantage? Are you thinking about using "Big Data?" Are you think-ing about changing your software or hardware to produce at higher speeds or with greater precision? Are you thinking about introducing new products and services to meet new demands in your market?

During the course of your career, you will face decisions to make changes that address obsolescence, best value, and competitive advantage. You may deal with them simultaneously or individually. Regardless of what type of change you make, be careful. Failure to manage change effectively may lead to breakdown of your security manage-ment program and increase your risk exposure as new vulnerabilities are introduced across IT, financial, and operational systems. Without an effective change management process that provides direct oversight and monitoring of change controls across every system, device, and application, organizations cannot adequately protect themselves from risks.

6.3 WHAT IS IMPACTED BY CHANGE?

People: Change impacts many things in an organization, but the most obvious and argu-ably the most important is the impact upon people. Recall our discussion of Edna, the elderly maintenance analyst in Section 4.3.3. Edna genuinely feared the proposed change to automate the maintenance system as she thought the effort presented an exis-tential threat not only to her job but perhaps to her very existence as "she lived for her job." Her concerns were legitimate. She was under great stress and she resisted the change. Any time you introduce change in an organization, expect your employees will be put under stress. Plan accordingly.

Changes also can affect how employees perceive the company and can influence their sense of loyalty. Loyal employees tend to work harder, create better products and services, and engender a positive attitude that permeates the organization. They also tend to retain loyal customers as well.[8] Employees expect that loyalty is a two-way street: they will be loyal to the organization and the organization will be loyal to them in turn. Change can introduce the stress mentioned above and may even cause some employees to question your loyalty toward them and their loyalty toward you and the organization. Some people may lose their jobs or have their jobs redefined as a result of the changes you direct. The key to maintaining employee loyalty is leadership. Be honest and truthful. Clearly communicate the "why" behind the change and define roles and responsibilities. Hold people accountable (including yourself!) Treat everyone with dignity and respect. Demonstrate through both your words and deeds that despite the impacts that the changes will have on people, including layoffs, that people are your most valued resource. Those who lose their jobs but are treated well as part of the change process are more likely to retain a measure of loyalty and fondness, are less likely to engage in acts of sabotage or malfeasance, and generally are more productive through the change process.

Service level agreements: Changes also affect your client and service provider relationships. Many organizations have service-level agreements (SLAs) that define specific performance levels that service providers provide to their customers. Typical example performance measures in SLAs include such items as minimum standards for network uptime and availability, help desk and maintenance response, and timelines for account creation and password resets. When you make a change to your computer systems, you may inadvertently cause an impact to your customers and business partners. Likewise, when they make changes to their environments, you and your organization may be affected. As you lead changes in your organization, always remember to ask, "Who does this affect both inside and outside of the organization?" Your staff may have blinders on that only allow them to see how changes affect their specific area of interest. Strive to broaden the aperture to specifically guarantee that any effects to your clients are properly identified and addressed. Likewise, be vigilant with your service providers to insure that your best interests are protected as they make changes. In fact, make it part of your SLA with them that any and all of their proposed changes affecting your service must be communicated well in advance so that you can conduct an appropriate risk analysis.

Contracts: Service-level agreements are contracts but they are not the only contracts that change can affect. Labor agreements, maintenance and service contracts, and acquisition and purchase arrangements should be reviewed for impacts before implementing any changes. You may find when you implement a change in product or procedure that you also have to change contracts. For example, an aerospace firm that transitioned from using specialty alloys to clad the wings of its planes had to buy out a long-term contract with its suppliers when the market demanded that it transition to composite materials; they no longer needed the specialty metals. In another example, when a firm converted from an Oracle to a SQL database, one of their costs was to terminate their Oracle support contract. A school district found after they purchased tablet computers for their students that their consumption of paper dropped precipitously.

[8] Domenico Azzarello and Ludovica Mottura, "*The Chemistry of Enthusiasm: How Engaged Employees Create Loyal Customers*," August 7, 2012, http://www.forbes.com/sites/baininsights/2012/08/07/the-chemistry-of-enthusiasm-how-engaged-employees-create-loyal-customers/. Accessed on October 13, 2013.

Although they had a long-term firm-fixed-price contract with a paper supplier, they were able to renegotiate the contract to accommodate the change in requirement. In a final example, when a manufacturer automated its assembly line, many employees were made redundant. The collective bargaining agreement signed with the employee's union called for company-funded retraining and preferential hiring into company vacancies. We recommend you, your general counsel, your business managers, and your contracting specialists carefully review change proposals to ensure there aren't any unexpected consequences hidden in any of your contracts, agreements, or purchasing arrangements. If so, that may increase both your risk and your costs.

Capacity: Changes also may impact your capacity. Most people implement changes to increase their capacity. Increased capacity generally equates with increased potential and increased profit. Many people view it as a good thing.

Nevertheless, be careful.

Increased capacity that is excess to your needs can be a bad thing for many organizations. While having the elasticity to expand as demand increases is a requirement in nearly all businesses, unused excess capacity is wasteful, adds drag to your organization, reduces profits, and can introduce increased risk. You wouldn't accept paying rental fees or positioning security guards at an old empty decrepit warehouse down by the river, would you? Should you do the same for the data warehouse sitting idle on the floor of your server room? Perhaps not. As cost models for data storage continue to drop per terabyte,[9] investing in huge storage devices may or may not be your best investment. In fact, purchasing storage from a cloud-based provider may present a better value.

Your excess capacity may also increase your risk. As an example, cybersecurity best practices call for a defense-in-depth approach. This means that rather than relying solely on the defenses at your outer network boundary, such as your firewalls and external router security configurations, every device on your network should have cybersecurity protections. We've already discussed many of them such as password protections, data encryption, and antivirus/antimalware software. (The following sections have discussions of these cybersecurity protections: 2.3, 2.4, 3.2, 3.3, 4.2, and **especially** 5.2 that focuses upon best practice policies. Also refer to the Index and Glossary.) Such protections have a cost in terms of time and treasure and should always be kept up to date.

The author is aware of an organization that had an unused server that was held in "strategic reserve" in the event that additional storage and computing power was needed in the organization. It had been a frontline server that had been replaced during a planned program upgrade (read that to mean "a change"), yet the network manager decided that since the server was still working well and was covered under the organization's enterprise maintenance and licensing agreement, he'd keep it "just in case." Over time, the server sat in a rack waiting for something to do yet was totally forgotten when the network manager left the organization. Neglected it sat. It was turned off yet sitting in its rack poised to spring into action to handle the next call to process information. Sadly, that day came when a new and very curious technician saw the idle server and decided to turn it on to see what would happen. The server, which hadn't received security patches in nearly two years, suddenly popped up on the network and was promptly

---

[9] 1 terabyte (TB) = 1,000,000,000,000 bytes = 10^{12} bytes = 1000 gb.

infected with malicious code that allowed a hacker to plant a RAT into the server and attack the network. The technical team spent several days fighting the effects of the hacker, and the resulting disruption to business operations was significant. If you are looking for a moral to the story, having a backup server to provide extra capacity is not necessarily a bad thing, but you need to have a change management plan when you introduce additional capacity. This organization didn't and it harmed them.

Changes can also decrease your capacity. If you have a smart phone, you may have already noticed that with every operating system "upgrade," your manufacturer offers to push more features to your device so that the available storage space for your applications, music, and photographs dwindles. The same thing happens with your business systems as new and improved software and patches are installed on your system. Beware of software bloating! Software bloating is a term that describes when successive versions of software become noticeably slower, use more memory or processing power, or have higher hardware requirements than the previous version while making only dubious user-perceptible improvements. Software bloating actually *decreases* your capacity and may hinder your performance, thereby increasing your risk. We recommend that anytime you make changes to your software through patches or upgrades, you thoroughly test it whenever possible in the event that it has negative effects on capacity and performance. If it doesn't work as planned, having a plan to return to previous successful versions is always an insurance policy you want to have.

Security: The final thing impacted by change is your security. Your cybersecurity posture will never be static. If your security officer is telling you that nothing ever changes, we submit maybe it is time to change your security officer. Of all the environments where your organization operates, the cyberspace environment arguably is the most dynamic. New threats and vulnerabilities emerge every day. As your systems and software age, hackers and other bad actors figure out new ways to exploit weaknesses. Mechanical parts in your computers sometimes malfunction too, causing a denial of service to those who have not adequately provisioned backups. Mechanical problems are not the only computer malady that can deny you access to your information. Your computer may even suffer the equivalent of a heart attack!

Many computers suffer from a condition called memory leakage. Memory leaks are caused by defects in software that incorrectly allocates memory. Picture your computer working by constantly shuffling data between the central processor and memory units. Over time, some programs do not release memory back to the system when they are done with their processing. This is like plaque accumulating in your arteries. Over time, unless you clean the memory by periodically restarting the computer (aka making "a change"), the memory will continue to fill up with remnants of previous processes until it doesn't have sufficient memory to process anymore and locks up. It has the equivalent of a heart attack! While you can do a reboot to restart the device after it locks up, you likely will lose all the information you've created since your last save, lose time, and are denied capabilities. While improvements in software have reduced memory leaks significantly, the risk is still there, and many network professionals will schedule reboots as part of a routine maintenance cycle.

A more common cybersecurity risk associated with change deals with patching. Some programs and operating systems have a series of patches and detailed instructions.

It is easy for both rookies and experienced technicians to make mistakes when patching systems, particularly when they are rushed. Three things can happen when patches are applied and two aren't good. The first thing is that all goes well and there are no problems. The second is that the patch does not go well and your technical team struggles to restore capability while your business suffers from the denial of service. The third thing (and arguably the worst) is that the patch doesn't work as planned and your environment is exposed to unacceptable risks.

Change impacts many things in your organization, principally your people, service level agreements (SLAs), contracts, capacity, and security. You should always plan changes to your information environment carefully to ensure you are conducting the due diligence and due care that your information deserves. Remember that due diligence refers to your activities to identify and understand the risks facing your organization. Due care demonstrates that you have acted in a prudent and appropriate manner to protect the organization, its resources (such as its information), and its people from possible threats. Always have a plan when implementing changes to your information environment and recognize that even seemingly small changes can have big effects.

6.4 CHANGE MANAGEMENT AND INTERNAL CONTROLS

Your internal control program is about managing and controlling risk. Internal controls usually are thought of as methods used by the CFO to safeguard organizational assets, protect the reliability and integrity of financial and accounting information, ensure compliance, promote effective and efficient operations, and achieve business goals and objectives. Many employees and perhaps even some managers will give you the "deer in the headlights" look when you ask them to explain how they contribute to the internal control program. While they may know there is a requirement to gather information used by senior managers, they may have no clue as to why or how the information is used. This is a sign that your risk management program is not sufficiently integrated throughout your organization. Organizations experiencing this type of disconnect between senior management's desire to have robust internal controls and lower levels not understanding those controls or their intent will have trouble on their hands. We contend that internal controls are not just the realm of the CFO; they apply to every manager and every employee in the organization.

Your internal controls should help you to manage and control change in your organization. While every organization is different, most use policies and procedures to guide actions to support the organization's strategic direction and plans. After reading Chapter 5.0, you have a good foundation to draw upon when creating plans, policies, and procedures to support your strategy. Incorporating change management processes into your internal controls program will bolster the effectiveness of your controls, help make your business more efficient as you improve management visibility into total cost of ownership, and enhance your security posture as you reduce risk presented by the unexpected consequences wrought by changes.

System configurations: We recommend you implement tight controls over how changes are made to your system configurations. The security of your information relies heavily on the proper configuration of your system. If your system is not properly configured, bad things can happen fast, causing huge problems for you and your business. Take, for example, what happened when a technician in Sweden misconfigured a device during routine maintenance. The technician introduced an error that caused all domain name system (DNS) lookups for the entire country of Sweden to fail. DNS is a terrific protocol that allows computers to convert a plain language address such as www.post-gazette. com to an IP address (108.166.31.131), which computers use to identify each other and exchange information. Without DNS capabilities, you would have to manually address every transaction using your destination's IP address, which is awfully inconvenient as most people do not maintain a contact list containing the IP address. Do you even know the IP address of your computer?[10] Most folks don't and when the Swedish technician misconfigured the DNS service, *the* **entire country** *effectively lost its Internet capability for an hour!*[11] During that hour, the Swedes lost their ability to surf the net, send and receive emails, conduct electronic business, use Internet-based media, and use Internet-based phone services. While the Swedish technician was performing his maintenance on a device serving the entire country, consider the consequences of system configuration mistakes created by your technicians. What would you do if one of your technicians disconnected your business for an hour or more? What would you do if one of your technicians disconnected *someone else's* **(how about everybody's)** business for an hour or more?

Software: Misconfiguration in software can be equally devastating and require thorough controls. Take, for example, the so-called misconfiguration of Facebook software that allowed bad actors to gain information from compromised email accounts harvested from exposed "Friends" lists to launch spear-phishing attacks.[12] Attacks like these are particularly nasty as they come in from the email accounts of your legitimate friends and usually contain attachments and links that contain or lead to sites that can poison your system or rob you blind. Because of the increased threat posed by these bad actors, as a policy, we refrain from clicking on links contained in emails and only go straight to the source to protect us from malicious sites. Perhaps you should too. What would happen to your business if *your* software was misconfigured? According to the NSA, misconfigured software is responsible for over 80% of cyber attacks they are sent in to clean up.[13] Not only do software misconfigurations expose you to possible attack and exploitation, they can take your business off-line. How much does it cost you when your systems are off-line? Do your customers leave you to go to another source? Do you have to compensate your partners when you are unable to fulfill your commitments? What happens when

---

[10] If you want to know your own IP address, you can check your own system settings or visit a web site such as http://www.find-ip-address.org, which will read you IP address and display it for you.

[11] CircleID Reporter, *"Misconfiguration Brings Down Entire .SE domain in Sweden,"* October 13, 2009, http://www.circleid.com/posts/misconfiguration_brings_down_entire_se_domain_in_sweden/. Accessed on October 13, 2013.

[12] David M. Ewalt, *"Facebook Says 'Misconfiguration' Allowed Spammers to Impersonate Users,"* October 29, 2012, http://www.forbes.com/sites/davidewalt/2012/08/29/facebook-spam-email-spear-phishing/. Accessed on October 13, 2013.

[13] Katherine McIntire Peters, *"Misconfigured Software Poses Significant Security Challenges,"* June 21, 2011, http://www.nextgov.com/cybersecurity/2011/06/misconfigured-software-poses-significant-security-challenges/49273/. Accessed on October 13, 2013.

your software at home is not configured properly? Does your home computer contain valuable information you want protected? Are you exposing more than just your family photos to potential bad actors? You have a lot at stake when making changes to your software. You need to protect yourself and your business by making sure that software changes are done properly and will not negatively affect your business.

Web pages: Don't forget web pages. Changes in web pages also have to be tightly controlled through your change management process to control your risk. Your web pages may be your principal means of communication with your clients, a dispersed workforce, and even your suppliers and partners. They likely are a major source of revenue as they may be the primary source of your sales through e-commerce methods. You have to tightly control any changes to your web pages.

Picture this scenario: you go to the web site of your favorite news source, the ZNews Network.[14] You visit their web site at www.znewsnetwork.com to catch up on all the news of the day. ZNews has stories and images from their own sources as well as from partners and "stringers."[15] You like to visit the weird and odd news section every day for its hilarious and preposterous stories. For example, today, there is an interestingly titled story called "Alien reveals secret UFO diet that turns Elvis impersonator into a JFK clone." Wait! What's this you see at the bottom of the page? ZNews Network is now selling its own line of clothing. You can get an official ZNews golf shirt to wear around the office on casual Fridays or when you finally get press credentials to serve as a ZNews stringer at the Super Bowl! You click on the link that takes you to a page that asks you to fill out a form so they can send you a catalog and keep you informed via emails whenever they have new products. You are delighted and fill out the form so that you can get your new ZNews golf shirt before the Super Bowl.

Meanwhile, a hacker from New Jersey named Bob also is reading the ZNews Network stories and sees the clothing offer too. Instead of seeing an opportunity to get a golf shirt, he sees an opportunity to cause a little trouble and make a little profit. He goes to the form and uploads a script containing Java instructions designed to steal a cookie[16] from everyone who visits the store. Bob the hacker will then use the information from the cookies exchanged between the sites to impersonate the victims and access their accounts in the ZNews store. He'll then harvest their stored credit card numbers, usernames, and passwords (which he'll try to use at other sites because despite warnings many people recycle usernames and passwords). Bob's style of attack, called Cross-Site Scripting (aka XSS), is very similar to SQL injection. But instead of attacking the database, XSS compromises the relationship between the user and the web site. It is what Symantec Corporation calls the most common vulnerability on the web today.[17]

---

[14] Modeled after the Zulu News Network (ZNN) that is featured on the hit CBS television show "NCIS."

[15] "Stringers" are independent correspondents.

[16] "Cookies" are small pieces of information sent to your browser by the web sites you visit and enable the web site to effectively manage the session between you and the web site. Cookies can also store passwords and forms a user has entered previously, such as a credit card number or an address, so many people make it a practice to routinely delete or block cookies when they are not needed. This is a good policy to have at home and the office.

[17] Symantec Corporation, *Internet Security Report 2013*, Volume 18, p. 29, http://www.symantec.com/security_response/publications/. Accessed on February 28, 2014.

How do you prevent changes in your web page from introducing an XSS vulnerability? What if your web site has an XSS vulnerability? What is your liability to your customers? What if an attacker executes a script that compromises your web server, reveals valid user credentials, and gains entry into your corporate network? Is your network sufficiently segmented to protect your intellectual property and trade secrets even if your web server is compromised? Even if your web site is secure now, could a patching change or misconfiguration suddenly expose you to risk? Your change management process needs to ensure that any changes to your system preserve the integrity of the system to protect against attack and exploitation.

Not only should your change management process preserve the integrity of your system's security, but also it needs to preserve the integrity of the information presented.

Your web site is your digital storefront. We are biased when we visit web sites rife with typographical and syntax errors. If a web site owner can't deliver proper grammar, how can we expect they deliver effective security? These are the types of businesses we pass by and move on to those who demonstrate professionalism from the moment we visit their digital storefront until the time we leave it. Maintaining the integrity of information should be a top priority.

United Airlines recently found out the hard way that maintaining the integrity of information is important. United customers were delighted to find that United had made a change to its web site that permitted them to book round-trip transcontinental trips for US $0–$10. The change, which occurred around 2:30 p.m. on the east coast, put travelers into a frenzy as word spread of the amazing deals and people booked trips and quickly printed boarding passes. United soon noticed the rush of customers flocking to take advantage of the deals and stopped online booking for an hour while they corrected the error.[18] In the aftermath of the misconfiguration, United announced they would honor tickets sold during the unintended "fire sale," most likely on the advice of their lawyers and public relations specialists.

Is the United experience unique in the world of web page-based e-commerce? Sadly, no. In another not uncommon example, Zappos lost US $1.6 million when an error at its www.6pm.com site capped all transactions at US $49.95.[19] Because 6pm.com emphasizes luxury items, capping prices at US $49.95 was a great deal for consumers, but not what management had in mind. Whether you are an executive in a large corporation like United Airlines or a small- to medium-sized business, if you do business through your web page, you have to protect yourself and your business by ensuring that all changes to your web pages deliver the results you expect.

Electronic exchanges: We've identified that you need to tightly control changes to your system configurations, software, and web pages. You also need to control the data

---

[18] Kent German, "*Error on United Airlines Web Site Results in Free Fares*," September 13, 2013, http://news.cnet.com/8301-1023_3-57602742-93/error-on-united-airlines-web-site-results-in-free-fares/. Accessed on October 13, 2013.

[19] Dan Nosowitz, "*Zappos Loses $1.6 Million in Six Hour Pricing Screw-Up*," May 24, 2010, http://www.fastcompany.com/1651302/zappos-loses-16-million-six-hour-pricing-screw. Accessed on October 13, 2013.

exchanges you make to preserve the integrity of the data you share with partners and that which you receive.

Many businesses, particularly large businesses and those specializing in logistics functions, use electronic data interchange (EDI) to accelerate the velocity and precision of their business. EDI eliminates many cumbersome manual processes as businesses share electronic documents, such as purchase orders, invoices, and shipping information, with their business partners. This type of electronic information management is proven to reduce business costs as information is standardized and the handling of forms and reports is automated. EDI vendors are proud of the fact that human error is minimized as the computers do all the work in processing routine forms and reports. They are also proud that electronic business transactions are more secure than normal paper transactions because they are protected using data encryption. Encryption requires the business partners to agree, as part of their contracts, upon common encryption techniques that enable them to freely exchange the information. Use of encryption is now so common that any EDI transaction not encrypted is viewed as suspect and rejected.

Because so much business now flows using data exchanges, your business likely uses EDI or other electronic exchange of information as a key component of your business partnership game plan. Your business relies on this flow of information to earn profits and maintain your competitive edge. You need to safeguard your information against inadvertent or deliberate changes that could compromise your vital information.

What could happen if a change occurs involving your data exchanges? How could it affect your business? What is the risk to you and your information?

Many third-party logistics providers (3PL) want your purchase order information so they can package goods. They want this information electronically as it saves them considerable time, eliminates errors from data entry, and speeds product delivery. But what happens if you make a change that shares *too much* information with your 3PL partner? What happens if your 3PL finds out from your information that your profit margin is US $100 million while you are only paying them 10 cents per box? Do you think they will demand higher compensation from you and your business if they know what your costs and profits are? What happens if you make a change that exposes financial information such as bank account and routing information? Are you exposing your business, its clients, and partners to the risk of theft or litigation? Do you have sufficient data segregation and controls to ensure that your partners in Plieno Steel Corporation cannot see what business you are doing with their competitors at Jeklo Steel? What happens if one of your technicians makes a change to your system and accidentally turns off your encryption? Will your partner automatically stop accepting your orders or proposals? They should! How much will the delay to find and fix the problem cost you? What happens if that same technician accidentally sends your sensitive transactions between Partner A and Partner B and vice versa? When you make uncontrolled changes to your data exchanges, you expose yourself and your business to significant risks.

The previous discussion on electronic data exchange focused on routing of information. What if your change affected the *content* of the information that you share between partners via EDI? We keenly are sensitive to the "garbage in, garbage out" principle of data processing, which states that if you input bad data into a process, no matter how good that process is, you will get bad information out of it. Recall the United Airlines web page example where a mistaken change in their web page permitted customers to buy flights at cut-rate prices, driving losses to the airline. United was able to detect the mistake relatively quickly and take action to correct it, but what would happen if the information in your EDI was tainted? What if a change was made and the bill you send or receive for US $100 had an extra zero appended to it, making it now US $1000? Do you have controls in place that would quickly catch and correct that mistake, or does that bill get paid automatically because it is between two "trusted" sources? What if the change is made on your partner's side of the transaction? How do you detect errors generated by your partners? What does your contract say? How long would it take to find and fix tainted information, and what would it cost to repair the damage? What would it cost if you didn't detect the tainted information? Are you paying too much for something *and don't even know it*? Your electronic data exchanges are vital to modern business yet need to be tightly controlled to preserve the integrity, accuracy, and security of your information.

Internal control policies that specifically address how you manage and control all changes to system configurations, software, web pages, and data exchanges will keep you out of trouble. Remember, next to your people, your most valuable asset is your information. Make it your policy that anything associated with the access, presentation, processing, storage, or transmission of your information is **tightly** controlled.

6.5 CHANGE MANAGEMENT AS A PROCESS

Change management is a critical component of your risk management program. Executives need to maintain positive control over how plans are executed and policies are followed. Whether you are an executive working in business, in academia, in government, or in a nonprofit organization, your information is a valuable asset that needs to be protected. You need to control all changes that can affect your information.

Cybersecurity professionals will tell you that the fundamentals of cybersecurity are controls that preserve the confidentiality, integrity, and availability of your information. These controls can come in the form of policies, procedures, or technical measures. They are there to enhance your business and its ability to meet your business objectives such as safely creating products that deliver value to your clients and earn profits for your ownership. Many managers look at cybersecurity controls as technical measures to be turned over to the "technical experts." To do so would be a huge mistake. Don't surrender management and oversight of those controls to your technical staff.

Managing change is a fundamental responsibility of management, while executing changes largely is a tactical-level activity conducted by employees under the oversight of management. Many organizations have little visibility into the effectiveness of their change management controls across the IT infrastructure. This lack of visibility can be devastating when changes are not managed and monitored effectively. Reduced availability, compromised information, and lost trust in the integrity of your information, and the systems that process it, could ruin your organization.

You can prevent catastrophe by doing what Kotter calls "Leading Change."[20] While Kotter's work centers on leading transformational change, he proposes that change is a process that can be led and controlled. We agree. When it comes to change, even at the operational and tactical levels, executives need to be both effective managers and great leaders, guiding and controlling the process that leads to change.

6.5.1 The Touhill Change Management Process

We submit that change is best effected when it is accomplished as the product of a well-managed and deliberate process executed by a properly trained and motivated workforce. Executives who lead these changes take proactive measures to make sure that the process delivers the desired effects in a manner that is predictable, reliable, and repeatable.

Do you have a change management process that is predictable, reliable, and repeatable? You may be surprised to find that many people do not have a formal change management process for their IT systems. Many organizations continue to install patches on an inconsistent basis (if at all), send untrained and inexperience technicians to perform tasks they have never done before, never test software before installation on production systems, and permit technicians to input instructions into critical systems without the requisite checklists or procedures that can insulate the organization from a potential loss of information. They do not have effective management controls over change.[21]

Does that describe your business?

In this section, we are going to share what we believe is the best practice for managing change in an information ecosystem: our process for managing and controlling change. You may look at it and say: "That's nothing new. We already use a process like that."

We hope you do.

Nevertheless, there are many executives who use process management techniques to manage risk at the strategic and operational levels but still neglect to use these same techniques to manage tactical actions. Such neglect can have profound adverse strategic effects. As you've seen from the anecdotes shared earlier in this chapter, tactical actions by your employees can have dramatic strategic effects, and you need to manage their activities as part of your risk management construct.

Figure 6.1 shows how we envision the change management process and the key steps involved. Frankly, it's simply a graphical expression of common sense, but we believe that you, as practicing executives, will relate to it.

The change management process presented should not be a startling revelation for you. We hope that you already are using a process like this to manage change in your organization. However, we wouldn't be surprised if you are not using such a process to

---

[20] John P. Kotter, *Leading Change*, Harvard Business School Press, Boston, MA, 1996.

[21] How would you like to fly in a commercial airplane where the pilots and mechanics are untrained and haven't tested their computer and radar systems and where nobody, especially pilots, has a safety inspection or flight checklist? We are reminded of Geena Davis' famous line in the movie *The Fly*: "Be afraid. Be very afraid!"

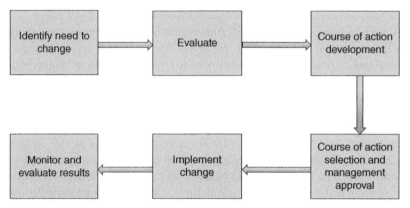

Figure 6.1. The Touhill Change Management Process.

manage changes to your hardware systems, software patches and upgrades, web pages, and data exchanges. You should use your version of the change management process to control change and manage risk.

6.5.2 Following the Process

If you want reliable, repeatable, and predictable results, you need to follow a process. It doesn't matter if you are manufacturing chemicals, creating machinery, delivering a service, or even implementing cybersecurity controls. Processes are the deliberate steps the user follows to accomplish tasks and achieve desired results. They are the procedures that implement your plans created to complement your strategy.

Let's walk through a common example that shows how to use a formal process, even when time is of the essence. It is Friday morning, and your network manager calls you out of a boring meeting going over TPS reports to alert you she just heard from her counterpart at another business that they were just hit by a new zero-day exploit that has taken down their computer systems and possibly corrupted their data. The initial damage report sounds really bad.

You say to her (while trying to look relatively intelligent), "Remind me again; what's a 'zero-day exploit?'" She reminds you that a zero-day exploit refers to an attack where there is no warning of the vulnerability; there are "zero days" to prepare to patch the vulnerability.

She tells you the manufacturer just released a patch this morning and the attacker hit your counterpart's business shortly thereafter. You now recall that's how many zero-day exploits occur; as soon as manufacturers announce a patch to some previously unknown vulnerability, some knucklehead decides to try to exploit the vulnerability before it gets patched. She says her counterpart at your competitor called her a few minutes ago to ask for her advice as they had problems installing the patch and their systems are down and will be for awhile while they try to reconstruct their configurations from backups. You secretly (and shamefully) are delighted and already are thinking how their loss may translate into opportunity for your business when she drops a whopper on you: your operating system is vulnerable to the same kind of attack. Your network manager wants to install the patch right now. Do you tell her, "What are you waiting for? There is a threat out there. Fix it now!" or do you follow your process to determine your best next steps?

The first step in the change management process is identifying what needs to change and why. In this case, you know that there is a vulnerability, a demonstrated threat, and a potential fix. Rather than jumping foolishly in reaction to the issue, you wisely follow the corporate change management process.

You move to the next step and evaluate the situation. What do you know about the attacker? Were they targeting your competitor for any specific reason? Who are they and what were they trying to accomplish? What do you know about the proposed patch? Will this patch prevent the attacker from harming your company? Hopefully, it will, but what other impacts are there? Will the patch operate in conflict with any of your other software products? Does your staff have experience installing patches like this? What happens if there are any problems in the patch installation? Does the software manufacturer offer technical assistance if you run into problems? You ask your staff how long it would take to develop an installation checklist, train your network personnel, and test the patch and checklist in your test environment. You need facts.

Step three is determining your possible courses of action (COAs). We always like to give our bosses a minimum three COAs, with all of them being feasible, acceptable, suitable, and affordable. After presenting the alternatives, we tell the boss what our first choice is and why. The first COA you may consider is to do nothing. As with all your COAs, you evaluate the pros and cons of the COA. Pros include you don't have make an unplanned change to your IT environment. Cons include the potential damage that could occur if your system is unpatched and attacked. Another COA is to patch immediately. Pros include that your system is patched right away, lessening your exposure to potential attack. Cons include that you don't have time to test the patch, train your staff, or check to see that the patch doesn't cause some collateral damage that negatively affects an application or key process. A third COA is to continue to operate your IT systems but disconnect from the Internet, where you believe an attacker may be lurking. This essentially would make your network an intranet and not accessible to outsiders. It also disconnects you from your customers. This may be an acceptable option if you don't rely on that connection for e-commerce and other revenue generation. It also may be acceptable if the duration of the outage is brief and your risk appetite allows for losses accumulated during the service interruption. A fourth COA is to accept the risk of potential attack, yet patch as soon as possible after creating, testing, and rehearsing a patching procedure in your testing environment. Your network manager believes it will take her and her staff approximately three hours to create, test, and rehearse the procedure. Usually, her estimates are pretty good.

Step four calls for COA selection and management approval. Because there is a serious threat to your organization, you don't delay and notify your chain of command. You inform your boss of the current situation and outline potential COAs. Many companies, including yours, use a Change Management Board to control changes to their IT environments. The Change Management Board is comprised of key stakeholders in the business and includes empowered representatives of key business functions; the general counsel; the CIO, CISO, and CRO; the financial department; marketing; PR; and the IT staff. The board reviews any proposed changes to ensure that sufficient controls have been employed to protect the business and its information from damage. You invite your boss to join you at the emergency Change Management Board meeting to be conducted in 30 minutes, where you will recommend COA 4 (accept the risk of potential attack, yet patch as soon as possible after creating, testing, and rehearsing a patching procedure in your testing environment).

Your boss approves and directs you to make similar calls to the key stakeholders on the board so they know what is going on and are better prepared to arrive ready to make a decision. When the board convenes, the facts are presented and the COAs debated. Insightful and probing questions are asked, and occasionally, a staffer is dispatched to get an answer if it is not immediately known. The culture and values of the organization contribute heavily to the decisions of the board. So does the risk appetite established by senior leadership. Because there is great confidence in the IT staff due to their proven track record of excellence, the board votes in favor of COA 4 and assigns the network manager to implement the change, noting that since the competition supposedly was devastated by an attempt to patch and your business is similarly at risk, the board believes that the CEO should make the risk decision whether to go with COA 4. The board normally has authority to approve changes; however, it is not unusual for the board to refer decisions involving high level of risks to senior management for final decision. The board chairman, who in this case is the CIO, calls the CEO, briefs him on the issues, summarizes the situation and board deliberations, and asks for CEO concurrence. Because the CEO has knowledge of and confidence in the process, he approves the board's recommendation and directs COA 4 be implemented immediately.

Fresh from COA selection and management approval, you are ready to move forward with COA 4 with **step 5: implementing the change**. Your PR specialist follows corporate policy by sending out a corporate-wide email notice of the proposed emergency patch along with information provided by the technical staff that identifies what will happen, why it is necessary to patch, when it will occur, what the expected results are, and what the staff should do if they detect any adverse effects. Similarly, the sales and marketing team is calling key business partners to give them a heads-up of the proposed change. They use a script that has been coordinated through PR, the technical staff, and general counsel to give just the right amount of accurate and informative information. Too much would be overkill and potentially shake the confidence of your partners. In fact, the courtesy call to the partners reinforces in their mind that your company is professionally managed and that you called to inform them of any actions that could possibly adversely affect them. They greatly appreciate the sharing of information and are now taking action to patch and protect their systems. They promise to share with your company any issues they may see. Good news arrives with a call from your network manager, who reports that her team created a checklist based on the manufacturer-provided patch instructions and her team successfully installed the patch in your test environment. They tested each application in your software inventory to double check they were not negatively affected by the patch and found no problems. You tell her the board and CEO have given approval to install the patch upon successful testing, so she needs to get on with the change, but you add, "Don't rush. Make sure you and your team do it right. Getting it right is more important than making a mistake, opening a bigger vulnerability, or inadvertently causing damage to our information." She acknowledges and executes the change flawlessly.

Once the change is made, you move to the **final step, monitoring and evaluating the results**. It is the end of the day and your environment continues to operate well. Your web page continues to operate at normal levels, and business is good with orders flowing in from an unusually high number of customers. You think many of them are coming to your business rather than to your competitor. Before you leave for the day, the CISO drops by to tell you that initial vulnerability scanning of the network indicates that the change did not appear to introduce any new vulnerabilities to the business, although it will be a couple of days for the full scanning procedures to run their course. He mentions that he likes to run penetration tests after significant software upgrades and special circumstances. He says over the weekend he'll decide whether this falls into one of those special circumstances where he'll ask senior management to authorize a "pen-test." He wishes you a great weekend and promises to call you right away if the cybersecurity team discovers any problems.

You prepare to go home ready to enjoy the weekend because your systems and their information are fully operational, are generating lots of business, and are secure. You have confidence they are effective, efficient, and secure. Meanwhile, your counterpart at your competitor is valiantly trying to recover from a change gone bad. As you log off your computer and clean your desk during your end-of-day security procedures, you think, "Whew! Sure glad we had a change management process and got that patch on line correctly."

If you and your business were confronted with a similar situation, how would it be handled? Do you have a well-documented and rehearsed process or do you rely on *ad hoc* procedures? Who approves changes and how is risk management incorporated into those decisions? When changes are made, is it well known who is in charge and responsible? In organizations that do not have a well-defined and disciplined process, the only way you find out who is responsible is to count whoever has the most fingers pointed at them. We prefer to know roles and responsibilities before changes occur. We also insist on communicating changes well in advance. Your process should inform who needs to be notified of any changes, including those internal and external to the organization. After your make the change, ensure that you assess how well you did. Conduct vulnerability scans to see if any new vulnerabilities have emerged. Depending on the magnitude of the change and the value of your information, consider investing in penetration testing or even red teams. You may want to even consider a full audit of your plans, policies, and procedures to make sure you are well postured to protect your shareholder's interests and that of the company.

Using a process to manage change in your organization has many benefits that we believe outweigh any costs. While many argue that processes increase bureaucracy, well-managed processes are inherently more effective and efficient than *ad hoc* procedures that are not reliable, repeatable, and predictable. They also better protect you from risk by delivering more secure solutions to your most important problems. If you don't have a change management process, now is a great time to invest in one.

6.5.3 Have a Plan B, Plan C, and maybe a Plan D

Sometimes, you encounter situations that you didn't expect. Do you sit gob-smacked or do you have a backup plan? What happens if your best laid plans go awry? Recall our previous advice: "Be Prepared. Have a plan and do things!" We believe it is essential that you always prepare contingency plans as a normal COA. The unexpected happens. Be prepared.

What Plan B actions would you have planned in the Section 6.5.2 example? What would you have done if the patch failed to load properly on the production system despite having worked well on the test system? Do you have a Plan B to "back out" of the installation without harm to the system and its information? Who makes the decision to go to Plan B? When do they make that decision? What information do they need and how do they get it? Does your Plan B permit you to return to the state or condition you were in before you started the change? If you have to switch to Plan B, who needs to know? What do you tell them? Who tells them? How are they informed (e.g., face-to-face, email, or phone call)? Be prepared.

Be prepared by asking the right questions. What do you do when Plan B to gracefully "back out" of the installation fails? Do you have a Plan C? Does your plan let you know how long it would take to reload the operating system if Plans A and B fail and your operating system was corrupted by the patch? Do you have the resources to reload the operating system in the event you face this unfortunate situation? Do you need any special reinforcements or technical help? If so, how much would that cost? How much downtime should you expect? What kind of loss are you facing? As with Plans A and B, who needs to know when you are moving to Plan C? Are you prepared?

You should always be ready for the unexpected. Be prepared.

6.6 BEST PRACTICES IN CHANGE MANAGEMENT

In our combined 80+ years of executive experience, we have been involved in the formation and execution of numerous change management plans, policies, and procedures. We have been successful yet both acknowledge early failures that turned into valuable learning experiences. Fortunately, our failures didn't leave a mark and enabled us to improve on our subsequent efforts leading to very successful careers.

Many of our clients ask us to share some of our change management lessons learned and best practices from these experiences to better posture themselves for success and minimize risk. We present them here to assist you as you develop your own change management plans and processes:

Touhill's Best Practices in Change Management

1. <u>Communicate early and often</u>: Nobody likes surprises. Every time you make a change to an information system, an application, a web page, or data exchanges, make sure you inform those affected by the change well before you make

the change.[22] Anticipate resistance and address fears by clearly articulating why the change is needed and how it will affect the individual. Seek to involve people in the change as opposed to imposing it upon them. Make sure you invest in two-way communication throughout all stages of your change management process.

2. <u>Don't rush to field crummy products</u>: If you rush a change that produces crummy results, you are rushing to fail. Crummy results erode confidence and make next steps even more difficult. Carefully balance the demands for speedy delivery of products with the demands for quality products as part of your risk management process. Decisions regarding timelines and quality are management-level decisions. Make sure that communication is strong between all parties to make certain that risk is appropriately identified and decisions are made with the right information. We've often found that adequate results are better than gold-plated results and always better than crummy products that leave everyone disappointed.

3. <u>Timing is everything</u>: Have you ever had a system administrator take down a service such as your email during the height of the workday because it was convenient for the administrator? We hate that and won't tolerate that in our organizations. Changes need to be accomplished for the betterment of the organization and its objectives. In his latest CIO position, the author was challenged by a fellow managing director as to why the IT staff performed maintenance on a key transportation system at 1:30 in the afternoon on Tuesdays. I told him we found that well over 90% of the users of the system were in Afghanistan, and when we asked them when would be a good time to do our maintenance so that it wouldn't interfere with their operations, they told us midnight leading into Wednesday was best. Because Afghanistan is 10.5 hours ahead of us in time zones, we did our maintenance at their convenience, during our afternoon. We recommend that when you introduce changes, you do so at the convenience of those affected by the changes. Not only is it common courtesy, but also it is good business.

4. <u>Change only what needs to be changed</u>: What would you do if you brought your car to the garage for an oil change and when you pick it up the mechanic tried shaking you down for extra money by saying he noticed some other things that needed to be fixed and did the work without your permission? You probably would blow a gasket and never return to that garage. Why should your network and IT systems be any different? Having a disciplined change management process prevents well-intentioned or plain-old-ignorant technicians from introducing unplanned changes that can actually cause more harm than good. Make sure you have positive control over all changes to minimize the risk of surprises and exposure to threats.

5. <u>Don't be afraid to ask for help</u>: Asking for help is difficult for many people. They mistakenly believe that it will be seen as a sign of weakness, whereas it actually is a sign of wisdom. In today's increasingly complex cyberspace environment, asking for help often is not an option; it is a requirement. One of the things that

---

[22] Just like our old friend Edna, nobody likes change.

executives do is build effective teams. When confronted by vexing problems that are beyond the skill or expertise of your team, don't hesitate to bring in expert consultants to assist. You also can use expert consultants to review (or even create) your plans and procedures. Ensure you protect your intellectual property and trade secrets with outside consultants through nondisclosure agreements and other measures whenever you consult for help.

6. Ensure everyone knows what is going on and why: Ignorance may be bliss for some people,[23] but it can lead to trouble when it comes to change. As an example, coordinating your maintenance schedule for your automated systems is an essential best practice of change management. It is very important to make sure that your personnel know when you will remove their systems from service or perform online maintenance. First, if your proposed maintenance window conflicts with vital business operations, the conflict can be identified and resolved before it becomes a problem. Second, if a problem arises as a result of the change, informed users are more alert to any issues and can dispatch maintenance personnel faster. Third, knowing what is going on and why permits employees to appropriately schedule their work around the period of maintenance, resulting in higher productivity and less frustration.

The author has a unique example of a change where making sure that everyone knew what was going on and why was critically important. When deployed overseas in combat operations, the author was responsible for all communications and computer systems supporting allied air forces. Anytime one of our troops was killed in combat, we wanted to make sure that appropriate next-of-kin notifications were made. Face-to-face contact is appropriate. An email is not. Early in combat operations, we found that emails from the frontline units were beating officers and chaplains to the doorsteps of grieving families. An email as simple as "Honey, I'm fine but Sergeant M didn't make it. He was killed in an attack this morning. I think it would be great if you and the other spouses would stop by Sergeant M's house to see if the family needs anything" dispatched concerned military spouses who would descend upon an unaware spouse while officials deliberately and diligently did the work to confirm the death, describe its circumstances, and gather the appropriate counselors to make official notification. In more than one occasion, these unofficial initial reports proved incorrect, and the wrong spouse was told of a death that didn't occur. To better serve our grieving families and avoid leaks and inaccurate reports, our higher headquarters made it policy to suspend all but mission-critical communication from the unit until the family had been properly notified. Web browsing and email services for everyone except the command post were disabled. Morale phone calls were suspended. We went "comm out" until we received word from the home unit that the family of our fallen comrade had been properly notified. Loss of communications in the military is serious business, and the troops hate any disruptions. Nonetheless, because they understood what we were doing and why, there was no griping, just the solemn recognition that the inconvenience

---

[23] One time long ago, we had a boss who sagely said, "Ignorance is bliss only for the ignorant!"

was part of a more important mission. Maintaining your cybersecurity posture to protect your vital intellectual property and trade secrets or operation of critical infrastructure may be your higher priority mission. The lesson to remember is to make sure everyone on your team and all your key stakeholder and partners knows what you are doing and why.

7. Monitor implementation closely: Anytime you make a change, you need to make sure the results are what you want and expect. Do you think someone at United Airlines should have checked their web page before they offered fares ranging from free to ten dollars? How do you check to make sure your changes are effective? Do you have internal controls that call for all changes to be "independently checked" before acceptance and completion? What happens if something happens in the middle of your change process that indicates that the proposed change is a dud and you need to restore to your previous state? How do you know? Do you have a process for detecting implementation problems? Do you have a Plan B, Plan C, and maybe a Plan D? Successful executives do not rely on autopilot to guide their operations. Monitor your changes closely to ensure they are effective, efficient, and secure.

8. Have a backup and a back-out plan: Helmuth von Moltke the Elder said, "No plan survives contact with the enemy." He was correct. The unexpected will happen. Be ready for it with a plan to address when your change doesn't go as planned. We suggest one of the key questions you should ask when presented with a change proposal is, "What do we do if this doesn't work?" Insist that your employees, especially your IT staff, create backup plans ahead of time to appropriately address situations when changes don't go as planned. Whenever possible, have a plan to back out of your proposed plan when it appears that the change is not going as intended. For example, you may find in step four of a twelve-step process that the installation of a new software application is not going well. The application may not be able to link with the database. Rather than trying to struggle through with the installation, you decide to stop and back out of the installation. You go back through steps 3, 2, and 1 to restore the system to its previous state. Once you have a stable environment, you can regroup and take the appropriate next steps. When proposed changes don't go as planned, don't stand there with your hands in your pockets wondering what to do next. Take proactive steps now to prepare to address the unexpected. Have a plan.

9. Make sure your plan doesn't break anything else: This is critical. The author is aware of certain software patches that wipe out previous security configuration settings; when you install the patch, you have to reset your security settings to protect your information. Likewise, you may find that blocking a port or protocol with the intent to better protect your information may actually deny a vital business function access to an important information source or client that can generate positive benefit for your organization. Your proposed change may have unintended consequences. That's why it is important for you to coordinate your plan carefully throughout your organization to minimize the chance that you'll interrupt an important process, deny an important information source, or break

something. Using a Change Management Board helps immeasurably to make sure that all stakeholders are involved in the process of developing the change and will help find any weaknesses in the change that could inadvertently cause it to break something.

10. <u>Be flexible</u>: Most people have pride in ownership. That can be a good thing when it means they feel a sense of responsibility and commitment. Those attributes are what we look for in the leaders who we assign to manage important processes and tasks. Regrettably, some people take pride in ownership too far and aren't willing to entertain or accept suggestions for improvements. These are people we pass over when it comes to assigning leadership responsibilities. Flexibility allows you to respond quickly when Plan A doesn't work and you have to shift to a contingency plan. Flexibility leads you to welcome suggestions and find better ways of doing business. Flexibility means you are more likely to remain calm and collected when confronted by the unexpected. Change introduces stress, uncertainty, and fear for many people. Don't be one of those people. Be flexible, embrace change, and lead others to do the same.

6.7 SUMMARY

- Change management is part of your risk management program.
- Change is inevitable and can be a very good thing when it is managed properly. Change brings you new capabilities, better efficiencies, and creative new ways of doing things. Change erases poor processes rife with wasteful steps, eliminates toxic leadership, and retires substandard products.
- Change can also introduce significant risk to you and your organization. In fact, periods of change are where most risk is introduced. Changes in personnel, process, and products represent great risk to you and your business. You need to manage change as part of your risk management process.
- At least 65% of all cyber attacks exploit misconfigured systems.
- Most downtime is found to be a self-inflicted wound. According to the Yankee Group, an IT research organization, over 62% of all network downtime is caused due to configuration errors.
- Changes are made to avoid obsolescence, to obtain best value, and to achieve and maintain competitive advantage.
- Changes not only affect your people, but also they affect your service level agreements, contracts, capacity, and security. You need to carefully manage change to protect against adverse impacts that can affect any of these items.
- Internal control policies that specifically address how you manage and control all changes to system configurations, software, web pages, and data exchanges will keep you out of trouble. Ensure that you have adequate controls in place to monitor and manage the change process for each of these important cyber-based capabilities.

- Change should be managed and controlled as a process to minimize risk while maximizing benefits. Your change management process should yield results that are effective, efficient, and secure.
- Change is best effected when it is accomplished as the product of a well-managed and deliberate process executed by a properly trained and motivated workforce. Executives who lead these changes take proactive measures to make sure that the process delivers the desired effects in a manner that is predictable, reliable, and repeatable.
- Managing change is a fundamental responsibility of management, while executing changes largely is a tactical-level activity conducted by employees under the oversight of management.
- The Touhill Change Management process is shown in Figure 6.1.
- The first step in the change management process is identifying what needs to change and why.
- Evaluate the situation thoroughly in step 2 of your change management process. Ask the right questions (and lots of them) to ensure you have a complete understanding of the situation. Evaluate whether you want or need to make a change.
- In step 3 of your change management process, identify potential courses of action. They should always be feasible, acceptable, suitable, and affordable. As a general practice, we always recommend giving your boss at least three COAs to consider.
- Step 4 of your change management process features management deciding which COA to pursue. Convening a Change Management Board as an integral part of your change management process is a best practice that makes sure that all vested stakeholders participate in the change process. It is important that people feel they are involved in the change rather than having it imposed on them.
- Implementing the change is step 5 in the process. Providing as much advance warning of the change to those affected by change and key stakeholder is essential. Don't forget to coordinate with key partners. Make your message clear by following an approved script that conveys the essential message but does not reveal sensitive information. Assign a manager to lead the change and assign specific responsibilities for all change activities. Monitor implementation progress closely and be prepared to back out or implement a backup plan in the event the proposed change does not go as planned.
- After you make a change, make sure that it delivers the desired effects. Always conduct vulnerability scans after making changes to your system configurations, software, web pages, and data exchanges. Consider conducting penetration testing, red teaming, and audits after significant upgrades and special circumstances.
- Be prepared for the unexpected. Have a Plan B, Plan C, and maybe a Plan D ready to execute in the event your Plan A fails or doesn't go as expected.
- There are several best practices for change management that you should keep in mind whenever you make changes to your information environment:

Touhill's Best Practices in Change Management

1. Communicate early and often.
2. Don't rush to field crummy products.
3. Timing is everything.
4. Change only what needs to be changed.
5. Don't be afraid to ask for help.
6. Ensure everyone knows what is going on and why.
7. Monitor implementation closely.
8. Have a backup and a back-out plan.
9. Make sure your plan doesn't break anything else.
10. Be flexible.

7.0

PERSONNEL MANAGEMENT

If you think it's expensive to hire a professional
to do the job, wait until you hire an amateur.

Red Adair[1]

7.1 FINDING THE RIGHT FIT

Having a great strategy complemented by great plans, policies, and procedures gets you absolutely nowhere without the right people, properly trained, with the right attitude, in the right positions doing the right jobs.

As an executive, one of your principal responsibilities is to select and align the right talent to execute your organization's mission. Throughout your career, you likely have seen the negative effects wrought by ill-prepared employees or ill-fitting personalities being overmatched by difficult tasks. In today's highly competitive market where organizations like yours rely on cyberspace capabilities to gain and maintain competitive

---

[1] Dick Ghiselin, *Red Adair, Remembering a Legend*, October 4, 2004, http://www.epmag.com/EP-Magazine/ archive/Red-Adair-Remembering-Legend_2318. Accessed on October 25, 2013.

Cybersecurity for Executives: A Practical Guide, First Edition. Gregory J. Touhill and C. Joseph Touhill.
© 2014 The American Institute of Chemical Engineers, Inc. Published 2014 by John Wiley & Sons, Inc.

advantage, you can't afford to have ill-prepared employees or ill-fitting personalities. You have to build a team that delivers results that are effective, efficient, and **secure**.

Cybersecurity is no longer just an IT issue handled by your technical staff. As earlier chapters in this book demonstrated, cybersecurity is now a key concern that affects *everyone* in your business, both at home and in the office. Cybersecurity issues at home can spill over to your business and vice versa. It touches every business function and activity and presents very real (and potentially existential) threats to your business and your livelihood. Cybersecurity must be a key component of your personal and professional risk management programs.

In earlier chapters, we recommended following Sun Tzu's adage of "Know Your Enemy and Know Yourself." After having read the previous chapters, we hope you agree. Understanding both your threats and your vulnerabilities to those threats is key to understanding your risks and developing an effective game plan to manage them. As you review your threats and vulnerabilities, recall our key point from Section 2.4.3: *"we contend a poorly trained work force presents the greatest cybersecurity threat to you and your business."* Fortunately, you can minimize your risk by investing wisely to inform, educate, and train your workforce to be what we call "cyber smart."

This chapter addresses that greatest cybersecurity threat: you and your people. The great American comic strip "Pogo" coined the phrase, "We have met the enemy and he is us."[2] When it comes to cybersecurity, that may well be the case. As an executive, you have to inform your employees of cyber threats and vulnerabilities, educate them on the risks, and train them to act in accordance with your policies and procedures to minimize risk.

Further, you have to make sure that you align the right talent to perform the right jobs. For example, too many times, we have seen organizations delegate information system administration duties to individuals who did not have the technical skills to handle the assignment adequately. Often, the systems were misconfigured, resulting in frustrating downtime and outages affecting the organization's effectiveness and efficiency. Many times, the systems were not kept up to date with current patches, exposing the organization to exploitation by bad actors. In both cases, the poorly managed systems exposed the organization to unacceptable risk. Misaligning talent will jeopardize your ability to be effective, efficient, and secure. It will cost you time and money.

We believe managers manage "things" while leaders lead people. As an executive, you need to be *both* a highly skilled and capable manager *and* a dynamic and proactive leader. In today's information-enabled environment, the challenges of knowing who to hire, who to fire, what jobs to assign, and who to assign them to become increasingly complex. You need to continually invest in technical training and education for both yourself and your employees to stay relevant and best postured to succeed. You need to know how to organize your team to best manage and protect its information. You need to *lead* your team to incorporate cybersecurity into its strategy, plans, policies, and

---

[2] Marilyn White, *"We have met the enemy … and he is us,"* http://www.igopogo.com/we_have_met.htm. Accessed on October 25, 2013. Pogo was a popular Walt Kelly comic strip. Walt Kelly first used the quote "We Have Met The Enemy and He Is Us" on a poster for Earth Day in 1970. In 1971, he did a two-panel version published in newspapers with Pogo and Porky in a trash-filled swamp. The author remembers them very well from his school days as Pogo's adage took its place next to Smokey the Bear's "Only You can prevent forest fires" and Woodsy the Owl's "Give a Hoot, Don't Pollute!" in promoting responsible protection of the environment.

procedures. You need to *lead* your team to do the right things to make your organization effective, efficient, and secure.

This chapter will give you the tools you need as you determine who and what are the "right fits" for your business. We will present you with information regarding the assignment of roles and responsibilities you should consider when determining organizational structures. It will provide you with the information you need to educate and train your workforce to make sure they are hardened against cyber threats. We'll also remind you of cybersecurity concerns that are of special importance to executives in general and detailed considerations for executives operating and managing critical infrastructure. The knowledge conveyed will better posture you and your team to be "cyber smart."

7.2 CREATING THE TEAM

When people are asked about teams and teamwork, their minds frequently gravitate to the world of sports. Ask someone to tell you about a great team and they likely will tell you a story about a sports team. They will tell you about a group of different individuals from different communities and circumstances who came together, blending their skills to achieve a common purpose. They may even tell you about those on the team who didn't get the spotlight as they subordinated their personal opportunities for glory in exchange for the team's success. We bet you are thinking of a couple examples of great sports teams right now.

Would you describe your business and its employees this same way? Is your organization the first great team that comes to your mind when people ask you to identify great teams? We submit that every business, whether it is a multinational conglomerate, a small- to medium-sized business, a partnership, or even a proprietorship, relies on teamwork to succeed. Isn't your business comprised of a variety of different people with different backgrounds and skill levels all contributing toward a common purpose? Don't you have some people who selflessly sacrifice for organizational success? Teams are especially important in today's highly complex information-enabled environment. If you don't think of your business and its employees as a team, perhaps it is time to change your thinking and do something about it.

Just like sports teams, your business has specialists who contribute their unique skills and talents to make your business succeed. For example, have you ever noticed that American football teams have a specialist called a "long snapper" to snap the ball during punts? It used to be that teams would put any lineman into the position. After many teams experienced crushing defeats due to the linemen errantly snapping the ball over the heads of punters, teams found they needed to hire somebody who would get the snap right every time. They needed a specialist. Teams value these specialists because the risk of a botched snap could be the difference between pinning your opponent deep into his territory or deep into yours. How well you execute your punt could be the decisive difference in a game. Do you view any of your employees as specialists with unique and indispensable skills that could make the difference in your game plan? We bet you have that special attorney that you go to handle difficult litigation. Likewise, you probably have your go-to accountant who handles your end of fiscal year closeout procedures. You may even have your "Top Gun" salesperson that you dispatch to handle your most important clients. What about your IT staff? Are they the long snappers in your organization? Perhaps, they aren't. But we submit they indeed are specialists, they aren't your "long snappers" who just come for a couple of plays. Instead, they are in on *each and every* play and, if they are successful in their jobs, they remain invisible and anonymous, much like Jon Kolb.

Do you know who Jon Kolb[3] is? Kolb was the offensive left tackle for the championship Pittsburgh Steelers in the 1970s. Labeled the National Football League's "Strongest Man" during that period, he protected Terry Bradshaw's blind side and earned four Super Bowl rings. Jon Kolb remains largely anonymous to most people because even though he did his job superbly, the position he played called for steady, secure, and stable (unspectacular) play. His role was not to pass like Terry Bradshaw, run like Franco Harris, or catch the ball like Lynn Swann or John Stallworth. His job was to make the blocks that enabled these teammates do their job so the team would win. He didn't receive a lot of notice or accolades, yet he was instrumental in the team turning from a league laughingstock into a powerful and perennial champion. We submit you need people like Jon Kolb throughout your organization in order to field a championship team. Is your IT staff like Jon Kolb, exceedingly powerful and opening opportunities for your business functions to exploit, or do they open up holes that hackers and others use to blindside you? In order to achieve your goals and objectives, you must recruit and align the right talent to execute the tasks that make your business prosperous and successful. You need people that like Jon Kolb will have an incredible positive impact to make your team better. When you do that, you have the start of a great team.

7.2.1 Picking the Right Leaders

You may have a great group of individuals and even have the best strategy, but if you don't have the right leaders, you are destined to fail. During your professional career, you probably have seen some very successful and profitable companies suddenly take a nosedive and flounder when they put the wrong person in charge. We have too. The person may not be bad, but if they are not the right fit, not qualified for the position, and do not possess the right leadership skills, their failure is almost certainly guaranteed. As the boss, one of the most important assignments you will have is picking the managers and executives to execute the corporate strategy and grooming those who ultimately will take your place.

With your business's reliance on information growing every day, hiring executives who do not understand nor appreciate IT is a losing proposition. With e-commerce, telework, office automation, mobile computing, robotics, computer-controlled manufacturing,

---

[3] Jon Kolb was drafted by the Steelers in the third round of the 1969 draft, Chuck Noll's first year as head coach. Oklahoma State University's center during his college days, Kolb did not start any games his first two years, yet in 1971, Noll shifted him to left tackle, where his incredible strength and skill proved to be invaluable in protecting the Steelers backfield. Some hard-core "Stiller" fans will recognize that the Steelers play-off teams started with Kolb's shift to left tackle. Coach Noll knew how to find the right talent and put it in the right places to win!

and a host of other information-enabled activities emerging at the forefront of business today, your C-suite and subordinate executives must have a thorough understanding of not only technology but also how to leverage it to create stunning victories every day.

Do all of your executives need to be technical experts? Of course not. In fact, we submit that sometimes it is better to hire a great executive with proven leadership skills who can be taught to understand information technology than it is to hire a technical expert and expect to transform him into a great executive and leader.[4] Nonetheless, you can't afford to hire executives who fail to show the interest in or appreciation of information technology. Likewise, you want to steer clear of those who are dismissive of cybersecurity; you can't afford them in your ranks, let alone permit them to be in charge.

Your executives not only have to be great managers, but also they have to be great leaders. Leaders and their attitudes set the tone for the organization. If your strategy calls for your organization to be information enabled and values cybersecurity to protect its vital information, you cannot afford to have leaders who do not embrace that strategy wholeheartedly.[5] We recommend you include focused questions on information and cybersecurity as you interview candidates for your executive positions. Does the candidate make unsolicited mention of cybersecurity as being part of their management and leadership philosophy? Does the candidate view cybersecurity practices, training, and tools to be unreasonable burdens or normal and prudent costs of doing business? Does the candidate believe information is an asset with an intrinsic value? Does the candidate believe cybersecurity is a "must-pay" priority investment or a discretionary expenditure? Is the candidate able to give examples of how they have incorporated cybersecurity into their business practices and that of the projects they have managed? Does the candidate demonstrate that they practice secure computing in both their home and office lives? With information and information technologies continuing to take a predominant role in business, you need executives who *lead* efforts to safeguard your information.

Your executives need to be savvy in their professional and personal lives. You would shy away from selecting executives who have a bad reputation for unacceptable behavior such as excessive drinking, gambling, and even caustic personalities. Now, you may include how they conduct themselves in their online activities. Many companies now comb the Internet and social media sites to vet their prospective executives and employees to see whether they have any embarrassing or inappropriate information on the net. While some people see this as an invasion of privacy, we view it as an essential business hiring practice. You don't want nor can you afford to hire executives whose web presence serves as a potential embarrassment to your organization. Individuals who protect their personal identities and sensitive information are more likely to protect your business's vital information too. Safeguard your organization by selecting executives who not only incorporate cybersecurity into their normal business and leadership processes,

[4] Chuck Noll, the coach who moved Jon Kolb to offensive tackle, believed that it was crucial to draft "the best athlete available."

[5] We are reminded of a CEO of a medium-sized but well-known Baltimore engineering firm that we interviewed in the early 1970s in conjunction with a survey of computer use in engineering who said: "If I find some fuzzy-headed jerk in my company who insists on using computers instead of slide rules to design facilities, I'll fire him!" Fortunately, the old man was ousted in a coup by younger and smarter partners, but the company to this day is haunted by his ignorance.

but also select those who demonstrate that they use strong cybersecurity practices away from work as well.

An essential skill for executives in today's information-enabled market is the ability to recognize and quickly act upon opportunities presented by information and IT. Consider businesses that decided to migrate to web pages as their digital storefront, offering lower-cost offerings available anytime, around the world. They recognized the value proposition that such e-commerce presents as traditional brick-and-mortar store-fronts with expensive staffing, facility expenses, and other overhead costs could be replaced by a web presence that is always open, always available, no matter where the customer is located. Companies such as Amazon and eBay were early adopters of e-commerce, while those who lagged and relied on traditional stores such as CompUSA and Circuit City did not survive. Do you want executives who are visionary and seek opportunities to leverage the power of IT to gain competitive advantage? Does your candidate have the proven ability to translate great vision into practical success? Is your candidate a technology innovator or troglodyte?

We recommend you make continuing professional education a priority for your executives, and when selecting the skills you want to enhance and nurture, make sure you include courses that emphasize cybersecurity. There are many excellent courses and seminars run by major universities and professional organizations that are now begin-ning to recognize the importance of information assurance, information security, and other monikers that apply to the cybersecurity realm. We recommend you invest in annual focused cybersecurity awareness training for your executive team to ensure they are well equipped to make decisions that keep your information and business interests secure in today's ever-contested cyberspace environment.

How do you know which course is best to meet your needs? First, determine whether the course content is aligned with your business strategy. If the course espouses principles that are consistent with your strategy *and* your core values, it is a contender. Second, consider how the course addresses information. Does it present information as a valued asset that should be managed, controlled, and protected like other valued corporate assets? If so, it remains a contender. Third, does the course have a specific block of instruction dedicated to cybersecurity? If it does, it sounds promising. Fourth, is the course a good value that fits into your budget and promises a good return on investment? If all four criteria are met, you likely have a winner worthy of your investment.

Finally, you need executives who can tell the difference between a slick salesman's snake oil story and ground truth. Sadly, there are many people in the IT business who overpromise and underdeliver. Select executives who can pick the winners and dismiss the losers. Successful executives are curious and ask the right questions. They do their homework and research alternatives and options before making decisions. They col-laborate with experts when the topic exceeds their area of expertise. They seek other opinions and recommendations from others. They seek to understand and appreciate technology before they make commitments to it. They are cautious when asked to be first adopters of technology and processes. They aren't afraid to say, "Prove it," when presented promises of fabulous returns. Successful executives can discern what is truly within the realm of the possible from proposals based on science fiction. You need to choose your executives with the same care and attention as you would a prospective son-in-law. Your C-suite and subordinate executives need to be "cyber smart." Your business is at stake. Choose wisely.

7.2.2 Your Cybersecurity Leaders

One of the things that you need to do as a leader is to remind your employees that cybersecurity is *everybody's* responsibility. This is at the core of building a corporate culture that values cybersecurity. Your cybersecurity programs, training, tools, and procedures do not belong just to one individual or department; they are the responsibility of everybody, including you!

Nevertheless, many organizations assign executives responsibilities for activities that cut across multiple product and business lines. These executives are responsible for policies that govern functional activities and provide oversight that makes sure the policies are appropriately followed in execution by personnel across the organization. These executives provide sponsorship and ownership of crosscutting activities. Financial, security, and risk management are examples of these types of crosscutting activities. Cybersecurity is another.

There is no single best practice organizational structure determined to optimize cybersecurity in business. While many organizations align sponsorship of cybersecurity programs under the chief information officer (CIO), we've seen others who place it under the auspice of the chief information security officer (CISO), the chief security officer (CSO), or the chief risk officer (CRO). We've even seen several companies place it under the COO or the CFO. The type of business you run and the corporate culture of your organization will guide your selection as to which officer is best suited to sponsor the cybersecurity program in your business.

We recommend you consider aligning management of your cybersecurity programs to your CIO with your CISO serving as a direct report. CIOs are responsible for the information of the business. If they do their job correctly, they are thinking and acting beyond the IT systems; they are focused on the process that creates, consumes, manages, stores, and protects the information that is such a valuable part of your business. Too many times, we have seen CIOs who become bogged down with the acquisition or management of software and IT systems while losing sight of the fact that these are complementary tasks supporting the management of information, which is truly the heart of every CIO's job. Your CIO should be responsible for managing the entire life cycle of your business's information, from creation to destruction, including its protection.

Because the CIO is responsible for the effectiveness, efficiency, and security of information throughout its life cycle, we recommend you consider aligning the CISO as a direct report to the CIO. The CISO provides the full-time focus to manage the programs and operational activities required to protect your information properly. We view this as a subordinate role to the CIO, who manages the entire spectrum of activities governing your information. A CIO without a CISO is a person ill equipped to properly do their job. Similarly, a CISO not working in concert with the CIO often is viewed as a competitor, resulting in needless friction that adds drag to your organizational processes and production. You don't need that kind of heartache.[6]

As senior executives, we make a point of thinking strategically, including with succession planning. We look for promising young executives who we can groom to take broader and more important roles in the future. We are looking for someone who has the skills and experience to prove themselves worthy to take our place when we decide to step aside (or move up). We believe that there is great opportunity to grow future CIOs from your CISOs. In fact, given the increased importance of information in your business, we believe that CIOs not possessing a thorough grasp of the skills required of CISOs are not adequately prepared to manage the full spectrum of information management successfully. We recommend you consider placing your potential future CIOs into CISO positions, where they will gain the necessary mastery of cybersecurity practices, policies, and procedures that complement your business and protect your information. These are essential skills that a CIO must have. If the candidates are unsuccessful as CISOs, they definitely will be unsuccessful as CIOs.

CIOs and CISOs share common technical education and training requirements and experiences, yet the CIO's experience base is arguably much broader than information security. CIOs must master information management, information security, software and hardware operations, system acquisition, architectures, and a host of other technical and managerial functions. This places CIOs in an ideal position to serve as not only the CISO's boss but also as a mentor. Too many times, we have seen CISOs evolve into the "Just Say No" security roadblock. Sadly, they put themselves in the position where they are viewed by other executives as the folks who say "no" to innovative ways of doing things rather than helping others find innovative methods with cybersecurity "baked in" to the process as a valued protective measure. CIOs typically have a broader view and can help provide the leadership to focus the CISO on providing value-added and practical security constructs that are appreciated and adhered to.

Many of our clients ask what type of credentials and experiences we recommend that senior executives such as the CIO and CISO have to ensure they have the skills needed to lead contemporary IT organizations effectively. While there are a host of certification and credentialing programs now available,[7] we believe there are two cybersecurity credentials that are "must-haves" for both your potential CIO and CISO candidates, the Certified Information Systems Security Professional (CISSP) and Certified Information Systems Manager (CISM) certifications.

The first is the CISSP certification issued by the International Information Systems Security Consortium (ISC2) organization. It is considered as the more technical of the two certifications, and we regard it as a must-have for CISOs and highly desired for CIOs. Members who possess this certification have demonstrated a detailed understanding and experience base in the following ten cybersecurity domains.

---

[6] In small organizations, it simply may be too expensive to have both a CIO and a CISO. In such cases, responsibility may be vested in one person. That can work if the "combined" duties are defined clearly and the span of control is manageable.

[7] Arguably, the best CIO training and education program is conducted by the National Defense University. Sadly, their course is not readily available to individuals outside of the U.S. government.

ISC2's Ten Domains

Information security and risk management	Business continuity and disaster recovery	Security architecture and design
Access control	Physical and environmental security	Telecommunications and network security
Cryptography	Legal, regulations, compliance, and investigations	Application security
**************	Operations security	**************

Many senior executives look at the CISSP certification as they do a Professional Engineer certification. Members possessing the CISSP certification have undergone a grueling six-hour examination that tests the candidate's knowledge in all ten domains. CISSPs must demonstrate five or more years of experience in protecting information and information systems. They subscribe to a professional code of ethics that promotes the safe and ethical use of IT. Further, they must maintain their currency through continuing professional education. When you hire a CISSP, you can have confidence they have demonstrated technical expertise in cybersecurity.

The second certification, the CISM certification, is issued by the ISACA organization.[8] While the CISSP credential program is more technical in nature, the CISM program focuses on the effective management of information systems using contemporary security principles and best practices. Like CISSPs, people with the CISM certification pass a comprehensive test that ensures they have the requisite technical and managerial knowledge, have to demonstrate years of experience in professional management of IT systems, and have continuing professional education requirements. The CISM certification is highly recommended for CIOs and CISOs alike.

Is one better than the other? Do you need both? Do you need to even have a certification? Frankly, there are numerous CIOs and CISOs operating without these credentials, and many of them seem to be doing just fine. However, with the ever-increasing cybersecurity threats continuing to mount, are your CIOs and CISOs adequately equipped with the skills and experience to excel in the coming years? Do you want to hire *or retain* individuals who are not prepared to adequately manage and defend your information?

We believe the current and future threat environment demands that your CIOs and CISOs need to maintain a professional cybersecurity certification. If your CIO and CISO already have their credentials, terrific! Make sure they maintain their currency through continuing professional education. Invest in them by sending them to courses and seminars that enhance your business objectives. If they don't have their credentials, ask them, "why not?" Consider including achieving certification as a mandatory performance objective in this year's performance plan.

---

[8] Previously known as the Information Systems Auditing and Control Association, ISACA has over 110,000 members in 200 chapters located in 80 countries. In addition to the CISM certification, ISACA offers the CISA program, which is the gold standard for auditing in the cybersecurity realm.

One other thing about credentials is worth noting. Technical staffs tend to give their best effort and deliver terrific results when they are working for someone who knows and appreciates their business. Having a boss who maintains their technical credentials often is viewed as a point of pride within the technical staff. They identify with their leader as someone who has taken the time to understand their jobs and what it takes to do the job right. They value bosses who have "been there, done that, and have the T-shirt" over those who are just there to have their ticket punched on the way up the career ladder. Many members of your technical staff seek to advance in the organization, and when they see unqualified executives from elsewhere in the organization placed in demanding technical roles such as the CIO or CISO positions as "career-broadening" jobs, it sends a signal that they have a limited career path in your organization. They read these signals and many of your best and brightest jump ship to find better opportunities.

Be careful what signals you send in picking your CIO and CISO.

If you make technical credentialing and experience a requirement in selecting your senior technical business leaders, you also send a message that gives hope to your technical workforce that with the right combination of technical prowess fused with managerial and leadership excellence, they can someday rise to the C-suite. This helps keep your best technical talent, provides invaluable stability within your staff, and builds loyalty and trust. When you establish a career path for your staff that potentially leads to a C-suite position, you set a solid foundation that pays rich dividends in strengthening your cybersecurity program.

Picking the right CIO and CISO is critically important for your organization. Not only do they need to have the technical expertise to effectively perform their jobs, they have to be great partners and leaders. They need to understand the business processes across the organization and look for opportunities to enhance business functions by innovatively leveraging information and IT to gain or maintain your competitive advantage. Many of our clients ask us to identify attributes we look for in selecting potential CIOs and CISOs to be great "cyber leaders." Frankly, CIOs and CISOs are executives who have to have solid technical credentials yet possess the same talents and leadership skills we expect of any other executive. We seek those who have demonstrated attributes that include the following:

- The ability to lead people and manage projects
- Thorough understanding of the business and how technology enhances it
- The ability to plan and forecast strategically and tactically
- The ability to communicate clearly across many different audiences and cultures
- The ability to adapt and perform at high levels in a variety of new and different positions (e.g., career broadening)

We've spent considerable time discussing your CIO and CISO, whom we consider your principal cybersecurity leaders, but what about your other executives? Aren't they also cybersecurity leaders?

Absolutely!

In fact, many firms now require that their CRO and CSO have demonstrated cybersecurity experience and possess CISSP or CISM certifications as a prerequisite in their job advertisements. Cybersecurity is about risk management, and both the CRO and the CSO have a huge stake in maintaining a solid cybersecurity program. It makes sense that they too should have experience in protecting and defending information. Like your CIO and CISO, you should consider recurring cybersecurity training for your CRO and CSO. Perhaps more than other executives in your organization, their duties call for a broader and deeper understanding of threats, vulnerabilities, and the tactics, techniques, and procedures needed to minimize risk. While they don't need the detailed technical knowledge of your CIO and CISO, they need to be able to speak and understand the same language. We recommend you consider investing in specific cybersecurity training for your CRO and CSO.

Having leaders who are "cyber smart" helps your business immeasurably. Executives who are well informed and understand the risks and opportunities facing your organization are much better postured to make the right decisions that yield success. As you look to pick the talent for those who lead your cybersecurity efforts, we strongly encourage you not only to pick great executives with terrific leadership skills but also to pick the ones that are established technical leaders with the right certifications and qualifications.

7.3 ESTABLISHING PERFORMANCE STANDARDS

Many executives believe they are successful if they meet their performance targets for the year. Nothing gets your attention more than your performance standards. Failure to achieve established goals can result in a change of scenery for you and the organization, so there is strong motivation to do well and meet "your numbers."

Does your organization incorporate cybersecurity "numbers" into its performance standards? Is cybersecurity performance specifically addressed in the expectations your organization establishes for you and your subordinates? Certainly, many companies are known to have a "zero tolerance" program when it comes to information disclosure, leakage of secrets, and cyber events, which tends to result in several employees packing their bags as they seek new employment after a single noteworthy cyber incident.

This approach is not uncommon nor is it necessarily unreasonable. Maintaining accountability is an appropriate and highly desired management and leadership attribute. Perhaps you should expect to get fired if your actions (or lack thereof) cause significant damage to the organization. For example, as we discussed in Chapter 5.0, where appropriate, your policies should always spell out that failure to follow policy can lead to disciplinary action up to and including termination.

We advise you to exercise wisdom and sound judgment when your employees permit bad actors to penetrate your cybersecurity shield. Is this person one of your star performers? Was it a lapse in judgment and common sense, did it fit with a long-term pattern of carelessness, or was it an accident? Can you afford to live without this person? If you exhibit compassion for one instance, do you dispense justice fairly and evenly to all thereafter? Demonstrate wisdom to your employees. Show them that there are consequences for mistakes and hold them responsible, but don't let intransigence stand in the way of intelligence.

Many companies are now struggling on how to determine cybersecurity success. How do you factor that into your annual "numbers"? How do you create meaningful cybersecurity performance standards that can enhance your business?

Determining measures of success is essential for any job. Everyone wants to know what the boss wants and how they will be measured for their performance. If you don't tell your employees what you value and what you expect them to do, they will give you what they believe is best and expect to be rewarded based on their own criteria, not yours. That is unacceptable for all parties. Be clear that you expect certain cybersecurity-related performance measures to be included in their annual performance plan. As with most performance standards, make them feasible, achievable, suitable, *and measurable*. (We bet you thought we were going to say affordable. Stand by, that discussion is coming.)

You may be reading this and thinking, what kind of performance standards can I set with cybersecurity in mind?

Let's use the Plieno Steel Company[9] and its third-shift supervisor, Rocky, as an example. Rocky is responsible for the successful production of Plieno's specialty steel during his shift. He runs the mill and all activities during his shift. Rocky is a great leader and his shift regularly outperforms other shifts. Sporting his trademark #20 Steelers jersey, blue jeans, and black handlebar mustache, Rocky inspires his team to exceed all performance standards. They usually do and they love him for his leadership.

How do you bring cybersecurity performance standards to a line supervisor like Rocky?

Let's start with costs. What happens if Rocky or one of his team cause or contribute to a cybersecurity incident? You already know the costs of remediation and cleanup can be very expensive, but you may be wondering how a steelworker can cause a cybersecurity incident. Let's look at a couple of potential opportunities Rocky and his third-shift team have to create a cybersecurity incident.

The first one may appear rather obvious. Rocky or one of his team could introduce malicious code into the corporate intranet or extranet by bringing in a thumb drive or CD poisoned with malicious code such as a virus or an even more devastating remote access trojan (RAT). Impacts could include loss of your network while the damage is cleaned; compromised, damaged, or destroyed vital information; or interruption of key production processes. Incidents like these can be catastrophic for Plieno—or your business.

A second opportunity might be in disclosing Plieno's secret formula either in an unguarded email or via a phone conversation, both of which corporate spies would love to get their hands on. Recall that Plieno's executives consider the secret formula to be one of their most highly treasured assets.

A third less obvious opportunity may involve the industrial control systems (ICS) that control the smelting and casting process. Most manufacturing processes now rely on computers to precisely regulate and control the flow of materials, the rate of production, and safety controls, all of which are critical at Plieno Steel. Some ICS are connected to intranets, which transport the data from ICS to automate important performance and production metrics. Others are sealed and only accessible by direct physical access. What if Rocky or one of his team accidentally alters the code of an ICS that controls the mixture

---

[9] Remember you met Plieno Steel Company in Chapter 3.

of materials? Plieno's success is dependent on its secret formula. Don't follow the secret formula while making Plieno steel and you have big problems. Your steel will not meet the specifications required by the customer, and if you are fortunate to catch it before it leaves the mill, you'll just have to replace it. If it leaves the mill before discovery, you expose Plieno to potential litigation and probable damage to brand reputation. In a worst case, your substandard product may fail and jeopardize public health and safety, which is accompanied by its own set of problems.

Some people may dismiss the vulnerability of ICS to potential tampering or altering. That is a dangerous attitude and is not shared by the U.S. government or key manufacturers of ICS. According to the U.S. General Accounting Office's May 2004 report, *Cybersecurity for Critical Infrastructure Protection*, "there is a general consensus—and increasing concern—among government officials and experts on control systems about potential cyber threats to the control systems that govern our critical infrastructures."[10] All businesses, especially those like Plieno and those who operate critical infrastructure (e.g., those specializing in chemical, water, pharmaceutical, transportation, and power production), need to take special care that their ICS are protected from inadvertent as well as malicious alteration. The public's health and safety depends upon the integrity of those systems.

What if you made "Suffer no losses due to cybersecurity incidents attributed to your team" part of Rocky's performance standard? Do you believe that Rocky would be more inclined to lead his team to take cybersecurity practices seriously? Would it be reasonable to expect that he would take proactive and repeated leadership measures to ensure he and his team understood risks and how to prevent losses from cybersecurity incidents?

What if we considered schedule? Not only do cybersecurity incidents incur monetary costs as resources are expended to respond to incidents to contain, control, remedy, and remediate incidents, but also they have noteworthy indirect costs. For example, Plieno operates on a tight schedule. Business is so good right now that they are operating multiple shifts around the clock every day except on Sundays and holidays. Many of their customers expect on-time deliveries of Plieno specialty steel. Most contracts have delivery dates specified with discounts and penalties identified. Time truly is money for Plieno as it likely is for you and your business. What happens when a cybersecurity incident happens that can be attributed to Rocky and his team? Is there a schedule slip? Perhaps. If there is, then there is a cost. Customers don't get their steel on time. Customers aren't happy. Customers are less inclined to hire Plieno again if there are suitable substitutes.

Could you identify schedule performance standards for Rocky that would correlate to cybersecurity? You certainly can. For example, you could include something like the following in his annual performance numbers: "Maintain all production schedules free from interruption by team-induced cyber incidents." Would that get Rocky's attention? If his Christmas bonus and company-subsidized Steelers tickets were linked to his ability to maintain schedules free from team-induced cyber incidents, do you think he would invest his time and attention to cyber harden himself and his team? We do.

What about performance standards such as product output? In Rocky's case, it is measured in tons of finished steel, but in your company's case, it could be tons of a

---

[10] U.S. General Accounting Office, *Cybersecurity for Critical Infrastructure Protection*, May 2004, p. 36.

certain chemical safely manufactured, millions of gallons of drinking water produced, or units of a medicine delivered. We submit that even these standards can be linked to successful cybersecurity results. How do you think Rocky would respond to the following performance standard: "Produce XX million tons of steel in prescribed timelines without rework or interruption due to team-induced cyber incidents"?

We hope you get our point. We believe that you can sharpen your organization's focus toward cybersecurity by incorporating it into your performance standards. Focusing on cost, schedule, performance, and (especially) business effects can drive home the point that proper cybersecurity practices have value in nearly every process and product in which your company is involved.

Linking pay to performance multiplies this effect many times over. When people know they will be rewarded for cybersecurity success and face consequences for cybersecurity failures, chances are very good they will give you their best effort. As you seek to develop a corporate culture that values cybersecurity and cyber hardens your workforce, consider incorporating cybersecurity into performance standards and link pay to performance. We believe you will get people's attention and they will respond exceptionally well.

7.4 ORGANIZATIONAL CONSIDERATIONS

Who is responsible for cybersecurity in your organization?

If you didn't say *everyone*, you need to reread the first six chapters of this book!

Actually, if you asked employees in many organizations who run their corporate cybersecurity program, they may tell you the name of the corporate sponsor, that is, the executive who administers and manages corporate-wide activities that promote cybersecurity. In some companies, the program is managed by the CIO, while others delegate those responsibilities to the CISO. We are aware of several companies who assign cybersecurity programs to the CSO and a very few who assign it to the CRO. Each company is different and assigns roles and responsibilities based on several factors including strategy, corporate cultures and values, company size, organizational goals and objectives, and the capabilities and personalities of their executives.

All things being equal, we submit that in a perfect organization (perhaps yours?), the cybersecurity program would be "owned or sponsored" by the CEO and managed by the CIO. This is because cybersecurity touches each and every activity in your organization. When the CEO stands up in front of the employees and says, "This is so important that I personally am its champion," and backs their words up with leadership in the activity (including holding subordinate officers accountable for promoting the program), people take notice and respond accordingly. After all, people tend to work the boss's problems first. If your organization is trying to develop a culture of cybersecurity, we suggest no more powerful signal of the value the corporation places on the effort than the CEO adopting the program as his or her own and actively promoting it.[11]

---

[11] Once again, we remind you of the success and positive impact that "Safety First!" programs had in the workplace.

With the CEO as the champion of the cybersecurity program and every employee responsible for their piece of it, you still need someone to manage it. We believe the CIO, who has responsibility for shaping and controlling the corporate information environment, is the best qualified position to manage the corporate cybersecurity program. Through the creation of plans, policies, and procedures; architecture development; and the selection of tools that create, manage, monitor, control, and store information, the CIO is at the heart of nearly all business activities. Furthermore (as previously discussed), the ideal CIO has the managerial experience and technical credentials to manage a cybersecurity program effectively. Supported by a capable and credentialed CISO and the CEO serving as the organization's cybersecurity champion, the CIO leads a powerhouse team.

Managing an effective cybersecurity program involves more than publishing policies, conducting annual cybersecurity auditing, training, monitoring intrusion detection systems, managing firewalls, and sending out occasional email reminders about the importance of cybersecurity. Unfortunately, that's all a remarkable number of companies do when it comes to cybersecurity. With the number of cyber incidents continuing to rise in volume and severity, you and your organization need to invest wisely to ensure you best manage your cyber risk.

We already mentioned in Chapter 5.0 that an organizational best practice to manage risks introduced during change is to convene a Change Management Board, chaired by the CIO to control and manage the risks introduced by change. We submit that this is not the only governance structure your organization should consider to manage its cyber risk.

For example, consider your organizational finances. Many companies convene Financial Review Boards chaired by the CFO to review organizational expenses.[12] One of the more common agenda items is a review of whether items of expense produced their desired results.[13] Is cybersecurity an agenda item in these financial governance sessions? In the organizations we've been affiliated with, it often is but determining whether cybersecurity investments are effective often is difficult to grasp. Nevertheless, if the organization approaches cybersecurity much as it does insurance, these meetings tend to be more fruitful and yield better decisions.

Consider also how risk decisions are made. As previously discussed, many companies establish risk committees as part of their governance structure. These committees evaluate the risks facing the company and establish the risk appetite of the firm. Does your company have a risk committee? Is cybersecurity on the agenda during these committee meetings? Who makes the decision to accept cybersecurity risk in your organization? We recommend you make it perfectly clear how risk is managed in your organization, because no doubt the board of directors will be acutely interested in how you answer these questions.

---

[12] Typically, in order to convene the Financial Review Board, the amount of the expense has to exceed a predetermined level, usually established by the board or CEO or both.

[13] Recall Chris Loftis' "Is the juice worth the squeeze?" comment. This agenda item often asks that very question.

The type of organization, its culture, and its values are key factors in determining the governance structure for cybersecurity programs. Secure leadership/ownership at the highest levels is necessary to reinforce the importance for safeguarding your information. Make sure everyone knows that cybersecurity is *their* responsibility.

7.5 TRAINING FOR SUCCESS

Investing in cybersecurity training for your employees can save your business more than money. It can save your brand reputation. When employees are not properly trained, they often devise methods and procedures on their own and the results are not predictable, effective, efficient, and secure. When your employees don't understand the need to practice cybersecurity methods at home and the office, they are more likely to expose themselves and your organization to risks. Much is at stake. Cybersecurity training for all employees is a wise investment that you can't do without.

7.5.1 Information Every Employee Ought to Know

We've already shared a significant amount of information that you ought to include in your employee training program. It is important to train every employee on basic cybersecurity principles and techniques so that they have a solid understanding of the threats, vulnerabilities, and risks confronting them and your organization. They should know what they should do to protect the organization's information and thus their own vital interests. Demonstrating how the individual can be personally affected is a powerful technique to reinforce the importance of the subject. Therefore, we recommend you include a section in your corporate training on how to protect yourself at home as well as how to protect the organization. We believe that a strong cybersecurity posture in the employee's home computer system is essential to effectively managing your corporate risk. For example, if they are working for you in the evening on a project that you and they enjoy, that's a good thing. However, if they have been lax in protecting their home computer and have a device riddled with malware, then that spreadsheet they created at home and transferred to a thumb drive probably will be plugged into one of your "clean" computers and unwittingly spread throughout your network.

Hence, computer safety does begin at home. We suggest that you continually invest in training your personnel to be sensitive to cybersecurity and proper use of computers. We have found that training that applies well at home and the office positively reinforces learning. As an example, we've included in Appendix B a training guide on email etiquette that is equally applicable to home use as well as in the office. If employees understand their vulnerabilities at home, they will relate more easily to the need for attentiveness at work. As telecommuting and home commuting become more popular, the likelihood of a compromised home computer infecting the company network increases dramatically. Elsewhere in this book (principally in Chapter 5.0), we articulate policies and procedures that apply.

Don't make the mistake many organizations do by getting lazy at the top. Your training should apply to every employee, from the Chairman of the Board to the newest hire in the mail room. Every employee shares in the responsibility to maintain a solid cybersecurity posture because a mistake by someone most likely will be felt by all. Make it policy that cybersecurity training is mandatory for every employee. Lead by your example, and don't exempt yourself because of your busy schedule. Schedule your cybersecurity training as you would any other important meeting. Check to make sure your employees take their training as well. Follow-up when they don't. Make satisfactory completion of their training a performance standard.

You may be asking yourself what your cybersecurity training program should look like. You are not alone as many people ask us to help them establish their cybersecurity training programs. For example, we have a CEO client who recently asked us to give our cybersecurity training to his employees. We tailored our presentation for his organization as all training should be put into a meaningful context to deliver the maximum positive effects. Nevertheless, there are some common items of interest that we recommend you include in your training programs. Many of these items have been covered in detail elsewhere in this book, and we will try to avoid replowing that ground. Rather, we propose an outline you should follow as a basis for training your employees to properly understand cyber risks and equip them with the knowledge to respond appropriately. Follow this training program and you will have a workforce that is not only "cyber smart" but "cyber hardened" as well.

Cybersecurity Training Plan Outline[14]©

Purpose: Describe why your organization is investing in the training and why it is important that they need to pay attention:

- Your organization needs reliable, accurate, and accessible information.
- Your information has value and needs to be protected; it is essential to maintaining your competitive advantage.
- Bad actors, such as hackers, and even some employees pose potential threats to your information.
- Need to balance security and effective information access.

Cybersecurity and risk management: Describe the threats and vulnerabilities facing your organization. Emphasize how they create risk if you don't protect against them.

- **Threats**

    Natural: lightning, fire, hurricanes, earthquakes, tornados, floods, etc.

---

[14] This is the Cybersecurity Training Plan we use in our classes and with our clients.

Human:

> <u>Unintentional</u>: accidents, safety violations, poor security practices, carelessness, and ignorance
>
> <u>Deliberate</u>: hackers, spies, disgruntled employees, and social engineering

Social engineering

> Phishing, spear phishing, and whaling
>
> Dumpster diving

- **Vulnerabilities**
- **Information**
- **Information Systems**
- **Infrastructure**
- **Humans**

<u>Procedures</u>: Describe how you want your employees to protect your vital information. Inform them how you defend in depth.

- **Protecting Information**

  Security classification, data accuracy, data quality, timeliness, authoritative sources, user authentication, roles and permissions, and need to know

- **Protecting Systems**

  Passwords, email policy, backups, threat awareness, antimalware software, firewalls, encryption, network design, demilitarized zones (DMZs), access control lists, redundancy, and physical controls

- **Physical Security**

  Facility access, escort control, screen locking, clean desk, and equipment control

<u>Privacy</u>: Remind your employees of the importance to protect the privacy of clients and themselves. Don't forget to discuss the legal requirements and liability concerns:

- Personally identifiable information
- HIPAA and other regulations

<u>Foot stompers</u>: In the college setting, "foot stompers" are things that you need to pay attention to because you'll see them again on the test. They generally are the things your professor believes are so important that you can't get wrong and are repeated often. These are some of our preferred "foot stompers" for everyone, regardless of whether you are at business or at home:

- Acceptable use, employee monitoring, and content filtering
- Email rules and email etiquette
- Spam
- Web browsing

- Social media
  - Benefits
  - Security concerns[15]
  - Proper use
  - Protection

<u>Common mistakes</u>: The U.S. Department of Defense maintains a database called the Joint Universal Lessons Learned System (JULLS) that catalogs lessons learned during operations. One of the benefits that users discover is in finding out how people made mistakes and learned how to prevent or fix them. I find learning from other people's mistakes is very helpful, so I like to share some of the most common cybersecurity mistakes with you in the hope that if you are aware of them, you don't make them too:

- Failure to install and keep antivirus software current
- Opening unsolicited email attachments without verifying source and contents
- Executing games, music, videos, and programs from untrusted sources
- Failing to install security patches
- Not making and checking backups
- Not installing the security features on your computer and network
- Leaving default passwords on your computer and network devices

<u>How to protect yourself at home and the office</u>: Here are some "best practice" techniques that everyone should follow at home to protect themselves and their information:

- Safeguarding information.
- Regularly scan with up-to-date antivirus, antimalware, and antispyware.
- Scan all email attachments and downloads you get from the Internet.
- Update and patch your software regularly.
- Install and use a firewall when you are connected to the Internet.
- Turn off and disconnect your computer from the Internet when not in use.
- Back up important files.
- Use complex passwords and keep them secret.
- Don't click on untrusted links.
- Don't reply to spam *and don't send it.*

---

[15] In one class I taught, I told my students not to post vacation pictures while traveling on their social media accounts. They were incredulous and challenged me. When I shared that some creeps in Philadelphia had committed several burglaries after having seen on social media sites that the victims were traveling on vacation, they agreed that using social media to advertise their physical absence was a poor security practice similar to putting an out-of-office message on your email that advertises that you are away on vacation. Be careful, you may be putting out the welcome sign for burglars! One of the best things you can do when using social networking is to lock down your account through the privacy settings. Tightly control who has access to all of your information at all times.

- Don't send emails to people who don't need the information.
- Don't surrender your personal information (i.e., birth dates, birthplaces, social security numbers, mother's maiden name, etc.) to untrusted sources.
- Don't use the same username and password for multiple sites.

<u>Ethics</u>: While your acceptable use policy should address ethics, we believe it is important to reinforce the importance of ethical use of computers. Our lawyers do too. Because violation of ethical standards frequently is viewed as an offense worthy of termination, consider making ethics a special "foot stomper" in your training program:

- Ten Commandments of Computer Ethics (see Section 5.2.1.2)

<u>Training timelines for accomplishment</u>: If you offer your cybersecurity training through a web-based or network-based training method, establish a policy that addresses when the training must be accomplished. Most organizations require the training be accomplished before the employee is given an account on the network with annual keep-it-current training. We believe this is a reasonable training rhythm that you should adopt.

<u>Certification/decertification</u>: Many of our clients like to compare their cybersecurity training programs to a driver's test. Prove yourself capable of safe driving on the highway and you get your driver's license. Prove yourself capable of safe driving on the information superhighway and you get access to the corporate network appropriate to your job. Prove yourself incapable of safe driving and your license gets suspended. Many of our clients decertify their employees and suspend their network privileges when they exhibit poor cybersecurity practices and only renew their privileges after they are retrained and demonstrate they understand and will follow the corporate policies. We believe this is a best practice.

7.5.2 Special Training for Executives

With the emergence of IT as an integral part of every business, cybersecurity has become a critical concern for all executive activities. As a result, we recommend that you consider investing in special training to sharpen the focus of executives on how they should incorporate cybersecurity considerations into their decision-making processes.

Special training for executives should include the numerous legal considerations that may affect decisions involving cybersecurity matters. For example, executive responsibilities to uphold governance and reporting under the Sarbanes–Oxley Act directly have cybersecurity consequences. Similarly, cybersecurity requirements driven by GLBA, HIPAA, CF DG 2, and other laws and regulations must be considered as executives create strategy and policies, oversee financial and budgetary activities, manage processes and projects, and carry out daily business.

Consider a case where an executive has to make a decision regarding a proposed cybersecurity investment to protect personally identifiable information. Staff members may believe that the investment is discretionary, but the executive realizes that legal and regulatory requirements to protect information from unauthorized disclosure make the investment a "must-pay" cost to stay in compliance. Awareness of key legal and regulatory requirements and how they affect cybersecurity controls is an essential executive skill set that you must invest in and nurture.

Consider also investing in additional ethics training for executives. Your executives are responsible for not only serving as the epitome of your ethics program; they also are your principal enforcers. Many organizations invest in additional ethics training for executives not only to highlight ethical behavior and business practices but also to give them tools to recognize and deal with employees who violate ethical standards of behavior. Regrettably, executives often are the people in organizations who make the most egregious ethics mistakes and violations. We have seen numerous executives who were on the fast track toward the highest leadership levels in their organization yet stumbled due to unethical behavior. We have seen some engage in business transactions with IT firms that resulted in an unacceptable conflict of interest. We have seen others whose personal financial interests inappropriately guided their transactions such as steering business; for example, large computer and information service contracts, to companies that they owned stock in. Others had their careers derailed when they accepted special gifts such as free or "specially discounted" software and games from IT firms. You invest a lot to develop and grow your executives. Make sure you invest in their ethics training as well.

As the people who will enforce your acceptable use, employee use, and Internet monitoring programs, your executives also should be given special training regarding your policies and enforcement mechanisms. You do not want an uninformed executive jeopardizing due process or violating someone's rights when confronted by a violation of your policies. Executives should clearly understand their roles and company processes when handling disciplinary situations arising from violations of these policies. The general counsel should be a key player in making certain your executives are well trained to handle violations of your policies fairly, quickly, and decisively.

Computer crime is on the rise. It can come from inside as well as outside your organization. Your executives may be among the first supervisory levels notified of incidents of computer crime. Make sure they know what to do! Whether the crime involves physical or information theft, your executives need to know the proper process to preserve potential evidence, respect individual rights, alert law enforcement authorities, and protect your business. Invest in special training for your executives on how to handle instances of computer crime.

Privacy and anonymity are two increasingly hot topics for Internet users. In the aftermath of Edward Snowden's disclosure that the NSA had collected sensitive information from several companies, anxiety over the exposure of private information and assumed anonymity on the Internet has become front-page news. Your company likely is the custodian of sensitive information your clients, customers, and employees want protected. They expect you and your organization will protect information about them and not release that information to a third party without their permission.[16] This

---

[16] The Code of Fair Information Practices was proposed by the Advisory Committee on Automated Personal Data Systems at the Department of Health, Education, and Welfare in 1973. It served as the basis for the subsequent Privacy Act of 1974 and has served as the model for privacy legislation and regulations around the United States. Executives should understand the principles governing fair information practices and implement them as good and reasonable business practices.

expectation has serious implications for you and your business. Ensure your executives know what information is sensitive and requires special handing. They must understand your corporate policies regarding information handling; be prepared to articulate them accurately to employees, clients, and customers when queried; and enforce your information handling policies responsibly. Likewise, your executives need to be prepared to implement your policies like the employee use and Internet monitoring policy, which provides for surveillance of employees to safeguard your organization against improper use of company resources. Recall the case of *Michael A. Smyth v. Pillsbury*,[17] where an employee sued his employer for wrongful invasion of privacy when his email correspondence was used by the employer. You must have comprehensive acceptable use and employee use and Internet monitoring policies and have your employees sign an agreement acknowledging them. If you don't, you may face problems if your employees engage in inappropriate conduct while using organizational resources. Make sure your executives understand privacy and anonymity issues and are well prepared to implement your policies well.

Intellectual property and trade secrets are among your company's most valuable assets. Your executives must understand the value of these assets and how to protect them from inadvertent disclosure, theft, or tampering. Invest in training your executives so they know the proper procedures for handling your intellectual property and trade secrets. From handling of paper versions of the information to electronic versions, this vital information should be tightly controlled so that only authorized personnel have access to it. According to a report by noted psychologists Drs. Eric D. Shaw and Harley Stock, over 75% of those who steal intellectual property and trade secrets have authorized access to the information.[18] They identify several indicators that are common to those who engage in intellectual property and trade secret theft, including previous violations of rules, policies, practices, or law; personality and anger issues; and disgruntlement. Train your executives how to recognize telltale signals that your intellectual property and trade secrets are at risk. You'll be glad you did.

Globalization and the Internet are tightly coupled. Being able to sell to anyone anywhere in the world presents special issues about which your executives need to be aware. For example, every entity that uses the Internet to conduct its business is a global business. Every one. Now, a small quilt shop in Eureka Springs, Arkansas, can advertise its patterns to prospective customers around the world. They can sell to anyone with a PayPal or credit card account and use worldwide shipping services to deliver their products to the consumer quickly and efficiently. What does that company do about sales taxes at home and abroad? Each country has its own importation and taxation laws. Does the quilt shop need a lawyer who understands the laws of every customer's country

---

[17] http://www.loundy.com/CASES/Smyth_v_Pillsbury.html. Accessed on October 25, 2013.

[18] Eric D. Shaw and Harley V. Stock, *Behavioral Risk Indicators of Malicious Insider Theft of Intellectual Property: Misreading the Writing on the Wall*, Symantec White Paper, p. 4. The authors commend the reader to carefully review the high-risk indicators in Table 5 on pages 15 and 16. These are excellent samples of indicators every executive should be alert to. Dr. Shaw is an expert in psycholinguistics who has consulted for the FBI and other organizations. Dr. Stock is a certified forensic psychologist and managing partner with the Incident Management Group.

to ensure they stay on the right side of the law? Perhaps they do! Also, depending on what your company produces, your product (including information) may be subject to exportation controls; you may not be permitted to sell your products in that country. Your executives need to be aware of the impact of globalization and how it affects your business. Make sure your executives are properly trained and your general counsel is included in the conversation.

Your executives can make you a lot of money through their expertise, imagination, and managerial and leadership skills. They also can cost you a lot when they are ill prepared and make big mistakes. You cannot afford for your executives to be unprepared in a hotly contested cyberspace environment. Make sure you give them the training they need to best protect your vital information and business processes. It will be one of the best investments you make.

7.6 SPECIAL CONSIDERATIONS FOR CRITICAL INFRASTRUCTURE PROTECTION

If you are an executive in an organization identified as being part of the critical national infrastructure, you have special responsibilities above and beyond that of many of your peers. Those who operate and maintain critical national infrastructure are the custodians of special public trust.

Critical infrastructures such as defense, financial and banking, transportation, pharmaceuticals, water supply, and power production are heavily dependent upon computers and networks. Regrettably, many of those computers and networks and the software they depend upon were not designed with cybersecurity in mind. As a result, much of the world's critical infrastructure is exposed to numerous vulnerabilities that could permit bad actors to disrupt, disable, or destroy these vital infrastructures that our modern society relies upon.

Cybersecurity incidents involving critical infrastructure can have huge effects that threaten public health and safety, potentially damage the environment, and cause significant financial loss. Because of these effects, critical infrastructure is a high-visibility target for hackers, terrorists, and others who are intent on creating mayhem. It even is a target for insiders, some of whom deliberately launch cyber attacks on critical infrastructure, while others do so by accident. Regardless of the attack vector, executives charged with the management and control of critical national infrastructure have a special responsibility to protect the well-being of the public by guarding against cybersecurity incidents.

Cybersecurity experts often cite the computers that operate and control industrial systems as being an Achilles heel for many critical infrastructures.[19] We agree. Because many of these automated control systems were designed and fielded without cybersecurity

---

[19] People use different terms to describe the computers that control industrial systems, including industrial control system (ICS), supervisory control and data acquisition (SCADA), and Programmable Logic Controller (PLC). Each has a distinct meaning. For the purposes of this book, ICS are computer-enabled systems that monitor and control industrial processes. SCADA systems are large-scale ICS that can monitor and control multiple sites and devices. PLCs are the digital computers that actually control mechanical processes in the machinery and supply monitoring information to the SCADA and ICS. We will use the term ICS to encompass all controls of industrial systems.

controls, many companies are scrambling to retrofit or replace their control systems to protect the systems from attack and exploitation. These are the smart ones. They surveyed their automated control systems looking for vulnerabilities and are taking proactive measures to insure against threats. Regrettably, other companies are blithely unaware of the vulnerabilities their automated control systems may have and are doing nothing or taking inadequate measures to protect against accidents or attack. If you are in an organization that is part of the critical national infrastructure, which best describes you? Are you an organization that is proactive and is making sure your automated control systems are secure or do you trust the security that the manufacturer built into the system, so you believe there is nothing to fear? What's the worst that could happen? (Clearly, the answer is **plenty**!)

When thinking about worst-case scenarios (and as an executive in critical infrastructure sectors, you should know what your worst-case scenarios are and protect against them), two examples leap to mind. First, a cyber-based attack against your automated control systems could cause catastrophic results. Cyber attacks could alter your products, disable safety controls, or introduce dangerous flaws in products or processes. Secondly, misconfigurations or other accidental alterations of automated control systems can have equally catastrophic effects. You must defend against both threats.

Many executives mistakenly believe they are immune to cyber attacks. They cite isolation from the Internet as one of their "insurance policies" that allow them to consider themselves protected against cyber attack. Especially for those in critical infrastructures, this is a foolish belief. As was demonstrated with the Stuxnet virus incident, even systems isolated from the Internet are vulnerable to cyber attack when inadequate cybersecurity controls and procedures are not implemented, such as protecting against malicious code transported by contaminated USB thumb drives.[20] Are you one of these people who believe that you are immune from cyber attack because your ICS are not connected directly to the Internet? Do you think because you operate on an intranet that you are adequately isolated and protected? Isolating critical infrastructures from the Internet whenever possible is a very good cybersecurity step yet is not the *only* step you should take. Thorough employee training, strong policies, disciplined procedures, and rigorous testing (such as penetration and red team tests) ought to be part of your overall cybersecurity program to protect critical infrastructures. We recommend you invest in training your executives on cyber threats to ICS not directly connected to the Internet. The Stuxnet case study is a good starting point but is only one of many examples worthy of your consideration. Just because your systems may not be directly connected to the Internet doesn't make you bulletproof. Be prepared.

There are other executives who mistakenly believe they are immune to cyber attacks because they believe they are "too small a target" and that hackers are out looking for "the big targets." This too is a dangerous and foolish belief. As evidenced by the November 2009 hack into the South Houston, Texas, water utility's ICS by a hacker

---

[20] For example, David Kusher's *The Real Story of Stuxnet*, February 26, 2013, http://spectrum.ieee.org/telecom/security/the-real-story-of-stuxnet. Accessed on October 25, 2013, provides not only an interesting description of the Stuxnet attack; it also provides a discussion of other emerging malicious codes that present threats to unprotected critical infrastructure and even our home systems.

called "pr0f," failure to protect critical infrastructure against cyber attacks poses a danger to society. Fortunately, in the case of the South Houston utility attack, the hacker did not damage any systems; he claims he was only demonstrating how easy it was to gain control of the system and urged officials to take better measures to protect the water supply. How did he get in? According to him, the water utility only used a three-character password to protect its ICS, which were connected to the Internet.[21] "This required almost no skill and could be reproduced by a two year old with a basic knowledge of Simatic [the Siemens code that operates the system]," "pr0f" wrote.[22] Some report that the three-digit password was still set with the out-of-the-box default password, which is the first thing hackers check and is shamefully all too common. Are your ICS properly protected? Do you have any systems that still are protected by their default passwords? We hope not. When was the last time you conducted a penetration test or a red team analysis of your systems? You are never too small or large to be subject to a cyber attack.

There are other executives who mistakenly believe that the new systems they are procuring have cybersecurity "baked in" and they can trust the manufacturers of the systems to deliver systems free of vulnerabilities. For those who operate critical infrastructure, this too is a dangerous and foolish belief. Think this doesn't happen? Take the case of the new Automatic Dependent Surveillance–Broadcast (ADS-B) system, which the U.S. Federal Aviation Administration (FAA) plans to introduce in 2014. Designed to replace the 1950s-era radar systems that serve as the backbone of the domestic air traffic control system, the system relies heavily on satellite-based telemetry, and according to Andrei Costin, a researcher at Eurocom who warned an audience at Black Hat 2012 (an annual conference dedicated to information security and attended by many hackers, most of which we hope are ethical), there is evidence that the new system is susceptible to hacking, according to a report by CNN.[23]

Costin said that the ADS-B system can be hacked and tricked into seeing aircraft that are not actually there. This is known as a so-called "spoofing" attack. According to reports, the system transmits *unencrypted* and *unauthenticated* information, which would enable anyone with the right technology to identify a plane and see its location. We hope the reports are wrong and the system is indeed protected by sufficient measures such as encrypted and authenticated information to protect the safety of passengers as well as people on the ground. Imagine yourself enjoying an uneventful flight from New York to Chicago's O'Hare International Airport when suddenly the pilot executes violent evasive maneuvers to avoid a potential collision with an airplane that suddenly appears out of nowhere only to find that there really isn't an airplane there, it is a false reading inserted by a hacker intent on introducing dangerous chaos. Your executives need to ask the right questions when procuring new systems. They need the training to

[21] Paul Roberts, Was Three Character Password Used to Hack South Houston's Water Treatment Plant a Siemens Default?, November 22, 2011, http://threatpost.com/was-three-character-password-used-hack-south-houstons-water-treatment-plant-siemens-default-11. Accessed on October 25, 2013.

[22] Matt Liebowitz, Hacker Says He Breached Texas Water Plant Network, November 21, 2011, http://www.nbcnews.com/id/45394132/#.UmSGOsu9KSM. Accessed on October 25, 2013.

[23] Heather Kelly, *Researcher: New Air Traffic Control System is Hackable*, July 26, 2012, http://www.cnn.com/2012/07/26/tech/web/air-traffic-control-security/index.html. Accessed on October 25, 2013.

understand potential cybersecurity vulnerabilities and make your suppliers prove that your new (and existing) systems are hardened against cyber attacks. Because most systems cannot be fully bulletproof from cyber threats, executives who are well trained to look for potential vulnerabilities are in a much better position to recognize and manage the cyber risks that accompany the procurement of new technologies. For critical infrastructures, this is an indispensable skill.

While cyber attacks on ICS get a lot of press, misconfigurations or other accidental alterations of automated control systems are the more common cyber-related problems executives in critical infrastructure sectors should guard against as they can have equally catastrophic effects.

One of the best examples of how this type of cybersecurity incident could cause catastrophic effects to critical infrastructure is the August 2009 explosion at the Sayano–Shushenskaya hydroelectric power plant in south central Siberia.[24]

Sayano–Shushenskaya dam, built in 1978 by the Soviet Union, is Russia's largest dam. The dam bottles up the Yenisei River, its ten turbines generating enough power to feed a city of 3.8 million people. On the morning of August 17, 2009, workers at the dam heard an unusual thump and the entire facility began to shake violently. Suddenly, turbine number two, a 1500 ton unit, exploded through the floor, rose 50 feet into the air before crashing back down, and shot debris like shrapnel throughout the facility as it disintegrated.

The structural damage was enormous and would get worse quickly. Water that would normally feed the turbine and its power generation unit now shot through the facility as a high-speed water jet, its 67,600 gallons per second water stream cutting through the support beams and machinery in the facility like a hot knife through butter. The facility quickly became immersed in water, the roof partially collapsed in the 950-foot-long generator hall, and safety locks that would have stopped the other turbines failed as the electrical system shorted and arced in the flooded hall.

All looked lost and the surviving frightened workers scrambled for safety, fearing that the dam was lost and would collapse. Many called their families downstream to warn them of an anticipated dam breach and to get to higher ground. Fortunately, not everyone ran for cover. The hero of the day, a brave security guard named Kataytsev, gathered several other employees and led them through the physically arduous and now extremely dangerous task of manually closing the water intake valves that fed the dam and its power generation facility. They stopped the flow of water in time, saving the dam and all those downstream.

How was this a cybersecurity incident? As it turns out, the accident investigation traced the cause to a *new automated control system*. The new system had been installed during scheduled maintenance between January and March of that year. It was designed to regulate the speed of the turbines to generate the right amount of power to meet demand. When demand rose, the control unit would speed the turbine up, and when

---

[24] There are many terrific case studies regarding this dam incident. Among our favorite sources for information on this incident are the Engineering Failure Organization's report on the Sayano–Shushenskaya Hydroelectric Power Station Accident (http://engineeringfailures.org/?p=703) and Joe P. Hasler's "Investigating Russia's biggest dam explosion: What went wrong" in *Popular Mechanics*, http://www.popularmechanics.com/technology/engineering/gonzo/4344681. Accessed on October 25, 2013.

demand fell, the control unit would command the turbine to slow down. Unfortunately, when the newly refurbished turbine number two was brought back online in March, something was not right. The turbine vibrated unusually, so much so that it was taken off-line due to safety concerns. Only after a fire at another power generation facility caused power shortages, managers at the dam rushed the suspect turbine number two back into service to meet the heightened demand. The turbine couldn't take the strain of the vibrations introduced by the new automated control system and failed. Some people have pointed to the fact that the turbine was two months shy of its projected 30-year life span as a factor in the accident. Perhaps it was. Nonetheless, the evidence found by the accident investigators indicates that the new automated control system failed to properly control turbine number two. The aged turbine could not withstand the resulting vibrations, which reached four times the unit's maximum allowable limit. It ultimately spun out of control, exploded, and led to the death of 75 workers. As of this writing, the dam's power generation capability has not yet been restored.

Could this happen in your critical infrastructure ICS controls? If you said "no," how do you know? How would you know whether you are the victim of a Stuxnet-like attack or a failure of your ICS? Do you insist on thorough testing before accepting new automated systems? Do you have a means of measuring performance to ensure your systems stay within performance standards? Do you have adequate controls that detect failures and react with a fail-safe or a fail-secure procedure?[25] Do you believe the executives at the Sayano–Shushenskaya dam would have made the same decisions had they been trained to implement testing and acceptance procedures for new automated control systems? Do you believe they fully understood the risks associated with bringing turbine number two back online and under load? Executives of critical infrastructure systems make decisions that affect *everyone's* risk, including yours. As such, we believe it is essential they are fully aware of and understand cyber effects and impacts on the critical infrastructure in their care. Well-trained executives are best prepared to make the right decisions.

Critical infrastructures are subject to cybersecurity issues involving more than just ICS. As mentioned earlier, critical infrastructures are high-visibility targets for hackers, terrorists, and others who are intent on creating mayhem. Disruption of services provided by critical infrastructure utilities and services can have profound and wide-ranging impacts that bad actors may find appealing. For example, in August 2012, Saudi Arabia's national oil company, Aramco, was the victim of a cyber attack that caused extensive damage to over 30,000 company computers and shut down the corporate network for a week. Using the Shamoon virus, hackers wreaked havoc as the virus wiped hard drives clean as it spread through the Aramco corporate network. According to Abdullah al-Saadan, Aramco's vice president for corporate planning, "The main target in this attack

---

[25] Fail-safe means that when the system detects a situation that it considers unsafe, it automatically changes to a state that preserves safety. For example, if you have a facility where the doors are controlled by automated locks, you may have your locks configured to automatically unlock when there is a power failure. That way, employees can safely leave in emergencies and not be trapped in the facility. Fail-secure means that when the system detects a situation that it considers unsafe, it automatically changes to a more secure state for its self-preservation or protection. Banks and many other facilities use fail-secure controls in vaults and similar areas.

was to stop the flow of oil and gas to local and international markets and thank God they were not able to achieve their goals."[26] Fortunately, the Aramco computers that operate and control the production of oil are physically and logically separated from the administrative and business computers that were infected. Imagine the impact not only on Aramco but also on the world's economy had production of oil in Saudi Arabia been interrupted. Imagine the damage that could be done if any group of people with an axe to grind against your organization activates a similar attack against you. The success of Shamoon is sure to attract copycats. How are your executives prepared to prevent such an attack or respond to one like it? Are they trained to make the right decisions?

A more likely threat to critical infrastructure is the theft of intellectual property that can be used to craft a later more devastating attack or perhaps be used as a means of gaining a competitive advantage by the attacker. Examples of the latter abound. They include attacks against 48 chemical and defense firms originating from a Chinese source. According to a report from Symantec, labeled as the "Nitro" attacks, company computers were infected with malicious software dubbed "PoisonIvy," a backdoor Trojan that was used to gain control of the computers to steal information such as design documentation, formulas, and manufacturing process details.[27] They also include an increase in attacks directed toward biotechnology and pharmaceutical firms. According to Mandiant Corporation, the noted cybersecurity firm that revealed the activities of coordinated Chinese hacking in February 2013,[28] Chinese hackers have stepped up their efforts significantly to gain insight into the biotechnology and pharmaceutical industries.[29] This increase in targeted activity coincided with the Chinese making pharmaceuticals and health care a strategic priority in the national 12th Five Year Plan covering 2011–2015. Coveted information includes drug trial information, chemical formulas, and confidential data for all drugs sold in the U.S. market. In one case, the Chinese hackers reportedly stole as much as 6.5 terabytes of information from a single company over a 10-month period. Don't be fooled into thinking the Chinese are the only ones out there trying to acquire intellectual property for critical infrastructures. There are a host of bad actors from around the world (or even down the street) that seek to gain a competitive advantage by gaining access to your information. You must protect it.

Executives in critical infrastructure operations are the custodians of a special public trust. Public health and safety, environmental protection, and economic well-being all

[26] Reuters News Service as reported in the *New York Times*, "Aramco Says Cyberattack Was Aimed at Production," December 9, 2012, http://www.nytimes.com/2012/12/10/business/global/saudi-aramco-says-hackers-took-aim-at-its-production.html?_r=0. Accessed on October 25, 2013. Like the Sayano–Shushenskaya dam incident, there are a plethora of stories reporting the Aramco attack including some very entertaining conspiracy theory articles. We prefer to stick with reporting that is independently verifiable through multiple sources.

[27] Jim Finkle, New Cyber attack Targets Chemical Firms, http://www.reuters.com/article/2011/11/01/us-cyberattack-chemicals-idUSTRE79U4K920111101. Accessed on November 1, 2011. Also see Angela Moscaritolo's "Nitro" Attacks Target 29 Firms in Chemical Sector, November 1, 2011, http://www.scmagazine.com/nitro-attacks-target-29-firms-in-chemical-sector/article/215781/. Accessed on October 25, 2013.

[28] Mandiant, *APT 1: Exposing One of China's Cyber Espionage Units.* http://intelreport.mandiant.com/ Accessed on October 25, 2013.

[29] Shannon Ellis, *US Biopharma Firms Hit By Cyber Attacks from China*, September 2013, http://www.bioworld.com/content/us-biopharma-firms-hit-cyber-attacks-china-0. Accessed on October 25, 2013.

depend on effective, efficient, *and secure* critical infrastructures. Many people refer to the operation of critical infrastructures as "zero-defect" operations, that is, the consequences of failure are so severe to such a large segment of the population that it is unacceptable to endure a failure. Cybersecurity is a new element in the decision matrix for executives in critical infrastructures and needs to be incorporated into every training program, into how systems are monitored and controlled, into maintenance programs, and into procurement processes. Cybersecurity needs to be an integral part of critical infrastructure internal controls.

If you are an executive in critical infrastructures, what should you do to safeguard the systems and information under your control from threats? How do you control your risk?

These are good questions that executives in every sector ought to be asking. Many executives are quickly overcome by the breadth and depth of the cybersecurity issues. Our clients ask us, "Where do I begin?" We recommend they start by "Knowing Your Enemy and Knowing Yourself." You should too.

Know yourself by asking, "What am I protecting?" Surprisingly, many executives do not conduct a thorough analysis of what it is that they are actually trying to protect. Are you trying to protect public safety, intellectual property, or perhaps both? Are you trying to protect machinery such as valves or regulators from inadvertent changes that could result in harmful effects? Are you trying to prevent bad actors from altering processes or information that could result in catastrophic effects? Are you trying to preserve the integrity of your ICS to safeguard your critical infrastructure processes and products? Are you trying to protect intellectual property or trade secrets from falling into the wrong hands, from destruction, or from alteration?

Successful executives, particularly in critical infrastructure businesses, seek to know the second-, third-, and fourth-order effects of their decisions. Know yourself like these executives by asking deeper questions and seeking more specific answers. Know what your systems are connected to. This is critically important in critical infrastructures as more and more ICS are being connected to business and industrial safety systems that may be connected to the Internet or potentially may introduce malicious code in a Stuxnet-style attack. In order to best manage and control risk, you need to understand your systems, your processes, your people, *and* yourself. You need to know where you are vulnerable and what are your options to address each vulnerability.

You should also know your enemies. Your enemies include adversaries, competitors, and potentially employees who are disgruntled or incompetent. As you conduct your risk analysis, make sure you look at all possible threats and don't expect that the list will remain static every year. In fact, threats to critical infrastructures are growing every day. Stay abreast of the evolving threats to critical infrastructures and regularly schedule management-level risk reviews to ensure your risk management posture and internal controls remain potent to control the risks presented by contemporary threats.

Many of our clients ask for a prescriptive checklist they can follow to "achieve cybersecurity," especially those who operate critical infrastructure. We hate to disappoint them, but there is no singular checklist that applies to every organization; every organization should be analyzed and managed separately. Nonetheless, there are several best practices that every critical infrastructure organization (and many noncritical infrastructure organizations as well) should incorporate into its cybersecurity program:

- Make cybersecurity a stated organizational priority and act upon it. Because so many critical infrastructures are susceptible to cyber attacks and accidents, senior management needs to focus attention on actions to mitigate weaknesses that may be exploited. For example, you may be an executive in the oil, gas, or water industries. If so, your facilities may use wireless sensors to monitor and control the flow of oil, gas, or water products. As pointed out by Lucas Apa and Carlos Mario Penagos of security consulting firm IOActive at the 2013 Black Hat Conference, many wireless sensors do not properly encrypt their signals and are susceptible to hacking.[30] Apa and Penagos were able to demonstrate how they are able to hack into the devices from as far away as 39 miles and change the pressure and volume readings the sensor displayed. If the facility's control system would change in accordance with the inaccurate readings, a pipeline, pump, or even an entire plant would be disabled. The faulty sensors typically cost between US $1,000 and US $2,000 each, and hundreds or even thousands of them are used in oil, gas, and water facilities. Is cybersecurity a priority in your organization? Will you make the decision to replace the sensors? Does the sensor manufacturer have any liability? Are you subject to any regulatory controls that compel you to control this vulnerability? What do you do? What are your priorities? How much risk are you willing to accept?

- "Bake in" cybersecurity to everything you do. Your strategy, plans, policies, procedures, and training should all incorporate cybersecurity best practices. With critical infrastructures increasingly reliant on IT and automated control systems, the need to incorporate cybersecurity best practices into your organization is no longer an option; it is an imperative. Executives in critical infrastructures should make a point of making sure that all due diligence and due care is accomplished to ensure that appropriate cybersecurity controls are in place throughout the organization. Remember that cybersecurity is much more than technical controls. Cybersecurity is about risk management. Control your risks by addressing cybersecurity early and often in the creation of your key business and technical activities. It is much more expensive to add cybersecurity later than to "bake it in" from the very start.

- Don't buy anything without evaluating its cybersecurity risks. With the advent of automated industrial control and safety systems, cybersecurity risks to critical infrastructures have risen significantly. Many vendors may try to sell you automated controls that are susceptible to cyber attacks. Scrutinize every potential purchase through the eyes of a hacker. How could a hacker exploit this particular unit? Consider hiring a certified ethical hacker to evaluate the unit to give you information that will help you with your business case analysis. It may turn out

---

[30] Jeremy Kirk, Oil, Gas Field Sensors Vulnerable to Attack via Radio Waves, http://www.computerworld.com/s/article/9241109/Oil_gas_field_sensors_vulnerable_to_attack_via_radio_waves. Accessed on July 25, 2013. Also see: http://www.ioactive.com/news-events/ioactives_ICS_xperts_lucas_apa_and_arlos_penagos_to_demonstrate_new_research_at_energysec_security_summit.html and http://www.homelandsecuritynewswire.com/dr20130730-black-hat-event-highlights-vulnerability-of-u-s-critical-infrastructure. Apa's and Penagos' presentation is available at https://media.blackhat.com/us-13/US-13-Apa-Compromising-Industrial-Facilities-From-40-Miles-Away-Slides.pdf and is well worth reviewing. Accessed on October 25, 2013.

that the great deal that the vendor is giving you will quickly be offset by the costs associated with a cyber vulnerability that is exploited.

- Implement strong internal controls and tightly monitor. Executives in critical infrastructures need to safeguard their systems to ensure public health and safety, protect the environment, and maintain economic stability. You can't let your guard down and must maintain constant vigilance. Make sure you have strong internal controls that give you the ability to monitor and control your key processes and procedures. Don't solely rely on automated systems and their reports. Factor in human monitoring and control mechanisms too. Your internal controls should maintain defense-in-depth principles with checks from primary and secondary systems.

- Identify and have a plan to address all single points of failure. Single points of failure are items in a system where an item malfunction or failure could cause the entire system to fail. Systems that require high availability are frequently built with redundant components and subsystems to insure that the system continues to function in the event of a failure of a component or subsystem. A common single point of failure in your house is your electrical power. Many people mitigate this single point of failure by installing uninterruptible power units and generators to provide continuous power to their critical electronics such as medical devices, computers, and (sometimes) their televisions. Critical infrastructures typically can't afford to have single points of failure, which could have catastrophic effects. Primary and secondary systems are normal configurations. If you are involved in critical infrastructure management, make sure you avoid single points of failure. Conduct a failure modes and effects analysis of your systems. Identify all potential single points of failure and analyze your risk to determine whether you need to mitigate, accept, or ignore the risk of single points of failure.

- Train your personnel. Your people are your most valuable resource and are the key element in protecting your resources against cyber attacks and exploitation. A well-trained and focused workforce is best prepared to find and fix vulnerabilities quickly and efficiently. When your workforce is cyber hardened and "cyber smart," they are more likely to recognize unsafe practices and procedures and detect aberrations that could be the early signs of a cyber attack or probing of your systems. Invest in your workforce by training them well to understand cyber threats and vulnerabilities. Teach them the procedures to follow to safely and securely operate and control your systems. Make sure they understand what to do when things go wrong including who to notify and when to do it. Empower them to act like the security guard Kataytsev who understood the situation he faced and knew what had to be done. Train them well.

- Practice! Many critical infrastructure operators conduct training exercises to make sure employees, local authorities, and other key stakeholders are familiar with risks and what to do when "the unthinkable" happens. The authors have been involved in countless defense, nuclear, and industrial disaster preparedness exercises and drills that have honed the skills of mission partners across multiple sectors. Exercises help gauge the effectiveness of plans and training. We believe exercises are invaluable and best prepare people to perform at high levels when confronted by emergency or unplanned situations. We recommend you make exercises part of your operations yet caution that in critical infrastructures your exercises should be carefully choreographed to protect against undesired effects. At no time should exercise controllers ever allow a condition to occur that could possibly jeopardize safety. Executives at all levels should insist that all plans and personnel be tested regularly and safely.

- <u>Audit</u>. Trusted independent audits of your systems should be regular parts of your business rhythm when you operate critical infrastructures. Don't just audit your policies and business processes. Invest in penetration tests and red teams that probe your systems for cybersecurity weaknesses. It is important to find and fix problems before bad actors find and exploit them. Audits also may find bad actors in your own organization. Take the case of "Bob," a software programmer for a critical infrastructure company. Bob was a longtime employee of the organization and was widely regarded as one of the firm's best coders. Supervisors praised him in performance reviews for his "clean, well-written" coding. All the attaboys evaporated when security audits revealed that Bob's account had been repeatedly accessed from addresses in China. In the ensuing investigation, it was discovered that Bob had outsourced his own job to software programmers in China, paying them one-fifth of his six-figure salary to produce the software he was responsible for. He had his Chinese employees adjust their work schedule to coincide with him in the United States and sent his security token to them via FedEx so they could access his computer system to make it appear that he was doing the work. While they toiled away, Bob surfed the net, watched cat videos, updated his Facebook and LinkedIn profiles, and dutifully sent his supervisors an end-of-day report detailing all the good work he (more precisely, his Chinese employees) had accomplished during the day.[31] Bob no longer works for that critical infrastructure firm, thanks to that security audit. Does your organization audit for cybersecurity vulnerabilities? Do you include penetration and red team assessments of your systems as part of your risk management program? Critical infrastructure audits should be comprehensive and not just limited to business functions.

7.7 SUMMARY

This chapter focuses on **personnel management**. Having the right team focused on the right tasks at the right time yields optimum performance. We describe some best practice techniques that executives can include in their processes for recruiting, retaining, rewarding, and managing talent in the Cyber Age. We also give recommendations on how to apply that talent as you organize for success.

[31] Ramy Inocencio, "US Programmer Outsources Own Job to China, Surfs Cat Videos," January 17, 2013, http://www.cnn.com/2013/01/17/business/us-outsource-job-china/. Accessed on October 25, 2013. You just can't make these things up!

Of special interest is the "Touhill Cybersecurity Training Plan Outline" that provides considerable detail that executives should follow as a basis for training employees to properly understand cyber risks and equip them with the knowledge to respond appropriately. The intent of this training plan is to have a workforce that is not only "cyber smart" but "cyber hardened" as well.

Remember that executives need training and regular updates in order to keep them up to date and to emphasize for employees the importance of cybersecurity training for everybody.

Additionally, in this chapter, we address special considerations for protecting critical infrastructure.

There are several best practices that every critical infrastructure organization should incorporate into its cybersecurity program. These include the following:

- Make cybersecurity a *stated* organizational priority and act upon it.
- "Bake in" cybersecurity to everything you do.
- Don't buy anything without evaluating its cybersecurity risks.
- Implement strong internal controls and tightly monitor.
- Identify and have a plan to address all single points of failure.
- Train your personnel.
- Practice!
- Audit.

8.0

PERFORMANCE MEASURES

I often say that when you can measure what you are speaking about, and express
it in numbers, you know something about it; but when you cannot measure it,
when you cannot express it in numbers, your knowledge is of a meagre
and unsatisfactory kind....[1]

William Thomson (Lord Kelvin)

8.1 WHY MEASURE?

Are you an executive who agrees with Lord Kelvin? Do you believe that if you cannot measure something, you can't truly understand it?

If you do, you are not alone.

---

[1] William Thomson, also known as Lord Kelvin, was a revolutionary physicist and engineer. The man was widely known for correctly calculating the value of "absolute zero" to be $-273.15°C$; he also was instrumental in formulating the first and second laws of thermodynamics and atomic theory. This quote, while intended to refer to scientific method, is illustrative of the power of measurement to enhance knowledge. The quote appears in numerous publications and web sites, including this one from the University of Washington: www.atmos.washington.edu/~robwood/teaching//451/Lord_Kelvin_quote.pdf. Accessed on November 2, 2013.

Cybersecurity for Executives: A Practical Guide, First Edition. Gregory J. Touhill and C. Joseph Touhill.
© 2014 The American Institute of Chemical Engineers, Inc. Published 2014 by John Wiley & Sons, Inc.

Performance measures (also commonly called metrics[2]) are an important part of modern business. Most executives rely on metrics to aid them as they manage their organizations. Many have learned through years of trial and error what sets of information help them better understand their business and its performance, giving them the basis needed to make decisions. "You get what you measure" has become a mantra in many companies.[3]

What about cybersecurity? How do you measure success? Have you invested enough? Have you invested too much? Are you getting your money's worth? These are key questions that many executives wisely are asking today.

When it comes to cybersecurity, there is no silver bullet answer delivering a singular metric or universal set of metrics that will answer these questions in every organization, nor should there be. As Peter Drucker said, "No two executives organize information the same way."[4] Nonetheless, we believe that use of performance measures is an essential part of gathering the information you need to make informed decisions, especially when it comes to cybersecurity.

With today's reliance on information, it is easy to get swamped by metrics. Ask an IT specialist to produce some metrics, and they happily will bring you reams of paper or PowerPoint charts showing system performance, traffic volume, and a host of network, system, and device measurements. We've found well over 1000 discrete performance measures for IT systems. That's far too many to cogently evaluate and draw conclusions in a timely and efficient manner. As a result, most executives cope by doing what Drucker observed as "A good many executives have found that the one way of organizing information effectively is simply to organize one's being informed about the *unusual*."[5]

When it comes to cybersecurity, the *unusual* may be a key indicator that something is wrong. It may mean you are experiencing a malfunction or it may mean you are under cyber attack. Either way, a quick and proper action is required. You should be paying attention to the "unusual" and have a performance measures program that actively detects unusual events and provides timely and accurate alarms to the right people, including you.

Performance measures are key and essential to make sure you and your organization practice due diligence and due care in the management of your vital information. As a reminder, we define **due diligence** as taking the right actions to identify and understand risks, while we define **due care** as developing and implementing policies, procedures, and standards to protect the organization against risks.

Using performance measures to assess the effectiveness of your policies, procedures, and standards in the highly complex information ecosystem in which your organization operates is expected. If you do not use performance measures, you likely will be seen as not practicing due care and due diligence, exposing you and your organization to charges

[2] The authors will use the terms *performance measures* and *metrics* interchangeably in this text. While we recognize that some people will note that there may be subtle differences between the two, for purposes of this discussion, they are considered synonymous.

[3] Jeffrey K. Liker, *The Toyota Way, 14 Management Principles from the World's Greatest Manufacturer*, McGraw-Hill, New York, NY, 2004, p. 305.

[4] Peter F. Drucker, *Management Challenges for the 21st Century*, HarperBusiness, New York, NY, 1999, p. 126.

[5] Ibid., p. 128.

of negligence. If something unsatisfactory results, such as a major data loss, cyber attack, or major IT service interruption, you and your organization will be held accountable for any and all ramifications of that negligence. In today's hotly contested cyberspace environment, we do not believe you should ignore investing in an effective performance measurement system to monitor, assess, and control your cybersecurity posture.

Your performance measures should be used to monitor and control your information systems:

1. They should **measure those activities that you employ to protect against threats**, such as measuring performance of your boundary protection devices, intrusion detection and prevention systems, and antivirus software.
2. They should help you **detect flaws to reduce faults or vulnerabilities**, such as conducting regular scans to find and fix flaws in your cybersecurity posture.
3. They should help you to **assess the effectiveness of your countermeasures**, which include both technical and human countermeasures. (For example, your organization may employ spam filters that are tuned to detect and filter spam and spear-phishing emails. A useful metric used by most CISOs is to compare how many spam and spear-phishing messages were detected and filtered by automated means, how many got through the automated systems only to be detected by well-trained employees, and how many eluded both automated filters and the workforce to cause an incident.)
4. Finally, when using performance measures to monitor and control your information systems, **your metrics are essential in measuring costs**. Senior executives such as chief financial officers rely on information systems to derive total cost of ownership metrics, and cybersecurity costs should be a line item on every organization's budget. Information is a key resource for executives, and performance measures are essential to monitor and control the systems that manage your vital information.

In addition, many organizations must maintain cybersecurity performance measures to remain in compliance with legal or regulatory requirements. Earlier in this book, we discussed many of these requirements. They include the requirement for publicly traded companies to disclose cybersecurity-related information as specified in the Security and Exchange Commission's Corporate Finance Division's Cybersecurity Disclosure Guidance 2. They also include internal controls, which include those cybersecurity controls to protect critical corporate information as required under the Sarbanes–Oxley Act of 2002.

Protection of nonpublic personal information specified in the Gramm–Leach–Bliley Act of 1999 has significant cybersecurity requirements. There are many other national, state, and local laws and regulations that drive organizations to maintain robust cyberse-curity-related performance measures to ensure they remain in compliance. As you create or evaluate your performance measurement program, you should involve your general counsel to ensure that you indeed are measuring all the relevant items required to remain in compliance with all appropriate requirements.

Many companies now find they also have to gather specific cybersecurity performance measures as part of their contractual obligations with partners and clients. We believe it is an appropriate and recommended practice to require those with whom you share your valued information to provide you with reasonable evidence that your information is well protected while in their custody. You should be prepared to do the same when serving as the custodian of others' sensitive information.

Cybersecurity-related metrics are a common and acceptable means of demonstrating proper controls of sensitive information but often are neglected. For example, one of our clients brought us in to evaluate their cybersecurity controls. They were proud to point out they had reduced costs by storing their client's sensitive financial data in a cloud-based storage capability provided by a major vendor. When we asked what type of specific security controls the third-party vendor provided and how many times someone had tried to gain unauthorized access to the client information, we received a "deer in the headlights" look from our client. Our client had a "wake-up call" moment when he realized he was responsible to his clients to protect their information, yet he didn't have a clue as to how his third-party provider was protecting the information. He was not receiving any performance measures that would indicate their effectiveness. He just *assumed* that his provider was doing all the right things because they were a large name-brand firm.[6] We recommend you don't make such assumptions.

When you enter into agreements with partners and clients, consider incorporating specific mandatory reporting of cybersecurity-related performance measures into your contracts. Specify in your requirements the type of measure you want, how often you want it, and in what format you want it delivered. Make your requirements thoughtful and reasonable because excessive and unwarranted reporting of metrics that are not used to make value-added decisions is wasteful and drives up costs. Therefore, *safeguard your vital information even when it is in others' custody by making cybersecurity a contractual requirement and insist on measuring your service provider's performance.*

While many organizations maintain robust cybersecurity performance measurement systems for contractual and regulatory compliance, others maintain cybersecurity-related performance measures to maintain coveted certifications. For example, one of our clients is involved in the financial sector and is in the process of earning their International Standards Organization (ISO) 27001 (Information Security Management) certification. This certification along with others, including ISO 15408 (Common Criteria), requires performance measurements to ensure that appropriate standards are maintained. Our client believes that ISO 27001 certification gives his organization a competitive advantage by sending a clear signal to his current and prospective clients that their information will be properly managed and controlled throughout its life cycle. He believes that failure to maintain the certification is a red flag to his discerning customers in the financial sector who will view it as a deficiency. He believes they will take their money somewhere else where it will be considered better secured. Evidence indicates his concerns are well founded. While investing in recognized cybersecurity performance measures and certifications that require them may not have a direct and immediate impact on your bottom

---

[6] As a result of our discussion, our client immediately ordered his director of IT to work with the vendor to implement mandatory reporting of key cybersecurity metrics including those detailed in this chapter. When it comes to cybersecurity, where the risk to your information is high, we recommend you follow Ronald Reagan and adhere to the Russian adage: "Trust but verify."

line, failure to achieve and maintain these valued certifications increasingly cause prospective clients and partners to steer their business to those who do make the investment.

Organizations that maintain performance measurement systems do so to facilitate decision-making. We contend that performance measurement principally is about proper management of processes, procedures, and products. We believe that the primary reason managers use information derived from performance management systems is to guide decisions to meet organizational goals and objectives. We are not alone in this view. For example, BNY Mellon's executive vice president and CIO Suresh Kumar says, "We want to take advantage of technology and use data insight to make better decisions. I call it decision science."[7]

When you choose the right metrics aligned with your corporate strategy, you drive the right decisions, but be careful to make sure that you are selecting the right metrics. According to MIT's Hauser and Katz, "Metrics empower managers and employees to make the decisions and take the actions that they believe are the best decisions and actions to achieve their metrics. If the metrics are chosen carefully, then, in the process of achieving their metrics, managers and employees will make the right decisions and take the right actions that enable the firm to maximize its long-term profit."[8] Subtly implied is that if you select the wrong metrics, managers and employees still will make what they believe are the right decisions to achieve their metrics, yet the results may not be what you want.

Linking your metrics to your strategy is important. According to Paul Niven, choosing performance measures that don't have an impact on your strategy can lead to confusion and lack of clarity because employees devote precious resources to the pursuit of measures that don't influence the firm's overall goals.[9] Fortunately, this misalignment between metrics and strategy is avoidable. As you select your cybersecurity performance measures, make certain that you are measuring the right things to make the right decisions in support of your corporate strategy.

Many executives and employees use performance measures to improve performance and accountability. An approach to performance management that has gained a loyal following in business and government circles is the **Balanced Scorecard** concept developed by Kaplan and Norton.[10] Kaplan and Norton propose that by using proper measurements in categories such as financial perspective, customer perspective, internal business perspective, and learning and growth perspective, managers can better align efforts to guide both current as well as future performance and identify better processes for meeting customer needs and shareholder objectives.

You can use a Balanced Scorecard approach to address cybersecurity in your organization. Kaplan and Norton state, "If employees are to be effective in today's competitive

---

[7] Bryan Yurcan, "*BNY Mellon's Kumar Uses Metrics to Make Better Business Decisions. Bank and Systems Technology,*" October 7, 2013, http://www.banktech.com/management-strategies/bny-mellons-kumar-uses-metrics-to-make-b/240161978. Accessed on October 25, 2013.

[8] John R. Hauser and Gerald M. Katz, "Metrics: You Are What You Measure," April 1998, http://web.mit.edu/hauser/www/Papers/Hauser-Katz%20Measure%2004-98.pdf. Accessed on October 25, 2013

[9] Paul R. Niven, *Balanced Scorecard Step-by-Step: Maximizing Performance and Maintaining Results*, John Wiley & Sons, Hoboken, NJ, 2006, p. 162.

[10] Robert S. Kaplan and David P. Norton, *The Balanced Scorecard: Translating Strategy into Action*, Harvard Business School Press, Boston, MA, 1996.

environment, they need excellent information."[11] Protecting that information is a major focus of cybersecurity and you can easily incorporate cybersecurity measures into the Balanced Scorecard construct. For example, say that one of your strategic financial objectives is to reduce costs. Let's also say that your business operates its own IT infrastructure. Using the Balanced Scorecard approach to monitor and control your operations, you may determine that your large server infrastructure is increasingly expensive to operate and maintain. It also presents a large attack surface to a potential bad actor, which consumes significant man-hours to maintain and secure. Using the Balanced Scorecard approach, you may determine that executing a server consolidation plan may reduce your total cost of ownership (as you will have less hardware and software to maintain), increase customer satisfaction by providing better security and reliable computing; improve your cybersecurity posture (because you will reduce your attack surface), and make you a hero to your overtaxed technical workforce because you will reduce their workload, permitting them to focus their time and energy on higher priority tasks. The Balanced Scorecard method is one of many that use metrics to improve performance and accountability. We recommend you find the method that is a good fit for your organizational culture, and when you do, make sure you include cybersecurity performance measures in your calculus.

Finally, most organizations use performance measures to reward and discipline their employees. Maintaining standards of performance is an essential part of nearly every business activity, but especially in critical infrastructures. Take for example a chemical processing facility that is producing ammonium nitrate. Bob is the manager of the plant, and his inventory figures show discrepancies over the last quarter. While the production data indicates the plant produced its targeted amount of the chemical, the stock data indicates there is much less of the finished product on hand. Over two tons appears to be missing. You are Bob's boss and you are alarmed by the performance records. Ammonium nitrate in the wrong hands is a very bad thing. Are the records wrong? Could someone have tampered with the records? Were your computers hacked and records altered? Are there two tons of a potentially dangerous chemical unaccounted for that you have to report to your superiors and law enforcement officials? What impact will the loss of accountability of the chemical have on customer and stockholder confidence in the business? After you complete your phone calls to corporate headquarters to launch an investigation into the missing chemicals, are you planning to reward or discipline Bob, your plant manager? We anticipate that Bob's job is in jeopardy.

Do you agree with Lord Kelvin that performance measures are important? Do you use metrics in your business? Are you already using cybersecurity metrics to help guide decisions regarding the protection and management of your information? There are many reasons why you should measure performance, and cybersecurity is no different than any other business area. Executives have a rich set of cybersecurity metrics to choose from yet need to tailor their performance measures to align with their strategy, goals, and objectives. Because cybersecurity metrics are newcomers to the executive suites, integrating them into business management processes is a challenging but necessary endeavor.

[11] Ibid., p. 134.

8.2 WHAT TO MEASURE?

"What to measure?" is a difficult question to answer for many executives. For some executives, trying to define acceptable cybersecurity measures is reminiscent of Supreme Court Justice Potter Stewart's description of trying to define pornography, "Perhaps I could never succeed in intelligibly doing so [defining pornography]. But I know it when I see it."[12] Are you like Justice Stewart when it comes to cybersecurity metrics? Do you struggle to articulate what performance measures you want to monitor and control your cybersecurity posture? Do you have difficulty finding the right metrics to link your risk management effects and strategy? Will you even "know it when you see it?"

If this dilemma describes you, you aren't alone. In fact, there are no universally agreed-upon standard metrics that measure cybersecurity. Multiple measures are needed and have to be updated regularly. A metric that is meaningful and useful today may be worthless tomorrow as changes in technology, process, and even regulatory controls affect how you assess your cyber defenses. As a result, executives have to stay engaged and pick the right measures that best fit their management style, their processes, and their organizational environment. The selection of cybersecurity metrics is a challenge as there are a lot from which to choose. While there are over 1000 discrete performance measures that your IT and auditing staffs can use to assess your information systems, most executives don't have the time, resources, nor the interest in collecting, monitoring, and assessing that many metrics to make decisions. So what should you do? Which of the hundreds of performance measures out there are best aligned to your cybersecurity and risk management programs?

8.2.1 Business Drivers

Many businesses focus their metrics on business driver categories such as cost, risk, quality, return on investment, compliance, and safety. Each of these categories offers potential cybersecurity measures that may prove valuable to you.

Cost: We are disappointed when we have discussions with cybersecurity professionals who zealously discuss security controls without thought to cost-effectiveness. Cybersecurity should enhance your business rather than be its centerpiece. Too many times, we have seen security professionals lose sight of the fact that the primary mission of their company is to make money and earn a profit. Implementing expensive security controls that eat into earnings without returning a satisfactory return on investment is wasteful and harmful to your business. You need to measure your security controls to make sure they are not only effective, but also that they are a good value. As an example, while serving as the CIO of a large organization, the author was asked to make a decision whether to invest in an automated cybersecurity control system. It was an effective system that would allow the organization to manage and control its ability to certify and accredit the security controls of the organization's major information systems. There

---

[12] Jacobellis v. Ohio, 378 US 184 – U.S. Supreme Court 1964. Justice Stewart's concurring opinion is found in section 197 of the document, a copy of which is available at http://scholar.google.com/scholar_case?case=15356452945994377133&hl=en&as_sdt=6&as_vis=1&oi=scholarr. Accessed on October 25, 2013.

were many attractive tools available, the sales pitches were compelling, and the testimonials watered the eyes of my cybersecurity staff. They winnowed the list of candidates down to one product and recommended I authorize the purchase. I didn't. Why? Did we need the capability? Yes. Did we want the capability? Yes. Was it the best value? No! In looking at our performance measures, I found that hiring an additional two employees to conduct the process manually was less expensive and a better value than purchasing the system. Our performance measures gave me the information I needed to determine my certification and accreditation costs accurately and compare the manual to the proposed automated processes. The lesson? Don't rush to buy an automated cybersecurity control just because others have it. Sometimes, a manual process remains the best value. Your performance measures should help monitor and control costs to ensure that you are getting the best value for your investments.

Risk: Most companies use performance measures to monitor and control risks. Recall from Section 1.3 our focus upon risk management. Therefore, your cybersecurity performance measures are related to risk management in one manner or another. We recommend you impress upon your subordinates the importance of this linkage as well as the process of how risk is managed in your organization. Too often, well-intentioned cybersecurity professionals take a zero-risk approach to cybersecurity and use performance measures to guide their efforts to drive risk to zero. We believe this is impractical and a losing proposition. You can't afford to have a zero-risk cybersecurity posture. You have to carefully balance risk and cost, which most likely results in always retaining and managing some risk. Experienced managers refer to this mindset as the "sin of excess perfection," which is characterized by exponentially rising costs in the quest for perfection (which incidentally is never achieved).

To best manage and control this risk, you need cybersecurity performance measures that give you insight into the risks your organization faces, their magnitude, and where they exist. Armed with cybersecurity performance measures, you will be better prepared to decide whether to mitigate, avoid, transfer, or accept cybersecurity risks. For example, your performance measures should include information derived from vulnerability scans of your network. Your scans may have detected hundreds of potential vulnerabilities. But not all of them are equal. Some represent a higher risk, while others may not even be a concern.[13]

Your risk management process needs to use your performance measures wisely to cull the wheat from the chaff. As an example, the author had a new boss who got his hands on raw vulnerability scan data and demanded to know why we hadn't been fixing a certain category of vulnerabilities that our scanning tools detected. He was eager to make his mark in the organization and thought this was an area where he could effect some immediate change. He showed me the metrics that indicated over 100 potential vulnerabilities on our network that we had known about for several months yet had done

---

[13] If you are interested in finding out the severity of vulnerabilities, we recommend you look at the Common Vulnerability Scoring System (CVSS) v2, which is used in the National Vulnerability Database. Vulnerabilities are categorized by high, medium, and low ratings depending on their impact from a loss of confidentiality, integrity, or availability. These ratings are not definitive measures, yet they are a terrific starting point to help prioritize work and place detected vulnerabilities in context. For more information, visit the National Vulnerability Database at http://nvd.nist.gov. Accessed on October 25, 2013.

nothing to correct. He was livid and said he was contemplating replacing me because of my "obvious neglect." After I explained that the vulnerabilities identified in the data were for a type of computer system we did not have, did not plan on getting, and that the costs of mitigating the vulnerabilities were wasteful, he wisely changed his mind. The lesson? Your risk management program should use cybersecurity performance measures to help you assess your cybersecurity risks and make the right decisions on what risks to mitigate, accept, transfer, or avoid.

Quality: Many organizations, particularly manufacturing companies, use performance measures to help them manage and control the quality of their processes. Many are required by regulatory or industry standards to measure and verify product quality. As an example, pharmaceutical manufacturers use a comprehensive set of performance monitoring and measuring capabilities to guarantee that their products maintain quality standards. These measurements are used for internal control procedures, third-party audits, and government oversight. Consumers expect that quality standards are maintained using thorough performance measures and many firms advertise that their quality measures are a key factor in delivering better quality products than their competitors. While some may argue that cybersecurity metrics cannot measure quality, it is a process and processes can and should be measured and controlled, allowing conclusions about quality to be drawn through metrics. Common cybersecurity quality measures include the number of emails that contain spam, viruses, and malicious code that elude filters (evaluates effectiveness of email filters), number of occurrences of spear-phishing emails that make it through defenses and trick employees into clicking links or downloading malicious code (evaluates effectiveness of your training), and number of unplanned configuration changes (evaluates effectiveness of your change management process). Executives at every level should use performance measures to control the cybersecurity processes under their responsibility.

Return on investment: Maintaining a solid cybersecurity program is not free. Like any other investment, you need to make sure you get your money's worth and an appropriate return on investment. Metrics are an important part of the discussion of return on investment, especially when it comes to cybersecurity. Traditional return on investment calculations are cost–benefit analyses that compare costs to returns, which typically are expressed either in percentages or dollar values. In contrast, return on investment calculations for cybersecurity investments typically do not address profits but instead address prevented losses. In many ways, this measures reduction in risk.

When measuring your cybersecurity return on investment, there are many methods you can use to put your measurement into a context that translates into the language of your organization. Say for example that you are a small-to-medium business that has a significant investment in the intellectual property and trade secrets residing on your servers. You also are the custodians of sensitive customer information. You are concerned that an unauthorized person who gets their hands on one of your passwords could cause significant damage to your business. You estimate the average cost per incident is US $100,000. Based on historical information, surveys, industry comparisons, and other data sources, your risk management team calculates the chance that your business will suffer a cybersecurity incident due to a compromised password in any given year is 75%. This is unacceptable, and you talk with numerous experts who convince you that by implementing a thorough cybersecurity awareness training program, you can reduce that chance to 10% in any given year. Your cybersecurity training costs you US $10,000. Hence, by investing in a cybersecurity training program that reduces your chance of incident due to password compromise from 75% to 10%, you have reduced your expected loss exposure from US $75,000 (i.e., likelihood times cost per incident = 0.75 × US $100,000 = US $75,000) to US $10,000. Therefore, for an investment of US $10,000 (the cost of the training program), you may expect a US $65,000 reduction in expected loss per year, which is an attractive return on investment.

Compliance: Many organizations use performance measures to maintain compliance with regulatory controls or contract specifications. Cybersecurity performance measures often fall into this category, especially because many organizations outsource their IT and cybersecurity operations to third-party firms. For example, you may be an executive in a manufacturing firm that has outsourced network operations and maintenance to Magija Computers of Coraopolis Corporation (MC3). In your contract with MC3, your company stipulates a series of cybersecurity performance objectives that you expect MC3 to achieve on your behalf. You specify how you want the performance measured, when and how often you want it measured, and when and how you want it reported. Many organizations link financial rewards to performance, and your contract provides incentives as well as penalties for MC3 to make sure they meet your cybersecurity requirements. This outsourcing example is not unusual and is increasingly the norm rather than the exception. Many companies find hiring IT and cybersecurity professionals to operate and maintain their information environments is a better value than maintaining and training their own group of technical employees. Be careful to ensure that when you do outsource these functions, not only do you include cybersecurity performance measures into your contract but also you incorporate third-party auditing into your contract surveillance and oversight procedures. It is important to safeguard your information and your business by making provisions to have independent experts verify and validate that your contractors are compliant with the terms of their contract and doing what you want them to do to keep you and your business secure.

Safety: Some of the author's students question why we include safety into a discussion of business drivers and cybersecurity performance measures. Safety is an important part of any business and should always be considered when evaluating performance, even when considering cybersecurity. This is best exemplified when considering critical infrastructures and industrial control systems. As mentioned in Section 7.6, executives in critical infrastructure sectors maintain positions of special trust in protecting public safety and welfare. Whether you are manufacturing chemicals, operating a power production facility, developing pharmaceuticals, or engaged in any other critical infrastructure activity, maintaining safety for your employees and the public is essential. With the advent of computer-controlled ICS, the potential negative impacts on public safety due to cybersecurity incidents such as hacking, malware infections, or even human errors are magnified. As you create your cybersecurity performance measures, make sure that you deliberately include safety into your calculus. For example, you may determine that the risk to public safety is of such a magnitude that you cannot tolerate a cybersecurity incident in one of your control systems. You deliberately reduce the attack surface by segregating the system from the Internet and placing it on a protected corporate intranet. Despite these prudent actions, threats still exist. To monitor and control your risk mitigation measures, you increase the rate of vulnerability scanning of your critical controls and implement a real-time monitoring system that continually looks for *the unusual* (e.g., unauthorized changes to software, failed log-in attempts, addition of unauthorized files, system performance out of assigned standards, etc.) and notifies designated personnel when situations out of normal bounds occur. Safety is critically important, and your information systems are increasingly the key link in maintaining safe and effective operating environments. Make sure you have the right performance measures in place that will monitor and control your systems to safeguard the integrity and security of your key systems.

8.2.2 Types of Metrics

As you look to align your metrics with your strategy and risk management program, it is instructive to know what types of metrics are available.[14] Executives should pick a few from each category to maintain situational awareness and positive control over operational processes.

Organizational metrics (those used for oversight at the corporate level) are used to describe and track the effectiveness of organizational cybersecurity programs and processes. They generally are used to check compliance with your decisions on how corporate programs and processes are to be managed. Examples may include measuring whether your cybersecurity awareness training is being conducted (e.g., percentage of people up to date on corporate cybersecurity training), whether your configuration management program is effective (e.g., percentage of approved configuration changes), and whether your vulnerability management program is effective (e.g., percentage of successfully mitigated discovered vulnerabilities). You can apply cybersecurity organizational metrics not only to your organic in-house programs and processes but also to your vendors and service providers.[15] Most organizations use organizational metrics as their primary metrics.

Operational metrics are used to describe and manage the risk to operating environments. You probably already have a system of operational metrics that you use to monitor and help you control risk. For example, say that you have an intrusion detection system installed on your network. Its function is to inspect all traffic on the network and identify suspicious patterns that may indicate that someone is trying to attack or compromise your system. When the system detects the anomalous patterns, it sends an alert to an operator, who assesses the event to determine whether it is actually hostile activity

---

[14] The metrics construct mentioned here is derived from the ISO 27002 (Information Technology Security Techniques Code of Practice for Information Security Management) and ANSI 04 (Security Technologies for Manufacturing and Control Systems) standards. These standards often are used to create cybersecurity controls for ICS such as SCADA systems. We believe that they are relevant for executives seeking to measure performance and assess cyber risks as part of a broader risk management program.

[15] The National Institute of Standards and Technology's Special Publication 800-55 (available at http://csrc.nist.gov/publications/nistpubs/800-55-Rev1/SP800-55-rev1.pdf) has a host of cybersecurity metrics available for review. As you look to create your cybersecurity performance measurement program, we recommend you review this document. We believe you will find some valuable measures that will aid you in providing better management and oversight over your resources. Accessed on October 25, 2013.

or whether some other anomalous event is occurring. These alerts are examples of operational metrics, that is, measures directly related to risk management that have been established to control risks to the organization.

Technical metrics are used to describe and compare technical items such as your smart phone, desktop computer, or even whole networks to technical standards, such as design specifications, architectures, and algorithms. For example, let's say you want to expand your office to accommodate additional employees but you don't want to pay to string additional cables around to hook the new employees into the corporate network. Instead, you want to incorporate wireless networking technology into this new office space. You believe this will avoid expensive initial wiring costs but future costs as well since it seems that your employees often want to rearrange offices. You direct your technical staff to create a plan that will provide the necessary capacity and performance to support new employees while providing sufficient security to maintain alignment with the organization's cybersecurity goals and objectives. Chances are good that your technical staff will evaluate potential network gear against the 802.11n standard, which is an Institute of Electrical and Electronic Engineers (IEEE) industry standard for Wi-Fi wireless local network communications.[16] Because your requirement calls for Wi-Fi capability, your technical team recommends a technical architecture based on this standard, which was approved by your change control board. The performance measures of the equipment are the technical metrics used to compare against the 802.11n standards. Therefore, equipment that is not compliant with the standard will not be selected for purchase.

As you manage your business environment, you should use a combination of organizational, operational, and technical measures to ensure you have full situational awareness and control over key business functions. We recommend you select measures where data are easy to collect, are easy to understand, assist in making decisions, and are aligned with your business strategy. Remember, your cybersecurity metrics merely are *indicators* of your security posture. Therefore, the wise executive will collect measurement information across a wide range of organizational, operational, and technical areas to provide sufficient information to make proper judgments and decisions.

8.3 METRICS AND THE C-SUITE

Chances are good that if we interviewed your CEO, we would be told that cybersecurity is a priority. After all, there are so many risks associated with cyber-related issues such as hackers, insider threats, corporate espionage, and increased reliance on IT that it has become a "hot-button" issue that CEOs increasingly are concerned about. But when we ask CEOs to show us their key metrics and performance indicators, cybersecurity-related measures usually are absent. When we query them as to why, they often

---

[16] The IEEE is an association dedicated to advancing innovation and technological excellence for the benefit of humanity and is the world's largest technical professional society. IEEE has become a leading developer of industry standards in a variety of technical disciplines. When you see a standard with an IEEE certification, you can be confident that it has undergone a thorough scientific process, peer review, and open comments period before it has been published.

say they don't understand what they should measure or how they should measure it. In fact, they aren't even sure they would know a good cybersecurity measure if they saw it. This section is written with those executives in mind. (As an aside, there are many CEOs who are acutely aware of the broad scope of cybersecurity issues and have an impressive grasp of what to do and how to do it. For them, the following section should be viewed as an update and a review.)

8.3.1 Considerations for the C-Suite

Cybersecurity has evolved from a technical issue to an agenda item in the boardroom. Executives need to be prepared to understand cybersecurity issues and make the key decisions that prevent cyber issues from evolving into full-scale problems. Some executives have found that ignoring cybersecurity was a big mistake and cost them and their firms huge amounts to recover from cyber incidents. As an example, according to the Ponemon Institute, in 2013, the average cost in the United States to address a data breach was US $188 *per record*.[17] Recall from Section 3.3.2 that BigRX (a large company) has over 10 million records. Given the Ponemon Institute's research, the potential loss associated with a data breach at BigRX could approach nearly *US $2 billion*! How many records does your organization have and what is your risk exposure?

With reports of cyber incidents arriving daily, many executives are coming to realize that cyber threats are real. What many still do not realize is that new threats emerge *every* day. How many threats does your organization face? How vulnerable are you?

Some organizations are proud to tell their stockholders, partners, and prospective clients that they are compliant with certain standards or regulations. Compliance is a good thing, but *compliance merely means you are meeting the minimum standards*. When it comes to cybersecurity protecting your vital information, executives need to decide whether they want to meet the minimum standards of compliance or whether the threat and vulnerability environment warrants additional controls above and beyond minimum standards to achieve compliance. Is your organization compliant with all relevant standards and regulations?

Most executives with whom we discuss cybersecurity struggle to maintain a balance between strong cybersecurity and productivity. Not that the two are in conflict, but rather, as we have suggested elsewhere in this book, production is the reason you are in business, and you do not want efficient productivity strangled by overly aggressive cybersecurity protocols. Cost, performance, and ease of use are key attributes of an efficient and successful cybersecurity program. How do you know when you have achieved a good balance between cybersecurity controls and operational performance? When do you know that you've spent too little or too much for cybersecurity? Sadly for some, they find out they've spent too little after they've suffered a cyber incident. Do you struggle with this balance too?

These are the types of cybersecurity situations commonly confronting executives in every business sector. How do you establish a performance measurement program that can help executives answer the questions posed earlier? How do you create the metrics

---

[17] Ponemon Institute, *2013 Cost of Data Breach Study: Global Analysis*, May 2013, p. 1. Sponsored by Symantec Corporation, this report is widely regarded as one of the more fair, objective, and accurate reports of its kind.

that will give executives insight into key cybersecurity issues that will make the right decisions leap off the page? Every executive and every organization will have a unique performance measurement system to monitor and control their processes, products, and information. There is good news, though. The process of developing cybersecurity metrics is *exactly* the same as you would use to create metrics in any other phase of your business.

Metrics Decisions for Executives

Decide what it is you are going to measure: Many executives don't make the time to sit down and actually decide what is important enough to be measured. Deciding what to measure is the most difficult part of any metrics development process. Invest in some "deep thought time" with both your technical staff and your business function leaders to decide what cybersecurity items of interest you want (and need) to measure. Start by reviewing your strategy. Define your cybersecurity goals and objectives. Identify your key information and understand its value. Establish your priorities. When you define these key items, you are well postured to decide what your most important cybersecurity measures are.

Decide how you are going to measure: Working with your technical team, determine the best way to collect your cybersecurity performance measures. Decide who will collect the data, where it will be stored, and who is authorized to access it. Develop procedures that instill discipline and rigor to collect the data in a reliable, predictable, and consistent manner. Note that not all measurements are easy to gather and not all are precise. You may want measurement information that is unaffordable, impractical, or unavailable. Make sure that whatever you decide to measure is in fact measurable; it is practical and affordable to measure.

Decide how your metric information will be reported: Executives need to determine how and when they want information reported. Decide what items of interest are "wake me up in the middle of the night" events (e.g., a cybersecurity breach that takes down your e-commerce web site) and what items can wait for normal business process rhythms. Do you want cybersecurity performance metrics briefings? If so, when and how often? Do you want reports? What format do you want? Do you want it in paper copy or sent electronically? Do you want to be notified when something unusual happens? If so, when and how do you want to be notified? Do you want an email or a phone call? Don't keep your employees guessing about what you want. Tell them. If for some unusual reason your staff cannot reach you in an emergency, assure that there is a suitable alternate identified who can react appropriately.

Decide how your performance measurement program will be implemented: Some executives will test metric and data collection in small pilots before launching larger-scale implementations. We believe this is a very good idea. Doing small-scale tests of proposed metrics has proven a very effective method of evaluating data collection methods, information flow, and value of the information collected. We have participated in many such tests only to have the boss realize that he had us collecting information he really did not need. On more than one occasion, the boss was like Goldilocks, testing different metrics until he found the ones that "were just right." This is a wise approach that conserves resources, minimizes disruption to the organization, and typically produces satisfactory metrics faster than several full-scale successive iterations. Also, consider using focus groups from inside and outside your organization to see what metrics are valued by your employees, partners, and peers. You may find that through the resulting dialogue you will find better ideas that point you toward the best metrics to meet your strategy and objectives. You will also find that a team approach to implementation yields better acceptance and adoption of the program, leading to a "win–win" result.

Set performance targets and key performance indicators: Many executives use metrics to establish performance targets and key performance indicators. While this is reasonable and common for well-established, predictable, and reliable metrics, we advise caution against using nascent metrics to base performance targets and key performance indicators upon. Doing so encourages some people to craft their metrics carefully to make themselves look good and meet targets rather than using the information as intended to encourage the appropriate action. (This is widely referred to as "gaming the system.") We recommend that you avoid setting performance targets and key performance indicators based on metrics that have not been tested, employed, and audited through a minimum of two fiscal quarters.

Remember, as you create your metrics, they should be linked to your strategy. Businesses whose metrics program is linked in a strategy-to-task relationship often find and fix issues before they become problems. Also, be mindful that your metrics should be quantitative (i.e., easily measured in discrete terms), they should be timely and accessible, they should be easily understood, and they should be relevant. When your metrics meet these criteria, even Lord Kelvin would be proud of you.

8.3.2 Questions about Cybersecurity Executives Should Ask

Great executives understand they don't know everything and ask a lot of penetrating questions to help them zero in on issues so they are prepared to make the right decisions. You do not need to be an IT expert to successfully manage cybersecurity risks. You just have to understand your risks (i.e., your enemies) and your capabilities (i.e., yourself). Asking the right questions can give you the information you need to make right decisions.

Great executives also know that the answers to their questions need to be backed up by facts. Lord Kelvin would agree that numbers matter and clear and unambiguous metrics help immeasurably in the decision-making process. Quantifiable measures are defendable before audits, help you identify trends, and help build consensus and support toward decisions. While you and your general counsel naturally will wish to avoid any legal proceedings, should you become embroiled in a lawsuit, verifiable and reproducible data can be instrumental in shielding you from liability.

We have found that the following cybersecurity questions help executives as they seek answers on what to do to improve their cybersecurity posture and better manage their risk. They should form the basis of your executive-level cybersecurity metrics program, regardless of your business sector. Your cybersecurity performance measures should answer these questions:

1. What are our threats? Executives should be asking themselves this question continually as the threat environment is always changing. New threats and new vulnerabilities emerge everyday. You should understand the threats your organization faces as well as measuring your vulnerabilities to them. You can never let your guard down in the dynamic cyber environment.

2. How effective are our systems? Nearly every organization relies on information technologies to perform essential tasks such as produce their products, manage their people and finances, and support their clients. Most people cannot do their jobs without information technologies that are effective, efficient, *and secure*. Information systems that are properly configured, properly operated, and managed well are considered "cyber hardened" and are less likely to be susceptible to exploitation by cyber attack. Executives need to have trust that their systems are performing at optimal levels and that they are properly configured and are meeting acceptable standards of performance. Your performance measures should indicate whether your information systems indeed are "cyber hardened" and are properly configured and operating as expected.

3. How vulnerable are we? This is a critical performance measurement area. You may have a very capable information system that appears to serve you well. It may be properly configured, properly operated, and managed well, but if it presents an inviting and undefendable target to a potential adversary, then it changes from an asset to a liability. Recall our discussion in Section 6.2 regarding the trusty old Windows 95 machine you may have tucked away in your home's basement office. It may have served you exceptionally well and still runs like a champ, but once you connect it to the Internet, it has a proverbial "hack me first" target on it as the operating system is no longer supported and there are so many exploits available against it that a bad actor could quickly take control of it and gain access to any information on the computer **and especially to any device connected to it**. Your corporate systems face the same issues. Your performance measures should indicate what your vulnerabilities are to known threats. You should continually be scanning your own systems to "know yourself."

4. Do we have the right people, are they properly trained, and are they following proper procedures? People are your most important resource. They also may be your greatest weakness. Executives know that they need to have the right people assigned to the right jobs. They know their workforce needs to be properly trained. They also know that the workforce needs to follow proper procedures. These are especially critical in cybersecurity. Performance measures help executives determine whether the workforce is indeed properly aligned, trained, and following procedures.

5. Am I spending the right amount on cybersecurity? Every executive should be asking this question. The answer, however, will be different in every organization. That is because every organization faces different risks and has different risk appetites. Investing in cybersecurity training, security hardware and software,

and related products and services is somewhat like purchasing insurance for your business. How much you purchase depends on how much risk you want to assume. As you manage risk in your organization, you should carefully balance your cybersecurity investments against the threats you face and the vulnerabilities you possess to make the best investment to protect your business. Your performance measures should indicate how much you are spending, what potential losses you could face, and what type and magnitude of risk you face.

6. <u>How do we compare to others?</u> Have you ever heard the story about the two hunters who were in the woods? In one version of this oft-told joke, the rookie hunter turned to his experienced partner and asked what they should do if they were chased by a bear. The experienced one looks at the rookie incredulously and says that you run away. The rookie appears perplexed and asks, "How fast do you have to run?" The experienced hunter says "Faster than you!" In some regard, maintaining a strong cybersecurity posture is like outrunning the bear; you just have to be faster than the next guy. Hackers and bad actors often look for targets with the weakest defenses. The reason is simple. They have to expend less time and fewer resources against weaker targets; it is easier and more efficient to attack weak links. Understanding how your cybersecurity posture compares with your peers is important not only to understand whether you are investing properly but also in understanding your potential risks. We recommend you invest in measures that compare your posture against your peers.

8.4 THE EXECUTIVE CYBERSECURITY DASHBOARD

Many executives like to have their metrics presented in a dashboard format. Dashboards, like in your car or in a cockpit of a plane, succinctly inform the operator the status of key subsystems. Your vehicle most likely tells you how fast you are going, how hard your engine is working, how much fuel you have, whether your tires need air or your engine needs oil, how many miles you have traveled, or how many miles you have to go. Most cars now rely on computers to tell you key and essential information needed to operate the car safely. Why can't you use your computers to create a dashboard to help you operate your information systems safely too? Can you add some measures that help you gauge your cybersecurity? The answer is, of course!

We believe that you can create an executive-level dashboard of performance measures that will help you answer cybersecurity questions every executives should ask (see Section 8.3.2). No matter what business sector you are in, the following cybersecurity measures should be incorporated into your metrics as you manage and control risk in your organization.

8.4.1 How Vulnerable Are We?

- **Metric Category 1.0**. **Number of Threats Detected**
  - <u>Metric 1.0.1</u>: *How many times are we being "pinged" and "probed"?* This is a quantitative measure that lets you know the amount of scans and probing against

your network. This is the cyber equivalent of "how many times did someone come by your house and check the doorknob to see if the house is unlocked?" This helps you gauge what the threat level is and how many times someone is evaluating your defenses. It also is a measure that shows what traffic load your systems face. Too many "pings and probes" may overtax your systems and expose you to a denial-of-service attack. This measure is most effective when compared over time. We recommend at least a monthly sampling rate.

○ Metric 1.0.2: *How much spam is filtered?* Many organizations and application service providers (e.g., Yahoo, Gmail, Hotmail/Outlook, etc.) offer spam filters to filter unwanted and potentially malicious emails. Spam is a common threat vector for malicious code. You should be aware of not only how much spam is being detected and blocked but also how much is getting through so you can tune your filters as threats evolve. You may also find that your filters cannot keep up with the volume and need to be upgraded or replaced. This is a quantitative measure and we recommend a monthly sample rate.

○ Metric 1.0.3: *How many phishing messages are we receiving?* This too is a quantitative measure and is a valuable measure to gauge the threats against your organization. When measured over time, you may detect patterns that give you insights into your potential foes as well as their strategies. We have seen phishing levels increase in some businesses after they make press announcements of new products, groundbreaking research, or other innovative and coveted information. Many executives heighten defenses against phishing during periods when they anticipate greater volumes of potential attacks. Know your enemy, know how many phishing attacks you face, and be prepared to predict when you may face increases.

○ Metric 1.0.4: *Who is targeting us?* In some regard, this is more than a quantitative measure. Using sensors and log files, your technical staff may be able to tell you where scans, phishing, and other potentially hostile actions against your networks and systems originated. Be advised that attribution of attacks is exceptionally hard to do and even harder to prove in court. Nonetheless, knowing where potentially hostile activities originate from is instructive and can help you develop the tactics, techniques, and procedures to better defend your business and its information. For example, you may have your metrics team prepare two measures for your review every month. One is a pie chart that shows what countries or Internet service providers the scans, probes, phishing, or other potentially hostile activity originated from (see Figure 8.1). We are aware of several firms engaged in fierce competition with foreign firms who saw probes and scans against their networks increase from sources originating from their competitor's home country. They took quick and immediate action to block all traffic from those sources to better protect their intellectual property and trade secrets. Review this measure monthly yet devise a process to alert you when significant and unusual activity occurs inside the monthly cycle.

Another measure worthy of your attention is a chart of "repeat offenders," that is, those Internet addresses that repeatedly serve as the source of unwanted pings, probes, and other potentially malicious activity (see Table 8.1). Your routers and other network

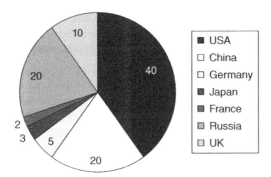

Figure 8.1. Sample chart indicating countries where probes originate.

TABLE 8.1. Sample "Repeat Offender" Table

IP Address	Date of Activity	Type of Activity	Source ID
10.0.19.255	10.31.13, 2353 hours	Probe	Alice's Restaurant, Barrington, MA, USA
172.16.19.18	10.31.13, 2354 hours	Ping	Lebowski Carpet Company, Los Angeles, CA, USA
192.168.20.04	10.31.13, 2359 hours	Probe	Telecom Italia, IT

devices have the capability to log addresses of every address that exchanges information with your systems. All IP addresses are registered and you can do what is called a "reverse lookup" to find out who the address is registered to. If there are some addresses that are a frequent source of bad activity, it behooves you to know who they are and block them. You may find that they are actually a business partner or friend who has been infected and doesn't know it (you can be a hero by letting them know!), a competitor trying to steal your information, or they may be a bad actor who is attempting something nefarious.

- **Metric Category 1.1. Number of Known Vulnerabilities**
  - Metric 1.1.1: *System vulnerabilities*. This is an important quantitative measure. Your technical staff should be scanning your networks regularly, both internally and externally, looking for known vulnerabilities. There are plenty of tools such as Nessus, Nmap, and Retina that can be used to identify weaknesses in your cybersecurity posture. We recommend at least monthly scans and prefer to have the identified vulnerabilities presented in the Common Vulnerability Scoring System (CVSS) format (i.e., high, medium, and low categories) to help prioritize potential next steps. The following are key metrics as part of this measurement:
    - Metric 1.1.1.1: Number of vulnerabilities discovered. This is important because it tells you the total number of vulnerabilities discovered on your systems during this scanning period. We recommend you compare your monthly data against historical data. If the number of vulnerabilities is

increasing beyond your staff's ability to control, perhaps you need more staff and better training, or there are other factors that warrant your attention and deeper action.

- Metric 1.1.1.2: Percentage of vulnerabilities mitigated in prescribed time frames. You may find a vulnerability, yet your staff may be able to mitigate it through a patch or other action. You should establish a standard for how fast you want detected vulnerabilities mitigated (e.g., one week from detection to correction) and measure against it. This measure is calculated by dividing the number of mitigated vulnerabilities by the number of detected vulnerabilities. The higher the percentage, the better. This is a quantitative measure that you should review monthly.

- Metric 1.1.1.3: Number of residual vulnerabilities. Often, you have vulnerabilities you cannot fix because software patches are not yet available from vendors, your staff may not have the requisite skills or sufficient resources to implement a mitigation, or the vulnerability is such a low priority that you decide not to act against it. Whatever the case may be, it is important to identify and understand the residual vulnerabilities remaining. This is a quantitative measure that you should review monthly.

○ Metric 1.1.2: *Other Vulnerabilities*. Your scans are prudent active measures that help identify known vulnerabilities. There are several other measures that you should use to monitor and control your risks.

- Metric 1.1.2.1: Percentage of systems and devices beyond projected life span. This is an important metric as older systems and devices have been found to be more susceptible to cyber attacks and exploitation than newer systems. Recall in Chapter 7.0 that we discussed how many older systems were not designed with cybersecurity in mind. As new products have been developed and fielded, security improvements have been incorporated in new designs. Understanding how many of your systems and devices are beyond their projected life span and therefore at greater risk of exploitation is a meaningful measure of vulnerability and may cue you to recapitalization requirements. This is a quantitative measure that you should review every quarter.

- Metric 1.1.2.2: Percentage of software beyond projected life span. Like systems and devices, your software has a projected life span that you should monitor. Recall our discussion in Chapter 6.0 that software vendors often terminate their support for older versions of their code, exposing those ancient versions to threats of exploitation. You likely know that today's Windows 8 is much more secure than its predecessors. Software can and does operate outside of its projected life span, but when security updates are no longer available to protect against cyber threats, your risk escalates precipitously; it is time to replace it. You need to keep an eye on your software and its cyber hardness. This is a quantitative measure you should review every quarter.

- **Metric Category 1.2. How Many Cybersecurity Incidents Have We Detected?**
  ○ Metric 1.2.1: *Number of cybersecurity incidents detected*. Past performance is an indicator of future activity. If someone has already breached your defenses,

your chance of being exploited again rises dramatically. Like sharks drawn to blood in the water, hackers and bad actors quickly find out who has weak defenses and will test their mettle against you, especially if they know that you have already been victimized. Measuring how many cybersecurity incidents you have detected is an essential measure. Cybersecurity incidents of all kinds, from password disclosures to hacks, should be included in this measurement. This measure should be reviewed by your security personnel at least monthly and by executives no less than quarterly.

○ Metric 1.2.2: *Number of detected cybersecurity incidents by category.* This metric is instructive to help you understand quickly which vulnerabilities are being exploited. For example, an incident may occur because of a weak technical control, a personnel error due to failing to follow a security control, or even a lack of a procedural or technical control. This is a quantitative measure that should be reviewed by your security personnel at least monthly and by executives no less than quarterly. An example method of displaying this information is included as Figure 8.2.

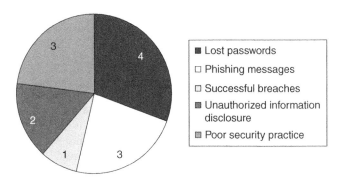

Figure 8.2. Detected cybersecurity incidents.

○ Metric 1.2.3: *Cost per incident.* This is an essential measure for several reasons. First, cybersecurity incidents cost time and money. As an executive, one of your primary responsibilities is to manage and control costs. What do you think a prospective investor would think if they asked you how much a recent hack into your database cost your company? Saying "nothing" would be a lie and "I don't know" is a signal of lack of control or awareness. Secondly, you should have a thorough understanding of the costs associated with cybersecurity incidents so that you can make the right decisions regarding proposed courses of action to address them and manage your risk. Remember that incidents are more than just breaches of your defenses. Other potential cybersecurity incidents include inadvertent exposure of personally identifiable information, phishing messages that breach your defenses, and viruses and other malicious code that breach your defenses and infect your network and devices. Each incident consumes different amounts of resources, yet each introduces a measure of risk. You should assess the costs of all cybersecurity incidents and make value-based judgments on whether to accept, mitigate, transfer, or avoid the risk. This is a quantitative measure that should be prepared by a team consisting of your technical, security, and cost accounting staffs. It can be measured discretely after each incident. It should be put into monetary terms and include all costs associated with incident response including labor, materials, and other resources that have been directly or indirectly allocated to handle the incident. You should review these costs (both individual and aggregate costs) no less than quarterly.

○ Metric 1.2.4: *Who is responsible for cybersecurity incidents*. When you know who is responsible for incidents, it is easier to determine your next steps. For example, if you see that business unit B has been responsible for several cybersecurity incidents over the past few quarters, you may have a special heart-to-heart discussion with the head of that unit. Remember that performance measures are an important part of your rewards and discipline program. As another example, you may find through this measure that many of your incidents are caused by your own technical staff. Further analysis may indicate that the technical staff is undermanned and overworked, causing them to introduce faults that result in incidents. We use this metric as a leading indicator that cues us to find and fix problems where they start rather than a stand-alone measure from which to draw conclusions. Review this measure quarterly. An example is provided in Figure 8.3.

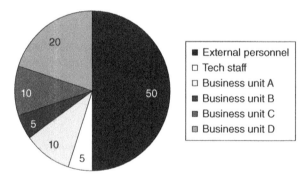

Figure 8.3. Sample "Who is Responsible for Cybersecurity Incidents?" chart.

8.4.2 How Effective Are Our Systems and Processes?

• **Metric Category 2.0. Network Performance Measures**

○ Metric 2.0.1: *Network performance measurement*. Your boss asks, "How well is our network performing?" You answer, "How well do you want it to perform?" We recommend you establish a reasonable standard of performance and availability for your network (e.g., 99.5% availability) based on your organizational goals and objectives, resource to sustain that performance, manage accordingly, and measure performance. A poorly managed network is a cybersecurity risk. Faults introduced by your own staff or bad actors can cause serious damage to your business and its information. Network performance is a sign of the technical health of your organization and also of the capability of your business to do its job. There are several ways you can measure your network performance. Most organizations choose to measure network availability (i.e., the percentage of time the network is available to users) as a measure of merit. Some organizations only use their networks during normal business hours; thus, outages during nonwork periods are not considered significant. We believe that with mobile computing, e-commerce, and a global marketplace, a more reasonable expectation is that your network will be available continuously with periodic planned maintenance periods that are programmed to occur during times where they will have the least impact. Therefore, we recommend you calculate your network availability based on a 24x7, around-the-clock, seven days a week, availability cycle. Measure and review network performance monthly.

○ Metric 2.0.2: *How does network performance compare to previous measurements?* It used to be that executives looked at the productivity of machinery and components over time and noticed that they deteriorated as they aged. This helped them make determinations when to recapitalize equipment. Deterioration sometimes happens with information technologies most noticeably with access and write times in older storage devices, but evaluation of network performance over time can expose more than just when you need to recapitalize equipment. It can highlight issues including capacity, training, productivity, or resource allocation. For example, if your standard is that the network needs to be available over 95% of the time and it is only available 90%, you may have to add resources (e.g., personnel or equipment) to meet your standard. Similarly, if your metrics indicate your network is performing at a 99.999% availability rate and your requirement is only 90%, perhaps you are allocating too many resources and can balance your resources to higher priority requirements. Compare network performance over time on a monthly basis as indicated in Figure 8.4.

○ Metric 2.0.3: *Percentage of devices with current security software.* Security software including antivirus and antimalware software, spam blockers, and other applications has proven very effective in providing defense against malicious code. All of your devices (e.g., desktops, mobile computers, and smart phones) should be updated with current security software. This measurement monitors

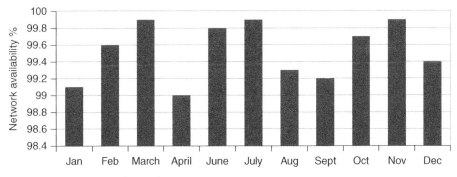

Figure 8.4. Network performance over time.

compliance with this standard. We consider it a network hygiene measure. While we trust our technical personnel to ensure that every device is properly patched and configured, we verify that it is done. This is a quantitative measure that should be collected and reviewed monthly.

- **Metric Category 2.1. Change Management**
  - Metric 2.1.1: *Number of unauthorized changes. Unauthorized changes to your systems are not good.* As indicated in Chapter 6.0, changes need to be the product of a disciplined and controlled process. Unauthorized changes often introduce vulnerabilities and flaws that lead to exploitation, can damage information and equipment, and can ruin your bottom line. You should strive to eliminate unauthorized changes to your information systems. By diligently reducing your unauthorized change rate toward zero, you will be in a better position to detect unauthorized changes that actually are the result of external actors tampering with your systems or information. The capability to detect such activity greatly improves your cybersecurity posture. Measure and review unauthorized changes monthly. This is another measurement that ought to be considered for coupling with your corporate rewards and discipline program.
  - Metric 2.1.2: *Percentage of maintenance successfully accomplished within schedule and budget.* Scotty from the Starship Enterprise said that he multiplied his maintenance projections by a factor of four so that he would maintain his reputation as a miracle worker. We seriously doubt there is any member of your IT staff who does not know Scotty's maintenance formula. Forecasting maintenance time is difficult, particularly with IT. Many organizations do not have the resources to have parallel systems or test environments and accept the risk of first-time installation and use on primary production systems. Often, this means that the first time a technician handles a patch or hardware installation is when the stakes are highest and when he alters your live system. Regardless of what circumstances you face, maintenance effectiveness is an important measure. Loss of capability beyond planned outage and budget forecasts costs you money, can damage your brand reputation, and introduces risks. We recommend that you consider measuring the percentage of maintenance successfully accomplished within schedule and budget. Measure it monthly and drive it toward 100%. Be mindful of Scotty's planning factor and continually seek to reduce the actual amount of time and money spent on maintenance.
- **Metric Category 2.2. Software Configuration Management**
  - Metric 2.2.1: *Percentage of software current with all known patches.* This is a critical cybersecurity measure. It makes sense to patch your software with all known patches and strive for 100% compliance, yet we recommend caution. Some patches may introduce issues with other pieces of code installed on your system. As mentioned in Chapter 6.0, whenever possible, ensure new software patches are tested before you load them onto your primary production systems. Your metrics should indicate you are very close to 100% compliance. If you have software that is not current, find out why and work to test and install the

patches quickly when it proves safe to do so. Make this measurement something you look at every month.

- Metric 2.2.2: *Number of unauthorized software and media detected on network and devices.* By now, we hope you realize that unauthorized software and media on your networks and devices is a bad thing. Unauthorized software and media expose you to malicious code, potential copyright violations (e.g., piracy involving bootleg applications, music, and video files remains a problem that you can and should prevent both in your home and office!), and potential interference between applications. You need to ensure that there is no unauthorized software and media on your systems. Review this metric every month.

- **Metric Category 2.3. Physical Security**
  - Metric 2.3.1: *Number of physical security incidents allowing unauthorized access into facilities.* This measures compliance with your physical security program, which directly complements your cybersecurity program. It is difficult to accurately measure because most employees are uncomfortable reporting when their colleagues fail to maintain proper control of those under their charge. Executives can promote their physical security program by creatively balancing rewards for detecting policy violations and correcting failure to comply. By instilling a sense of discipline and accountability, you can better protect and control your sensitive information and prevent it from walking out the door. We recommend you make this a measurement you review at least semiannually. For critical infrastructures, we recommend a higher frequency of sampling depending on the sector.
  - Metric 2.3.2: *Number of violations of clean desk policy.* This metric is a bit easier to gather when you implement an end-of-day security check. Many government and private sector organizations working with classified material governed by national security rules routinely maintain end-of-day security procedures that include cleaning desks of all materials and locking everything up in approved safes. For other organizations, this may represent a cultural change that takes some getting used to. Nonetheless, a clean desk policy has repeatedly shown to be effective in preventing theft and exposure of sensitive information. We have found that holding work teams or offices accountable to adhere to the clean desk policy heightens security awareness and ownership. Consider posting this security metric right next to safety metrics in the work area for all to see and reward excellent performance. This is a quantitative measure you should review monthly until the policy and culture matures. After you reach a steady state of compliance, consider switching to quarterly reviews.

- **Metric Category 2.4. Acquisition**
  - Metric 2.4.1. *Percentage of System and Service Contracts That Include Security Requirements and/or Specifications.* Cybersecurity is most effective when it permeates all of your activities. Make sure you include it in your system and service contracts. You do not want to buy any systems that have cybersecurity weaknesses nor do you want to hire contractors that introduce risks either. When purchasing hardware and software, deliberately consider cybersecurity in each

contract. Similarly, when procuring services, ensure that cybersecurity controls are incorporated into each contract. For example, if you have service contracts with companies that provide individuals who supplement your workforce with special skills that require access to your networks and information, make sure the contract clearly identifies their responsibilities to meet your corporate cybersecurity program requirements. As you negotiate new contracts, make a cybersecurity review by your CIO part of the contracting officer's checklist for every contract. Measure the percentage of contracts that have been reviewed and approved by the CIO for cybersecurity content and review semiannually.

8.4.3 Do We Have the Right People, Are They Properly Trained, and Are They Following Proper Procedures?

- **Metric 3.0. Percentage of Employees Who Have Current Cybersecurity Training**. An untrained workforce can harm your business. Make sure your employees receive current cybersecurity training. We recommend initial training for every employee and annual training thereafter. Monitor this metric every quarter.
- **Metric 3.1. Percent of Technical Staff with Current Certifications**. You would not want an unlicensed doctor doing surgery on you nor should you have individuals without the proper certifications operating your networks and protecting your information. We recommend that your technical staff maintain the appropriate certifications relevant for their duties. This should be reviewed every quarter.
- **Metric 3.2. Number of Users with System Administrator Privileges**. System administrators have a complete control over your network and the information contained on it. You need to tightly control who has system administrator privileges and ensure they are extremely well trained, are trustworthy and reliable, and are properly compensated. There will be many users who believe they should have system administrator privileges so they can install their own software, change configurations, establish user accounts, and perform similar functions. Resist the impulse to allow just anyone to have these privileges and only grant the minimum privileges to perform assigned tasks. Keep the number of those with system administrator privileges to a minimum. Review how many you have every quarter and ensure they have the right training and skills to accomplish these responsibilities successfully.
- **Metric 3.3. Number of Security Violations during Reporting Period**. This measure gives you an indication as to the effectiveness of your security program. The stronger your program, the fewer violations you should observe. Measure this quarterly.
- **Metric 3.4. Percentage of Security Incidents/Violations Reported within Required Timelines**. This is a compliance measurement to ensure that security incidents and violations of policy are reported properly and in a timely manner. This helps you enforce good order and discipline. Review this quarterly.

8.4.4 Am I Spending the Right Amount on Security?

- **Metric Category 4.0. Cybersecurity Costs**
    - ○ Metric 4.0.1: *Percentage of the IT budget devoted to cybersecurity*. You'll never know if you are spending the right amount on cybersecurity if you don't know what you are spending right now. It is instructive to know how much of your IT budget actually is allocated to cybersecurity. Consider including appliances such as firewalls and intrusion detection systems, tools such as network monitoring and scanning software, and cybersecurity personnel overhead costs. Review this during your annual budget allocation process.

    - ○ Metric 4.0.2: *Percentage of the organization budget devoted to cybersecurity*. You likely have cybersecurity expenses outside of the IT budget. For example, your employee cybersecurity awareness and training program may be a budget item in your human resources department. Comparing your aggregate cybersecurity expenses to your entire organizational budget is helpful in assessing whether your investments in cybersecurity are in alignment with your risk management program and organizational priorities. Review this during your annual budget allocation process.

    - ○ Metric 4.0.3: *Execution of current budget*. Managing your current budget is critically important. If you do not effectively and efficiently manage your budget, you will fail and senior management will never allocate more. Measure how your expenditures adhere to your budget and spend plans. Review this on a regular basis. We recommend you review your budget at least once a month. Be prepared to explain deviations from the budget to the board and CEO.

- **Metric Category 4.1. Value of Information**
    - ○ Metric 4.1.1: *Value of information*. This is the dollar value estimate of your information. Like any other asset, your information has value that should be defined and monitored. It is a quantitative measure put in dollar terms that considers the cost to acquire, maintain, and replace your information. As your information likely is continually changing in composition, volume, and value, we suggest that you conduct asset valuation of your information at the same time as your normal accounting cycle conducts valuation of other assets. Generally, this is done annually and in conjunction with financial audits.

- **Metric Category 4.2. Consequences of Information Loss, Tampering, or Destruction**
    - ○ Metric 4.2.1: *Cost to replace*. In the event that a cybersecurity incident causes irreparable damage to your information, you should understand the cost to replace it. Costs may range from relatively low if adequate backups are maintained to very expensive costs if the information must be reconstructed or likewise replaced. As your information likely is continually changing in composition, volume, and value, we suggest that you review the cost to replace your information at the same time as your normal accounting cycle conducts valuation of other assets. Generally, this is done annually and in conjunction with financial audits.

○ Metric 4.2.2: *Estimated costs associated with loss, tampering, or destruction of information.* You need to have a clear understanding of the consequences associated with your information being lost, tampered with, or destroyed by deliberate or inadvertent action. Legal fees, regulatory fines, costs associated with mandatory notifications, and anticipated losses all should be factored into your calculus. This metric should be expressed in monetary terms such as dollars per incident. Make this estimation annually during your normal accounting cycle.

○ Metric 4.2.3: *Estimated costs associated with regulatory fines for failing compliance.* This is a helpful metric to consider when determining whether to implement cybersecurity controls mandated by government regulations. Some organizations find the costs to implement controls far exceed the penalties of not complying with the government regulations. Do not neglect to factor in indirect costs such as loss of brand reputation or accreditation in the costs that will be incurred if you fail to maintain compliance with government regulations. This metric should be expressed in monetary terms such as dollars per incident and should be addressed at least annually through your normal accounting cycle.

- **Metric Category 4.3. Cybersecurity Risk Exposure**
  ○ Metric 4.3.1: *Cybersecurity risk.* As discussed in Section 3.3, there are several ways you can calculate your risk exposure. Understanding your annual loss expectancy, single loss expectancy, and annualized rate of occurrence is valuable in determining your risk exposure. We advise you calculate this measure annually or whenever you have a major change that is expected to drive a change in your risk profile.

8.4.5 How Do We Compare to Others?

As with any other issue in the competitive marketplace, you should keep a close eye on the competition. A strong cybersecurity posture may give you a competitive advantage, while a weak posture may put you at significant risk. Understanding the current contemporary standards used by others in your business sector (as well as others) exposes you to a wider sample size of cybersecurity practices that permit you to determine what best practices you want or need to incorporate into your organization. This can save you time and money as you learn from others what works and what doesn't rather than having to invest in your own trial-and-error efforts to find the best mix of cybersecurity practices, procedures, tools, and training that fits your organization. Be constantly on the lookout for potential best practices that you can incorporate into your operations, even when they come from your competitors.

Many peers are forthcoming when it comes to their cybersecurity metrics and will freely share their experiences on measures they value to assess their cybersecurity posture. We do not consider the measures themselves as sensitive material and recommend that you engage your peers to exchange information on what cybersecurity measures are best suited for your business sector. Don't be timid. Ask your peers what they do to measure their cybersecurity posture, what lessons they've learned, and how they use the measures to make decisions. For the price of a lunch, you may share a valuable set of measurement techniques that can significantly improve your ability to monitor and control your cybersecurity posture and improve your organization's risk management posture.

Another way to compare your performance measurements to those used by others is through industry groups and associations. Many host annual conferences and seminars that bring members from throughout the business sector together. Some even host seminars that are dedicated to performance measures. We encourage you to participate in these conference and industry-specific events that permit you to meet counterparts in your business sector. Ask the attendees what they do to discover cybersecurity performance measurement best practices. If the topic is not on the formal agenda, ask that it be added as it is a topic. With sufficient lead time, most conference organizers are pleased to include discussions of cybersecurity to the agenda of the conferences as it applies to every business sector. Conferences of like-minded individuals and industry experts are great places to compare what you are doing to that of your peers.

Another method is to consult with industry experts who specialize in cybersecurity and IT. Companies like Gartner, Forrester, and Wisegate are exposed to a wide variety of businesses across multiple various sectors and maintain libraries of best practices that they regularly update and keep current. Their partners can help you find where you stack up across not only your industry but across many others as well.

8.4.6 Creating Your Executive Cybersecurity Dashboard

Recall Drucker's observation that how many executives find a way to organize information effectively is to simply organize one's being informed about the *unusual*. That is the concept we use when implementing our Executive Cybersecurity Dashboard. Referring to the performance measure categories previously cited, you can establish control limits for each performance measure and instruct your staff who monitor these measures on a daily basis to give you updates when one of the measures exhibits unusual performance outside of control limits. For example, if you set control limits that define that no more than two incidents of maintenance overruns (Metric 2.1.2.) are acceptable, then on the third incident, you should be notified so that you can investigate and take proper action. The Executive Cybersecurity Dashboard can be configured to alert you to the *unusual*.

There are many ways to depict the Executive Cybersecurity Dashboard. Some executives use automated tools to monitor their performance measures, while others rely on their staffs to maintain watch over the measures and alert the executive to changes outside of control limits. We prefer a simple stoplight chart with green representing all is well, yellow indicating a warning, and red indicating a malfunction, serious incident, or alarm.

A simple representative sample of an Executive Cybersecurity Dashboard visual display we would receive every morning from our cyber operations team might look like as follows[18]:

---

[18] Our former work colleagues may find that this representation looks a lot like what we used when we served as a senior officer in the U.S. military.

Threats	System Performance	Vulnerabilities	Personnel and Training	Costs
Red	Green	Yellow	Yellow	Green

In this example, a new threat has emerged in the form of zero-day exploit that was announced this morning. Your cybersecurity and technical teams are doing the assessment to determine what its impact may be on your organization. Until their assessment is complete, you will remain in alarm status for threats.

System performance looks good this morning. All performance measures are within the standards you have established.

Vulnerabilities indicate warning (yellow) status. This is due to several high priority vulnerabilities (as defined by the CVSS database) that remain unpatched on your network. You have approval to patch through the Change Management Board, and the technical team will patch the vulnerabilities tomorrow night. In the meantime, your watch team is on heightened alert and conducting a combination of manual and automated monitoring of the affected system as a mitigation technique. Your senior leadership has approved this action and accepted the risk of waiting to install the patch until tomorrow.

Personnel and training also is in warning (yellow) status. Despite a series of countdown reminders, several employees have let their annual cybersecurity awareness training lapse, and three of your IT technicians do not have their current certifications. Per corporate policy, the employee's network access will be suspended until they complete the training. The IT technicians are on track to complete their certifications at the end of the month. Until that time, they are not permitted to hold system administrator privileges and will be assigned other tasks commensurate with their certified skill levels.

Finally, costs remain within expected control limits.

Your Executive Cybersecurity Dashboard relies on the decisions you make regarding your key performance indicators. Set your control limits to cue you as to what is unusual and warrants your attention and establish the processes to ensure you get accurate information in a timely manner. When you do that, you have the start of a good cybersecurity situational awareness tool.

Your aggregated performance measures should be considered confidential information not to be freely released outside of official organizational channels. They should only be shared on a need-to-know basis approved by management. Make sure that they are properly secured as they are stored, transmitted, and displayed.

While the sample Executive Cybersecurity Dashboard has worked well for the author, don't be constrained by its format in building your own. Use what works for you. As Peter Drucker has observed, "No two executives organize information the same way."[19] We like to follow Jim Collins's advice to "Try a lot and keep what works."[20]

Give cybersecurity performance measures a try in your organization and let us know what lessons you learn. We would love to hear from you.

[19] See Note 4.

[20] Jim Collins, *Built to Last, Successful Habits of Visionary Companies*, HarperBusiness, New York, NY, 1994, p. 140.

8.5 SUMMARY

- Use of performance measures is an essential part of gathering the information needed to make informed decisions, especially when it comes to cybersecurity.
- Many executives have found that an important way to organize information effectively is simply to be informed about the *unusual*.
- You should measure your cybersecurity posture as part of your efforts to practice due care and due diligence, monitor and control your information systems, maintain legal and regulatory compliance, meet contractual obligations, and maintain certifications.
- Cybersecurity performance measures facilitate decision-making, improve performance and accountability, and support efforts to reward and discipline your staff.
- There are no universally agreed-upon standard cybersecurity metrics for executives.
- Executives have several key decisions to make about their performance measures:
  - Decide what it is you are going to measure.
  - Decide how you are going to measure.
  - Decide how your metric information will be reported.
  - Decide how your performance measurement program will be implemented.
  - Set performance targets and key performance indicators.
- Every executive should seek to answer the following questions through their cybersecurity performance measures:
  - What are our threats?
  - How effective are our systems?
  - How vulnerable are we?
  - Do we have the right people, are they properly trained, and are they following the proper procedures?
  - Am I spending the right amount on cybersecurity?
  - How do we compare to others?
- You can create an executive-level dashboard of performance measures that will help you answer the cybersecurity questions every executives should ask.
- Recommended cybersecurity performance measures are provided to assist you create your Executive Cybersecurity Dashboard.
- Your Executive Cybersecurity Dashboard relies on the decisions you make regarding your key performance indicators.

9.0

WHAT TO DO WHEN YOU GET HACKED

Computer hacking really results in financial losses and hassles.[1]

Kevin Mitnick
The World's Most Wanted Hacker

9.1 HACKERS ALREADY HAVE YOU UNDER SURVEILLANCE

You will get hacked. Sooner or later, one way or another, someone will penetrate your defenses and gain access to your network and potentially to your most sensitive information. It may be a traditional hacker, who seeks a trophy by breaching your defenses and emerging with your valued code to show off to their buddies. It may be a hired-gun hacker, that is, someone who is paid by a competitor or adversary to penetrate your

---

[1] Verne Kopytoff, *Q & A/Kevin Mitnick/Ex-hacker shares secrets of deception/Mitnick says 'social engineers' play big role in cyber attacks*, October 28, 2002, http://www.sfgate.com/business/article/Q-A-Kevin-Mitnick-Ex-hacker-shares-secrets-2758743.php. Accessed on November 11, 2013. Mitnick has been called the world's best hacker. After serving nearly five years in prison for his hacking exploits, he now is one of the world's most renowned computer security consultants. We highly recommend his books including *Ghost in the Wires*, *The Art of Deception*, and *The Art of Intrusion* to those who seek to better understand the motivations, tactics, techniques, and procedures of hackers.

---

defenses to steal your information or damage your ability to compete in the market-place. It may be a nation state that is gathering intelligence by seeking access to your intellectual property or trade secrets to give them an advantage. It may even be one of your own employees, acting deliberately or accidentally to circumvent your security systems with the resulting unauthorized exposure, tampering, or destruction of your vital information. The threat is real and you face it right now.

Hackers are an enemy. They engage in criminal activity. They will seek as much information about you as they can before they even attempt to attack you. They have a game plan when they seek to breach your defenses, and it is important that you know as much about their plan so that you can best posture your organization to defend against it, to detect them when they try to execute it, and to recover in the event they are successful. You must "Know Your Enemy and Know Yourself."

Hackers know that the goal of cybersecurity is to protect the confidentiality, availability, and integrity of your information. Most will have knowledge equal to or better than your own cybersecurity staff. They will make an effort to understand your security policies and programs in an effort to find weaknesses or deficiencies they can leverage in order to gain unauthorized access to your systems. They always will seek the path of least resistance to your vital information.

Their first steps usually have nothing to do with probes or scans of your networks. Before they even attempt to penetrate your defenses to gain access to your information, many hackers will analyze your physical security to see how tightly you control physical access to your facilities and the devices that contain your information. They know that if they can gain physical access to your computer, they can gain complete control of the system and bypass all security measures. Weak physical security usually is an indicator of an overall weak security program and is considered an encouraging sign for hackers. On the other hand, a strong physical security program often is a potent deterrent as it is assumed such a program is an indicator that the organization has similarly invested in a strong cybersecurity program. While some dedicated hackers will take that as a challenge, the vast majority will move on to less secure and easier targets. Potential foes are observing how your facilities are secured (guarded), such as seeing (1) if you permit uncontrolled physical access to your computers and networks, (2) if your employees are lax in enforcing your physical security controls, or (3) if they can gain access to your information through weaker defenses at your off-site data storage facilities. They are watching.

Hackers are analyzing your network security. They know that if your computers are connected to the Internet, they have the potential to communicate with that device, gain control of it, and access your information. They know that if they are able to gain physical access to your network, they can install software or equipment to *sniff* all unencrypted network traffic, potentially giving them the ability to capture passwords they can use to gain control of your systems. They are looking to see if you allow dial-in access to your systems, which could permit them to use the phone system to gain access to your network. They are trying to map your attack surface by seeing how many potential access points you have. They will systematically probe each one to find the weakest link for their future planned attacks.

Hackers are trying to find what type of computers you have and how they are configured. They want to know what types of computers you operate, what operating systems and versions you employ, what skill levels your IT personnel have, who are your vendors, who does your maintenance, and what applications you use. They diligently do their homework to compare your profile against known vulnerabilities to help them identify weaknesses and craft their attack plans, many of which start by testing to see if you are using default passwords or even bother to activate many of the security features available in your systems.

Finally, hackers are watching you and your personnel. They know that people are both the strongest and weakest links in your cybersecurity program. They know that personnel weaknesses can negate great physical, network, and computer security programs. They are looking to see if you and your people follow your promulgated policies in a reliable and predictable manner. They will look to see when and how you make exceptions to policy and will look to exploit those circumstances when they can. They will look for opportunities to socially engineer your organization through your people. In a customer-focused organization, personnel are trained to be helpful. A hacker will plan to use this sense of helpfulness to their advantage as they seek to gain access to your information. They will gamble on the fact that your customer-focused workforce will be fixated on the customer aspects of their job and lose sight of the need to maintain a strong cybersecurity posture. They are watching what your people do and how they do it.

Chances are fairly good that sooner or later you will get hacked, but that does not necessarily mean that your business will face catastrophe and potential collapse. You should minimize the effects of hacking incidents by preparing in advance. Your best defense against hacking is not only to defend against it with a cyber-hardened environment and workforce but also to prepare for what to do when you do get hacked. Remember the words of Vegetius, who stated, "*Si vis pacem, para bellum*" (If you want peace, prepare for war). Hacking is cyber war on you and your organization. Be prepared.[2]

9.2 THINGS TO DO BEFORE IT'S TOO LATE: PREPARING FOR THE HACK

You likely have a last will and testament that details to your heirs, the government, lawyers, and other interested parties what you want done with your estate when you die. You don't prepare a will because you *think* you eventually will die; you prepare it because you *know* you will. While preparing for a hacking incident does not have the moribund sense of finality typical of a last will and testament, preparing for a potential hacking incident is a prudent investment that will protect you, your organization, and your vital information.

---

[2] This phrase is well known to those of us who have graduated from the U.S. military war colleges as it is woven into the curriculum in numerous courses. Some people wrongly attribute its English translation to George Washington, who almost certainly was familiar with the quote and concept, but did not originate it. For those who seek more information on the quote and its origin, we commend to you the following 1767 translation of Publius Flavius Vegetius Renatus' *De Re Militari* by British Army Lieutenant John Clarke: http://www.digitalattic.org/home/war/vegetius/. Accessed on November 11, 2013.

9.2.1 Back Up Your Information

A startling number of people and businesses do not back up their vital information properly. Despite calls from industry experts to make regular backups of critical information an important business ritual, even the most clever individuals and powerful companies stumble when it comes to maintaining adequate backups.

Your information has value and should be protected through the practice of regular and comprehensive backups, at work and at home.

A best practice in data backups involves analyzing and prioritizing your information. Not all of your information is of equal value. Not all of your information is collected at the same rate or in the same volume. It pays to understand what information you have, who needs it, how it is collected, and what its value is. Sometimes, it does not pay to back up every bit of information you have; you should only store and back up what you need. When you have a thorough understanding of your information, you are best prepared to make a cost–benefit analysis to decide what gets backed up, when it gets backed up, where the backups are stored, how you will recover backed-up information, and who is responsible to execute these tasks.

With the costs of data storage continuing to fall, some companies and individuals are finding they can afford to back up *all* of their information, especially when they use cloud storage services. This is a very attractive option for many. To protect their information that is under the control of third-party providers, most opt to employ the best practice of encrypting all the data they have in storage. This is a great idea and a prudent cybersecurity practice yet **key management** needs special attention. Imagine what would happen if you back up all your information into the cloud and encrypt it. Then, you are subject to a cybersecurity incident that corrupts or destroys your primary computer databases and storage. How do you recover from backup if the key to unlock the backup was destroyed or corrupted in the attack? Smart organizations and individuals place their decryption keys in a form of key escrow, where they will have access to the keys in the event of disaster.

As an executive, you should understand the value of your information and how that information is protected. Make sure that your important information is adequately backed up in a timely, effective, efficient, and secure manner. Have a plan on how to recover from a cyber incident using your backed-up information through scheduled testing of the plan. Include regular audits of your backup and recovery plan as part of your overall business continuity planning.

9.2.2 Baseline and Define What is Normal

People who specialize in computer forensics often are called in to assist investigators looking into suspected hacking and other cybersecurity incidents. One of the first questions the investigators will ask you and your IT team is for the documented baseline of your computers and network devices so they can compare the current presumed *hacked* state against the presumed *pristine* state of your computers and network devices. Sadly, not everyone or every organization maintains a documented baseline, but they should. You should too.

Your IT staff should maintain documentation that shows how your computer systems are configured and controlled. Your baseline documents should contain information including details on the type of equipment used and version control documents that indicate what authorized software and patches were installed. Proper maintenance and control of this documentation is expected of professionally managed information systems functions. You should make it a requirement for your IT staff, and your auditing function should be checking to make sure it is done properly.

Having an accurate and well-maintained baseline will make it a lot easier to determine whether you have been hacked or not. It will aid in diagnosing and troubleshooting problems. In the event that you have been the victim of a hacking incident, it can prove to be extremely helpful in recovering and returning your systems to full operational capability faster and more reliably. Maintaining your baseline is a great investment to protect you and your information in the event you are attacked.

9.2.3 Protect Yourself with Insurance

The author recalls a commercial on television whose tagline was, "The best time to insure for an accident is before it happens." That is good advice, especially when it comes to cybersecurity.

Cybersecurity insurance is a rapidly growing discipline within the insurance industry. Many brokers now offer a wide range of insurance options that cover many cybersecurity incidents including hacking and other forms of malicious activities all the way to inadvertent self-induced data loss.

Not everyone needs cybersecurity insurance nor does everyone want it. Like any other insurance, deciding whether to invest in cybersecurity insurance is a best-value business decision. Individuals and organizations that possess highly valued information such as intellectual property and trade secrets, personally identifiable information for a large number of clients, prized financial records, or medical records are typical investors in cybersecurity insurance. So are those businesses that rely on information systems to generate or promote their income; they can't afford losses associated with hacking and other activities that take their information systems off-line. Businesses that operate in critical infrastructure sectors are investing in cybersecurity insurance in ever-increasing numbers. They do so because their expected loss due to cybersecurity incidents could be significant due to incident damages and remediation, anticipated regulatory fines, and expected litigation. To them, insurance is a good investment (perhaps mandatory).

Regardless of your type of operation or business sector, evaluating your options regarding cybersecurity insurance is a worthy activity. Your decision whether to invest in cybersecurity insurance should be based on a cost–benefit analysis congruent with your organization's risk appetite, just as you would decide for any other insurance option. Because cybersecurity insurance is relatively new to the marketplace and actuarial tables influencing rates still are maturing, noticeable variations in rate structures exist between different firms. You should be a discerning shopper when considering cybersecurity insurance. Compare plans and rates from many brokers and don't be afraid to tailor your own package based on your requirements rather than accept a generic package that may not cover all your needs or may provide more coverage than you need.

If you and your business determine that you are at high risk of being hacked or are similarly susceptible to cybersecurity incidents, cybersecurity insurance may be an indispensable investment. Wishing that you had insurance after you've been hacked is unacceptable. Anticipate what will happen if you get hacked, what the likelihood of being hacked is, how much it will cost you, and compare that cost against the cost of insurance. You may find that you cannot afford not to have cybersecurity insurance.

9.2.4 Create Your Disaster Recovery and Business Continuity Plan

While serving in the military, I was exposed to and created countless plans. From the moment I became an officer, the importance of having a plan was continually reinforced. My units would create comprehensive plans that addressed what we would do in every conceivable situation. We would train everyone in the unit on what to do and would test the strength of the plans, our training, and our personnel through rigorous exercises where inspectors would deliberately throw us curveballs by introducing situations we didn't originally envision when we created the plan. These curveballs introduced stress but also promoted creativity. We would take our lessons learned from these training exercises and update our plans and training to make us even better prepared in the event we had to implement the plan for real. We always looked to be prepared for any contingency.

When we had to implement our plans, we had confidence that we were well prepared. We knew what our objectives were. We knew what everyone on the team was doing, from the commanding general down to the junior airmen. We knew that there were very few scenarios we had not anticipated, but if we were confronted by something new, we knew we could adapt because we had *our eyes on the ball* and knew what was important to our mission. We had confidence in each other, our leaders, and in our planning. As an example, as we concluded Operation DESERT STORM, where the author served as a squadron commander, one of my senior noncommissioned officers commented, "We weren't confronted by anything we hadn't been trained to do and that made what we did easier than we thought it would be. In fact, our training was more difficult than actually going out and doing it for real!" Despite being thrown numerous curveballs that weren't in our plans, we were prepared to hit them out of the park because we had a master plan that was flexible and accommodated changes. We were confident because we all knew our goals and objectives, we trained to handle multiple contingencies, and we worked as a team. You need to do the same with your **Disaster Recovery and Business Continuity Planning**.[3]

Having a comprehensive disaster recovery and business continuity plan is essential for every organization, regardless of business sector. It is a crucial part of your business continuity planning. You and your organization need to plan for what you will do to maintain the continuity of your business when disaster strikes regardless of whether they are acts of nature or man-made.

---

[3] The then-Master Sergeant Glyn Howells, Jr., said this in our after-action meeting after the cease fire order was given. Howells would conclude his active duty career as a chief master sergeant, the highest enlisted rank in the U.S. Air Force.

Natural disasters are a very real and unpredictable threat. For example, the earthquake and tsunami that pounded the coast of Japan in March 2011 had global effects as it interrupted numerous supply chains, including that of IT manufacturers; introduced harmful radiation into the atmosphere and Pacific Ocean; severed numerous undersea cables providing key Internet connectivity; and sent debris across the Pacific. Every geographic location around the world is susceptible to one or more potential natural disasters. Often a disaster in one area will have an effect elsewhere, as did the 2011 earthquake and tsunami in Japan. Although nobody knows when natural disasters will hit, you need to be prepared.

Recent terrorist activity highlights another potential source of disaster and reinforces the need to invest in business continuity planning. For example, in the aftermath of the 1993 terrorist bombing of the World Trade Center in New York, many New York-based firms recognized their vulnerabilities to having their IT infrastructure located in one place and quietly began to invest in backup capabilities in New Jersey that allowed them to maintain business continuity in the event that their primary facilities in New York were damaged or otherwise interrupted. Such capabilities proved to be invaluable for many firms in the days after the tragic attacks of September 11, 2001, and 2012 Hurricane Sandy and allowed them to continue their operations and ability to serve their clients. Proper planning and implementation led to superior results and effective business continuity.

Cybersecurity events can be a disaster for you and your business if you don't have a plan to address them. Planning in advance better prepares you to address all possible courses of action so that you are not distracted by *the fog of war* typically found during disasters or crises. When you are confronted by crises, you are easily distracted by heightened emotions, confusion, spotty information, and angst. So are your people. Making it up as you go along is a losing proposition too as you increase your chances of increasing confusion, losing unity of effort, and making mistakes.[4]

We discussed planning in detail in Chapter 5.0 and introduced numerous items of interest that you should consider when creating your plans to prevent cybersecurity incidents as well as what you should do when incidents occur. We do not intend to replow that ground again. Nonetheless, we strongly encourage you to *invest now* in creating a plan that addresses how you and your organization will react when confronted by a cybersecurity incident. Identify every likely scenario and build your plan of action to address what you will do when disaster strikes.

9.3 WHAT TO DO WHEN BAD THINGS HAPPEN: IMPLEMENTING YOUR PLAN

The plan you create to address a cybersecurity incident may not be perfect. In fact, you may find that it does not adequately address every facet of the incident you face. It is neither unusual nor unexpected to see plans go through several iterations and updates

---

[4] Carl von Clausewitz was a nineteenth-century professional German–Prussian soldier, whose book *On War* is considered one of the classics of military theory. In it, he introduced what he called *the fog of war*, referring to the uncertainties introduced by such things as erroneous and incomplete information, excitement, fear, and doubt. Even though he was referring to warfare, a *fog of war*-like phenomena is evident in disasters and other crises. Regardless of whether you are a captain in the military or a captain of industry, great leaders must be able to see through the *fog of war*, make the best decisions possible, and lead their troops to success.

over its life span as new technologies are introduced, as new threats and vulnerabilities emerge, and as exercises of the plan illuminate new and better ways to address situations. You should consider your plan a *living document* and prepare to update it regularly. While your plan may not be perfect, it is a rallying point upon which you and your team can begin the process of assessing the situation, determining your next steps, and working together in a well-coordinated manner congruent with your strategy to minimize risk and put you firmly in control.

We recommend that you adopt the following ten *things to do* checklist items to guide your activities when you are hacked. Incorporate these items into your cyber disaster response checklists that are part of your organization's cybersecurity incident disaster recovery and business continuity plan. You may find that your corporate culture, type of organization and products, and other factors may spur you to add to this list of things to do with additional steps tailored to your organization. You should consider writing the items down on a sheet of paper and store it in your wallet for quick reference as you may use them at home as part of your personal disaster recovery and business continuity plan.

You are going to be hacked. Here is our 10-item checklist indicating what you should do when a hacker breaches your defenses:

9.3.1 Item 1: Don't Panic

The Hitchhiker's Guide to the Galaxy has "DON'T PANIC" inscribed on its cover. Perhaps, this book should too. Nobody wants to follow a leader who panics at the first sign of trouble. When it comes to cyber incidents, your first step is to remain calm and not to panic.[5]

Leaders who remain calm and resolute can be the decisive factor in achieving success. For example, in the first Battle of Bull Run in July 1861, panic struck Confederate forces that were withering under attack by Union forces. Their lines were collapsing and many soldiers frantically were retreating. Trying to motivate his troops, Brigadier General Barnard Bee pointed to Brigadier General Thomas Jackson and his troops, who were anchoring the Confederate forces and standing their ground. "There stands Jackson like a stonewall! Rally behind the Virginians!" When confronted by a cyber incident, you must be like Thomas *Stonewall* Jackson and not panic. You must rally your troops to stay calm and deliberately take the right next steps.[6] Recall that eventually it was Union troops who retreated in panic (along with many spectators who blocked an orderly retreat).

Many people lose their cool when they detect telltale signs of hacking or are informed they have been hacked. Hacking is a crime and suddenly people feel the full weight of being a victim. For most, they feel violated as identities may be compromised, monies stolen, secrets revealed, or treasures destroyed. Some will feel personally responsible and professionally embarrassed. The emotional impact can be crushing to some and devastating to others. As an executive, you have a leadership responsibility to remain

---

[5] Douglas Adams, The Hitchhiker's Guide to the Galaxy, Harmony Books, New York, NY, 1979, p. 3.

[6] National Park Service, *The Battle of First Manassas (First Bull Run)*, http://www.nps.gov/mana/historyculture/first-manassas.htm. Accessed on November 11, 2013.

calm and not panic. Crisis is the time when your calm and deliberate leadership is most crucial. You must be the leader to calm people down and focus them on the task at hand, which is to gather the facts and resolve the situation according to plan. You can't do that if you aren't calm yourself. Be like *Stonewall* Jackson. Don't panic and be the leader your people can rally around to resolve the situation properly. *Ride to the sound of the guns* and prove yourself worthy.

9.3.2 Item 2: Make Sure You've Been Hacked

Brace yourself for the day you get a report that you've been hacked. Such reports can come from various sources. They may come from one of your own people who have detected anomalous behavior on your network or devices. It may come from your personal observation of poor system or network performance. It may even come from the hacker himself, who has advertised their supposed success. Regardless of your source, verify that you actually were hacked and what you are seeing is not the result of an accident, system or software misconfiguration, or malfunction.

There is a popular military saying, "First reports are always wrong." Verifying you actually have been hacked is critically important because you likely will expend significant resources in the event it's true that you have been hacked. You don't want to waste your precious resources chasing a bogus report or overreacting.

Some people have learned the hard way the importance of verifying whether they indeed were hacked, many with embarrassing results. Take, for example, the case of the U.S. Department of Commerce's Economic Development Administration. In early 2012, it was alerted that malicious code was detected on its network. In January, employees were disconnected from email and web sites when the chase for the malware was launched. By April, the employees finally were reconnected using alternative means while the hunt continued and agency incident responders destroyed numerous desktops, printers, cameras, mice, keyboards, and other devices believed to be infected. The agency then went on to spend more than half of its IT budget, over US $2.7 million dollars, chasing down what appeared to be a major malware infection. By August, they "had exhausted their funds and halted the destruction of its remaining IT components, valued at $3 million," wrote inspector general auditors assigned to investigate the facts and circumstances surrounding the incident. The auditors found that the person put in charge of the incident response was unqualified and untrained for the task and only after they ran out of money did they find that the infection was limited to just two devices.[7]

There are many other similar examples of similar *target fixations* and overreaction across each business sector, but not everyone is as forthcoming and transparent as the U.S. Department of Commerce in publicly admitting their mistakes. Nonetheless,

---

[7] We wished we had made this up but sadly we didn't. There are numerous public reports on this incident including Allya Sternstein's July 9, 2013, NextGov report, *"Commerce Trashes $170,000 Worth of Tech to Disinfect Imaginary Viruses,"* http://www.nextgov.com/cio-briefing/2013/07/commerce-trashes-170000-worth-tech-disinfect-imaginary-viruses/66248/ and Jacob Kastrenakes's July 8, 2013, breaking of the story on *The Verge*, "US Commerce Department destroyed $170,000 worth of TVs, mice, and more to root out malware," www.theverge.com/2013/7/8/4503946/commerce-department-unnecessary-cybersecurity-computer-destruction.

there is a valuable lesson to be learned from the fiasco at the Economic Development Administration: verify that you have indeed been hacked before you commit yourself to some expensive and potentially irreversible steps.

9.3.3 Item 3: Gain Control

Once you confirm that you indeed have been hacked, it is essential that you gain control of your computer or device. This is critical so that you can assess the situation and determine the next best steps for you and your organization.

Hackers may have control of your computers and are quietly using them to do such things as monitor your online activity, steal your information or that of your clients and partners, or even use your computers as tools in criminal activities. As an example, hackers continue to develop sophisticated programs that secretly are installed on as many computers as possible through use of poisoned email attachments, web sites, and other surreptitious means. These infected computers then are banded together like zombies, unwittingly committing criminal activity such as distributed denial of service attacks, fraud, and information theft. These bands of zombie computers, called *botnets*, are increasing in size, number, and severity on a daily basis and may expose individuals and businesses to liability concerns and damages. David Dagon, a Georgia Tech researcher specializing on botnets, estimates that a minimum 11% of the more than 650 million computers connected to the Internet are infested with botnet software. "It's the perfect crime, both low-risk and high-profit," said Gadi Evron, a computer security researcher for an Israeli-based firm, **Beyond Security**, who coordinates an international volunteer effort to fight botnets. You need to gain control of your system to stop hackers and their malicious code to reduce your liability and risk.[8]

How and when you do this is situationally dependent. If you are on your home computer, for example, you may simply disconnect from the Internet to deny the hacker access to your system while you do your analysis and remediation. In contrast, if your business systems are infected and under someone else's control, you may take different actions ranging from disconnecting from the Internet to immediately calling in law enforcement and computer forensic specialists.

Disconnecting your infected device from the Internet is a technique commonly referred to as *isolation*. Similar to techniques used in medicine, the infected device is isolated or quarantined from other devices. This ostensibly reduces the likelihood that any suspected infection will spread to other devices.

Many businesses find they need to have their systems online in order to generate revenue; thus, downtime to study suspected hacking activity is unacceptable. Firms like these often will do whatever they need to in order to restore their services to full operational capability as fast as they can, even when it means they accept the risk of destroying the potential evidence law enforcement personnel can use to track the perpetrators and bring them to justice. While many firms have redundant capabilities that allow them to switch to back up systems to preserve evidence on their infected systems, these cases

---

[8] John Markoff, "*Attack of the Zombie Computers Is a Growing Threat*," January 7, 2007, http://www.nytimes.com/2007/01/07/technology/07net.html?pagewanted=all&_r=0. Accessed on November 11, 2013.

are relatively rare. Some firms now attempt to take a *snapshot* or full copy of the computer's now-infected hard drive to preserve evidence for law enforcement and forensic investigators and then wipe and reload the device to return it to service as fast as possible. What actions you take depends on the level of risk you are willing to accept.

Regardless which avenue of approach you take, you should recognize that once your computers and network devices become victimized by a hacker, they may be considered evidence in the commission of a crime. As such, any action that you take may contaminate or erase potential evidence. The computer or devices are evidence that may be helpful to investigators from law enforcement, regulators, computer forensic professionals, and auditors. You should maintain detailed and accurate chain of custody records of every action that you take. They should document everything including who touched what device, why they did so, what they did, and how they did it. Maintaining a *snapshot* copy is very helpful to preserve the *crime scene*, especially the computer logs. You should be prepared to justify to your board, shareholders, regulators, and law enforcement officials every action you've taken in response to a hacking incident.

The best time to evaluate and decide what course of action to take to gain control in a hacking incident is before it happens. Think ahead and include in your plan of action what you would do to regain control of your systems in the event that a hacker successfully attacks you and when you would do it. There are several likely scenarios you may face including root level access, botnets, and other levels of control and infection. Consider in your calculus when (if at all) you would call in cyber forensic professionals and/or law enforcement officials. Include your public relations and general counsel personnel in the preparation as you will find their expertise indispensable. You should have a plan of action on how to deal with each likely scenario tailored to your risk appetite and business requirements.

9.3.4 Item 4: Reset All Passwords

This may appear to be a no-brainer; however, we continually find instances where individuals and organizations fail to change their passwords on a regular basis, let alone in the aftermath of a hacking incident.

If your business is the victim of hacking, change every password on *every* computer and device, *especially* system administrator passwords. We caution that you should be on heightened alert to the appearance of any new and unauthorized accounts during this period as they may be evidence that the hacker has established their own account (or accounts) on your network to permit them continued access. Hackers who have established what appear to be legitimate accounts will comply with your calls for all personnel to conduct mandatory password changes to retain their access. Therefore, you should combine your mandatory password resets with a 100% validation of accounts to ensure you deny hackers access to conduct further damage.

In the event that your business suffers a hacking incident, you should change your passwords on your home systems. Many people exchange information between their home and work computers. While this often increases productivity, it also increases exposure to cross-contamination; you may inadvertently bring a virus home from work that could open your home system to the hacker who attacked your work environment.

Hence, you should change your password (don't forget to use our recommended password rules found in Section 3.2.2) and thoroughly scan all your personal and home devices with current antivirus and antimalware software products to verify the integrity of your home systems.

When your home system is the victim of a hacker, you should move quickly to change *all* your passwords. Despite recommendations from security professionals to make sure your passwords aren't all the same, many people continue to recycle the same password, using it on multiple accounts because it is easy for them to remember. Hackers love this as it permits them to not only gain control of your computer but also of your bank accounts, web-based email, stock funds, Internet shopping accounts, and other sources that rely on your username and password. Don't make it easy for the bad guys. Change all your passwords whenever you find evidence that you have been hacked!

9.3.5 Item 5: Verify and Lock Down All Your External Links

Many people use applications that share information with other applications through various plug-ins and application programming interfaces (APIs). These relationships improve the user experience and permit information to flow seamlessly between different applications without you having to manually move files around to share them. It also provides a great capability for hackers to gain access to your information and exfiltrate it through these links.

As an example, you may use your phone to help you navigate while you drive. The application you use has to continually swap information with multiple servers. First, the phone has to determine where it is. This typically is managed by the operating system that uses telemetry from multiple satellite sources (such as the Global Positioning System [GPS]) or even cell phone towers to fix your position. Your phone then shares this information with an application server that calculates your optimum routing based upon your current position and target destination. This information is continually updated based on a predetermined data exchange rate managed by the application. There are many background processes quietly executing and exchanging information while you drive toward your destination.

What makes all these information exchanges work are the APIs. Their importance and capabilities make them inviting targets for hackers who see them as a means to both enter your system and take information from it. Many applications such as games, social media, and productivity tools (such as your online schedule) use this capability, so it pays to be cautious in the aftermath of a hacking incident to verify and lock down these pathways.

Additionally, your baseline should define your network's authorized ports and protocols. Unusual network traffic using previously unauthorized and unexpected ports and protocols is a telltale sign of malicious activity. Make sure it indeed is unauthorized, and if it is, shut it down and call in the professionals to investigate. Recall the *Deny All* policy first introduced in Section 3.3.3. You only should allow network traffic that you specifically authorized and block everything else. If you've established this policy and implemented it before the hacking incident, your technical staff should be able to compare the current configuration to your authorized baseline and close any holes a hacker has opened.[9]

9.3.6 Item 6: Update and Scan

When someone like a hacker *owns* your computer, you should consider wiping it (such as reformatting the hard drive several times), reloading it with up-to-date security software, scanning for any residual threats, and returning it to service as appropriate.

It is difficult to determine whether a hacker has left behind capabilities that will enable them to reenter your computer at another time of their choosing. Some hackers have the capability to even thwart reformatting of hard drives and maintain a persistent presence on computers. For example, a technique used by highly sophisticated hackers is to create what often is called a *stealth drive*. Using this method, the hacker segments a portion of your drive, tells your operating system to not consider it part of the working disk yet accept any commands emanating from it. This allows the hacker to maintain a persistent *backdoor* presence on the machine, even if you reformat the hard drive and reload the software, as the operating system will not reformat the segment created as the *stealth drive*. While this is a possible worst-case scenario of which you should be aware, the cost to detect a stealth drive and remediate it is noteworthy (i.e., expensive). Your risk appetite may lead you to accept the residual risk after wiping and reloading your computer.

Scanning the computer and device with current and updated antivirus and antimalware software should be a mandatory control on your checklist before permitting the infected device to return to operation on your network. We have seen many schools of thought on this with many companies changing their antivirus and antimalware software after incidents. Those who make the change state they lose confidence in their original vendor when they are the victim of an incident and believe that another vendor's product may provide better protection. While scanning for any residual risk should be a mandatory control, switching software packages may not be the right answer. Make your changes based on a capabilities-based cost–benefit analysis comparing candidate products.

9.3.7 Item 7: Assess the Damage

Not all hacking incidents will ruin your day because not all hacking and cybersecurity incidents have the same effects. You will allocate resources based on the type and severity of damage you face so it is important to assess the damage to provide you the information you need to best align your precious resources to the incident response.

As an example, one of the most common hacking incidents prevalent today is a web site defacement. Defacement is similar to someone spray-painting graffiti onto the front

---

[9] While this does not prevent a hacker from using legitimate authorized communication channels to gain or maintain access to your system and its information, experience has shown that searching for and closing unauthorized backdoor communication channels established by hackers is a best practice to follow.

windows of your business and may have similar effects. In this type of hack, the hacker gains access to your web server and changes or adds content to your web pages. The list of organizations that have fallen victim to this type of hack is long and includes the United Nations, Fox News, and even the NSA. If your web site gets hacked and is defaced, the damage depends on several factors including what the defacement says, how long it is posted before it is cleaned up, and who reads it. If someone hacks into your web site and defaces your page with a sign that says, "You've been pwned by Hacker Z,"[10] very few people visit the site before it is restored to its proper (baseline) configuration, and very little resources are expended to fix it, and then the damage from the hack may be considered negligible. In contrast, if your web site is used for electronic commerce and a hacker gains control of your site to change pricing, post comments critical of your business or products, or revealing customer information, the damage could be severe not only to remediate but also to your brand reputation. The severity of damage drives how you respond, governing the velocity of response actions, the expenditure of resources, the assignment of personnel, interaction with law enforcement and regulators, and public disclosures. You need to know how bad things really are (or aren't) so you know what resources to commit to the problem.

Assessing damage is part of your risk management program and should be measured in monetary terms. Understanding the cost to repair the damage and restore to the desired baseline is only one factor that ought to be considered. Damage to brand reputation is not to be ignored and must be addressed. Your disaster recovery and business continuity plan should include defining who will assess potential damage, what damage they should assess (e.g., physical damages, data integrity, brand reputation, legal and regulatory action exposure, and other monetary expenditures), and what timelines you expect for reporting.

Assessing damage should not be limited to organic in-house personnel. Consider as part of your plan bringing in third-party experts to help assess and define the damage wrought by hacking and other malicious activity. For example, many companies retain PR firms to help them rebuild to retain their brand reputation in times of crises or disasters. You may find that a major cyber incident calls for reinforcements to protect your brand reputation. Likewise, the hacker may have employed techniques that are beyond the capabilities of your in-house technical team. You should consider hiring a specialist such as a certified ethical hacker or certified information systems auditor with computer forensics experience to augment your team to find and fix the root cause of the incident. Third-party experts can be extremely helpful in determining the extent of damage and in recommending appropriate next steps to reduce your risk profile. They are an investment worth considering.

While it is important to understand the magnitude of the damage and its effects quickly, we recommend patience. It is not unusual to not receive a full accounting of the damages associated with a cyber incident until the thorough investigation is concluded. Nonetheless, it is important to get a rough order-of-magnitude assessment as soon as possible so that you can align resources appropriately. Include in your planning

---

[10] "Pwned" is hacker slang meaning "owned," which refers to that hacker having taken control of your system.

provisions for damage control parties, make sure your people understand what their roles are in the damage assessment process, and remember that first reports aren't always accurate.

9.3.8 Item 8: Make Appropriate Notifications

When you find out you have been hacked, one of the things you have to do is determine who needs to know. Many organizations do not want it known that they have been hacked. There are many very good reasons they want to keep notifications to a minimum including the fact that once it is public knowledge that an organization has been hacked, such information often attracts other hackers to attempt copycat attacks. Also, public disclosure that an organization has been hacked often is viewed negatively by prospective clients, partners, and investors, harming brand reputation and value. Hacking incidents also increase risk exposure to litigation as law suits alleging misconduct or malfeasance associated with insufficient due care and due diligence have increasingly been seen in the aftermath of major cyber incidents. With these reasons in mind, while you should retain an open and honest approach to your notification process, you should keep your notifications to a minimum; only notify those with a need to know.

Your leadership should be notified when you are the victim of hacking incidents. As an example, many boards of directors and their risk committees issue guidance directing what critical information they require and when they need it. Hacking and other cybersecurity incidents are increasingly finding their way to the top of their reporting requirements. Bad news does not get better with time, so make sure you keep your leadership informed with fact-based information in a timely manner. Be prepared to answer the 5 Ws: Who, What, When, Where, and hoW. Recognize that initial reports may be incomplete, so plan for regular updates and status reports. Always make sure that your boss is informed!

Besides your leadership team, there are several entities within your organization that need to know when you've been hacked or had a similar cybersecurity incident. The first is your general counsel. Your legal team should be fully informed and engaged as you respond to the incident. They should be responsible for identifying all legal requirements for any and all notifications and remediation actions. Your lawyers should give you sound advice on what you should do, what you have to do, and what you might do when confronted by a cybersecurity incident.

Another internal notification that should be made is to your PR staff. A cybersecurity incident can have significant negative effects on your organization if not managed well. Your PR staff should be charged with managing your information flow to the public. Nobody in your organization should be communicating about the event without the expressed approval of the PR staff. Make sure your PR staff is notified of cybersecurity incidents promptly and completely.

When you are making notifications internally about your cybersecurity incident, don't forget to notify your CFO. Cybersecurity incidents can result in significant unplanned expenditures that your CFO will have to find funding to cover. Moreover, CFOs also typically control many organizational auditing functions. As you respond to cybersecurity incidents, auditing is an important component of investigations and remediation activities, and the CFO ought to be fully engaged. They also need to be fully informed.

You may maintain partnerships that specify a contractual obligation that the parties must notify each other in the event that one is the victim of hacking or similar cybersecurity incidents. This is common for those businesses that share information through electronic means and those who maintain custodianship of information. Your contracts should spell out under what conditions notifications are to be made, to whom, when they are to be made, and in what manner. Partnerships can be poisoned by lack of transparency during cybersecurity incidents, so make sure you understand what notification requirements you have in your contracts and hold yourself and your partners accountable to meet obligations.

Your organization may be associated with critical infrastructures whose regulatory authorities mandate reporting of cyber incidents. Whenever you are subject to mandatory reporting requirements, you should comply with the requirements in accordance with the directives. Follow the instructions specified by the regulatory agency and provide the required information; but exercise caution. In the famous television drama *Dragnet*, the detective Sergeant Joe Friday would always tell those he was interviewing he wanted *just the facts*. You should follow this advice and provide just the facts as requested. Anything more or less may expose your organization to additional risk. Be cooperative with regulatory agencies, provide only the information they require, and comply with all control mechanisms.

Because hacking is a criminal activity, you should notify law enforcement officials. Effective prosecution of criminals such as hackers requires cooperation between the victim and law enforcement authorities, and if you do not report the crimes to law enforcement authorities, it is very likely the hackers will continue to victimize you and others. Moreover, as the FBI says, "In a digital world where evidence can disappear at the click of a mouse, swift investigation is often essential to successful ... prosecutions."[11]

Finally, public notification to shareholders and the public at large should be handled carefully. For publicly traded companies, the SEC's Corporate Finance Division's Cybersecurity Disclosure Guidance 2 (see Section 3.4.2) spells out requirements for public disclosure of cyber incidents. Most companies are very succinct in their reporting of cyber incidents in their disclosures and reveal the minimum amount of information to satisfy regulators. As of this writing, there are no similar regulatory controls for private companies, nor do we anticipate some anytime soon. Therefore, private companies are reticent to reveal any information regarding hacking and cybersecurity incidents perhaps under the belief that such information would have deleterious effects to their competitive advantage. Whether you are a publicly traded entity or a private concern, public disclosure that you've been hacked or subject to a cybersecurity incident can have second- and third-order effects on your bottom line. Determining what information you disclose, who releases it, and how and when it is released should be a leadership decision supported by your staff. When creating potential COAs to address notification to the public, make sure you include legal, marketing, PR, and technical personnel to the team creating the options for consideration.

---

[11] U.S. Department of Justice, Computer Crime and Intellectual Property Section, "Reporting Intellectual Property Crime: A Guide for Victims of Copyright Infringement, Trademark Counterfeiting, and Trade Secret Theft," March 2013, p. 6.

9.3.9 Item 9: Find Out Why It Happened and Who Did It

Your initial response to a hacking or cybersecurity incident is focused on what happened and how to fix it. Many cybersecurity professionals like to compare this to fighting a forest fire; you try to contain the damage and will worry about other things later. Just because you have extinguished the fires that erupted from the hacking or cybersecurity incident doesn't mean that the incident should be closed. In fact, we've found that lasting fixes come from understanding why the incident occurred and who was responsible.

Determining why a hacking or other cybersecurity incident occurred is helpful in making sure that your controls and defenses are adequate to prevent the attacker from successfully breaching your defenses again. Your investment in cybersecurity controls is not inconsequential and you want to make sure you have a good return on investment. Understanding why the perpetrator attacked may give you valuable clues as to whether you can expect them or others to return with other attacks and whether you need to invest in other more robust defensive controls to prevent further damage.

For example, consider the hacktivist group Anonymous first introduced in Section 2.4.2. Anonymous is a hacktivist organization that has been very effective in wreaking havoc on several organizations and individuals with whom their members have a disagreement. Their activities have been largely motivated by political or ethical concerns. Their attacks have sought to punish those responsible for certain actions; to coerce changes in policies; or as postulated by some, to intimidate people and organizations. Anonymous can bring powerful forces to bear on you and your organization through their diffuse crowdsourcing method of attacks. Defending against one of their attacks may prove to be more expensive and difficult than if you are being attacked by a disgruntled employee.

Many organizations find they have been attacked by disgruntled employees or former employees who feel the organization has wronged them. Often, it is difficult to determine whether your attacker is inside or outside the organization. Many attackers who hold a grudge or are disgruntled like to advertise why they are attacking their victim. This information frequently leads to clues that reveals who the attacker is, points to how to prevent them from successfully attacking again, and leads to their ultimate apprehension and arrest.

Finding out why you were attacked and who did it also comes into play in assigning liability. Sad as it may be, hacking and other cybersecurity incidents drive financial losses and hassles, and there will be great calls to assign accountability and liability; somebody has to pay for the incident. Understanding why and who attacked may be extremely helpful during adjudication of culpability. For example, let's say that your firm does business in Africa and was recently hacked, causing significant losses. The investigation found that the attack came from one of your IT staff, a contractor from another firm, who was offended by your company's business dealings in one of the African countries and sought to prevent your company from successfully doing business there. The other firm may be liable for the damages caused by their employee, but this information is valuable in other ways. Alerted by the contractor's motivations, as part of your risk management program, you may actively search for others who share this motivation and posture yourself to mitigate risks accordingly.

Finding out why you were hacked and who did it may take a while. In fact, you may never find out who attacked you and why. Nevertheless, you should investigate the facts and circumstances behind the attack so that you can take prudent and reasonable actions to prevent the next one. Be patient yet know when to quit. If your investigation is not showing signs of producing the evidence that leads to answers, don't waste time, effort, and money. Instead, take the information you have and adjust your defenses.

9.3.10 Item 10: Adjust Your Defenses

If you've been hacked, chances are excellent that word is out where you are vulnerable and other hackers will try their hand to penetrate your defenses too. In some regard, hackers are like sharks in the water; when they smell blood, they all flock to the kill. That's why is it essential that you adjust your defenses to remedy the deficiencies that enabled the initial successful hack into your systems and prevent further losses.

Hackers are extremely smart and talented people. Many seek affirmation of their talents and skills through boasts of their conquests in hacker forums and web pages. Some even believe they are performing a public service by exposing your weaknesses to compel you to improve your cybersecurity posture. Many will return to the scene of the crime to see if you've *learned anything* from their initial attack. They will try the same exploits to see if you've taken appropriate measures to remedy the conditions that allowed them to attack you in the first place. If not, they will hit you again and may try to inflict even greater damage.

Your defenses should never remain static regardless of whether you have been hacked or not. Technology changes continually as do the tactics, techniques, and procedures used by hackers and other bad actors who seek to gain access to your information. Your cybersecurity staff should be keeping up to date on the latest threats, vulnerabilities, and defensive techniques to best posture you and your organization to protect your information. Be prepared to continually adjust your defenses.

9.4 FOOT STOMPERS

As previously discussed in Section 7.3.1's "Cybersecurity Training Plan Outline," we like to tell our students to pay attention in class because there will be some foot stomper information that will be on the test. You should pay attention to these following foot stompers regarding issues you will have to address when you get hacked.

9.4.1 The Importance of Public Relations

When you've been hacked, your PR team is needed more than ever. In the aftermath of a hacking incident, confidence in your organization's ability to adequately protect information plummets. Clients, partners, and some shareholders start consulting with their attorneys to discuss potential litigation. Potential investors begin to look at alternative investments. You need to be very careful and deliberate in your public messaging during this period.

Maintaining the goodwill and confidence of your employees, customers, partners, stockholders, and potential investors is critical when you've been hacked. How you lead and manage during this period will be put under the microscope, and people will want to see that you are doing the right things at the right times for the right reasons. Stakeholders at all levels want to see evidence that you are protecting their best interests. They expect you to demonstrate competency and professionalism in dealing with a difficult situation, expect you to demonstrate that leadership *owns* the issue and is fully engaged, and expect you to keep them informed. This is where your PR team earns their keep.

Unfortunately, many companies, even great ones, drop the ball when responding to hacking and cybersecurity incidents. As you build your disaster recovery and business continuity plan to hacking and cybersecurity incidents, you should carefully study the lessons learned from other companies that already have experienced hacking and cyber incidents so that you are well prepared to have the right public message to maintain confidence in your organization.

The 2011 hack of the Sony Corporation's PlayStation Network serves as an example of the importance of timely PR that you should consider as you develop your plan.

On April 19, 2011, Sony learned there was an intrusion on the PlayStation Network used by gamers around the world and shut down the popular system so that it could conduct a thorough investigation. On April 20, they issued the following statement: "We're aware certain functions of PlayStation Network are down. We will report back here as soon as we can with more information." On April 21, a posting on the Sony Europe PlayStation blog suggested the network had been attacked, but the posting was quickly removed. The next day, in response to the hue and cry of numerous gamers and media, Sony issued a short statement acknowledging an external intrusion but offered no details nor information on when the service would be restored. The hacker group, Anonymous, quickly issued its own press release, stating that the group had nothing to do with the attack. On April 25, the computer security experts called in by Sony concluded that consumer data indeed had been breached, but the company decided to hold off on making a public announcement until the next day. When Kaz Hirai, head of Sony's gaming division appeared the next day at a Tokyo news conference to unveil the company's tablet PCs, he expressed condolences and support for the victims of the March earthquake and tsunami, talked about the new tablets, but failed to mention problems with the PlayStation Network. He left the stage without taking the scheduled questions from the press. About 12 hours later, Sony released a statement that confirmed that consumer personal information including names, addresses, birth dates, and other personal information had been compromised. "While there is no evidence at this time that credit card data was taken, we cannot rule out the possibility."[12]

PlayStation Network subscribers were outraged. Not only were they angry over the potential compromise of their personal and financial information, they were furious over how Sony handled the incident. For a week, wild rumors swept the gamer community as the 77 million PlayStation Network subscribers speculated why the network was down, when it would return to service, and whether a hack indeed occurred and how

---

[12] Martyn Williams, "*PlayStation Network Hack Timeline*," May 1, 2011, http://www.pcworld.com/article/226802/playstation_network_hack_timeline.html. Accessed on November 19, 2013.

it may affect their personal and financial information. On the day that the breach was announced, legal action quickly was discussed. "Many states have laws that require notification to individuals if the individuals' information is hacked (and each state's law is slightly different about the how, when, and what of the notification, as well as the effect for failure to notify)," said Andrew Ehmke, an attorney from Texas. "Another place that people may look is the terms of use and privacy policy and whether those were actually complied with by Sony."[13] Even politicians got involved including Senator Richard Blumenthal of Connecticut, who sent a letter to Sony calling for the company to provide network users with free financial data security services, including two years of credit reporting services. Blumenthal declared, "When a data breach occurs, it is essential that customers be immediately notified about whether and to what extent their personal and financial information has been compromised." He criticized Sony for what he called a "troubling lack of notification from Sony about the nature of the data breach," stating, "Although the breach occurred nearly a week ago, Sony has not notified customers of the intrusion, or provided information that is vital to allowing individuals to protect themselves from identity theft, such as informing users whether their personal or financial information may have been compromised."[14]

Midlevel executives at Sony tried spinning their story well, stating that they didn't release the hacking information until they were positive that they indeed had been hacked but the damage was done. Senior-level executives remained noticeably silent and seemingly disengaged, with the senior director of corporate communications and social media being the principal link between Sony and the public. Users were already sensitive to an April 4, 2011, anonymous-affiliated DDoS attack of the PlayStation Network (just three weeks before!) in retaliation for Sony's lawsuit against hacker George Hotz, who discovered the internal password to *jailbreak*[15] the PlayStation 3 and posted the password online. Imagine the impact on consumer confidence a week later on May 2, 2011 when Sony pulled the plug on the network linking players on the massive multiplayer games when they found that it too had been breached, raising the total number of compromised accounts to over 102 million.[16]

Sony found itself wearing a cyber *Kick Me* sign as hackers circled the Sony networks like sharks smelling blood, yet on June 1, 2011, Sony relaunched the PlayStation Network to a suspicious consumer base eager to resume their gaming. Offering free games, movies, and other incentives to woo back subscribers who were leaving in droves, Sony looked forward to resuming full operations. Unfortunately, later that day, it was discovered that the Sony Pictures web site had been hacked by the LulzSec hacker

---

[13] Patrick Klepek, "*PSN hacked: What Sony's Security Breach Means for You (And What Comes Next)*," April 27, 2011, http://www.giantbomb.com/articles/psn-hacked-what-sonys-security-breach-means-for-yo/1100-3092/. Accessed on November 11, 2013.

[14] Jason Schreier, "*PlayStation Network Hack Leaves Credit Card Info at Risk*," April 26, 2011, http://www. wired.com/gamelife/2011/04/playstation-network-hacked/. Accessed on November 11, 2013.

[15] Jailbreak is a term commonly used to describe modifying the operating system of a device to remove restrictions installed by the manufacturer.

[16] Matt Liebowitz, "*Cybercrime Blotter: High-Profile Hacks of 2011*," February 24, 2012, http://www.technewsdaily. com/6644-websites-hacked-government-commercial-cybercrime-2011.html (posted with a typo incorrectly listing the publication date as February 24, 2011). Accessed on November 11, 2013.

group, who bragged how easy it was to access over a million usernames and passwords in the Sony system, calling Sony's security *disgraceful* and *insecure*. As a result of this and previous episodes, the U.S. House of Representatives Energy and Commerce Committee launched an investigation into Sony's data security and criticized Sony for waiting a week before they confirmed the PlayStation hack to the public in April.[17]

The hacking had profound impact on Sony and its bottom line. Sony executives acknowledged that the hack cost the company an estimated US $170 million. The hack also shook the confidence of investors, dramatically affecting Sony's competitiveness in the market, as evidenced by their stock prices. On April 1, 2011, days before the first incident, shares of Sony stock were trading at US $31.87 per share. By the end of 2011, shares were traded at US $18.04 per share, a 43% drop in value. As of this writing, Sony has yet to recover its previous value, and on the eve of the release of the PlayStation 4 gaming console, the stock is trading at US $16.74 per share, a 47% drop in value from the prehack levels.[18]

The gaming community, which represents hundreds of millions of people around the world, continues to cite Sony's response to the April 2011 hack as an example of how not to respond to a hacking incident. Sony lost consumer confidence as many gamers felt that Sony was placing profit above loyalty to their subscribers, and their PR efforts were too slow, incomplete, and perceived as disingenuous. It remains to be seen whether the launch of the new PlayStation 4 will help the company rebound.

While Sony's hacking experience touched subscribers around the globe, even local companies can feel the sting of hackers. The 2013 hacking experience of Schnucks, a St. Louis, Missouri-based grocery chain, also gives valuable lessons about the importance of maintaining good PR during a hacking incident.

On March 15, 2013, Schnucks was notified by a major credit card firm that it noticed that 12 cards used at Schnucks stores had been used to commit fraud. An internal Schnucks investigation was launched and quickly ruled out employee misconduct and any tampering of point-of-sale devices. Within days, more incidents of fraud were reported, and Schnucks created an incident response team on March 19. They hired Mandiant to perform a forensic investigation and contacted law enforcement. By March 28, Mandiant had identified the malware used by the attackers, and two days later, Schnucks had contained the malware and publicly admitted the attack with a statement by Schnucks CEO Scott Schnuck, who apologized for the incident, stating in a press release that Schnuck's would be *relentless* in maintaining security.[19]

Like the Sony subscribers, Schnucks customers were concerned about the integrity of their personal and financial information and upset over the perceived lack of transparency and slow notification by Schnucks that customer information may have been

---

[17] Investopedia, "*Most Costly Computer Hacks of All Time*," July 20, 2011, http://www.investopedia.com/financial-edge/0711/most-costly-computer-hacks-of-all-time.aspx. Also see Ed Oswald's PCWorld report, "Sony Gets Hacked Again and Again, Pilfered Data Released," June 6, 2011, http://www.pcworld.com/article/229520/sony_gets_hacked_again_and_again_pilfered_data_released.html. Accessed on November 11, 2013.

[18] Ibid.

[19] Brian Prince, "Schnuck's Grocery Store Hack Details Emerge," April 17, 2013, http://www.darkreading.com/attacks-breaches/schnucks-grocery-store-hack-details-emer/240153091. Accessed on November 11, 2013.

compromised. Like the Sony customers (many of whom shop at Schnucks), many felt they should have been notified earlier so they could have taken proactive steps to protect their personal and financial information.

The incident got widespread attention in the Schnucks market area around St. Louis. Local television personality and radio host Charles Jaco noted that by mid-April that Schnucks had refused requests to make executives available to the press for comment and had communicated solely through emails and press releases. When pressed for information on the depth and breadth of the attack, Schnucks' response was, "no comment." Jaco's comments were emblematic of the consumer thoughts, "Schnucks responses seem to be orchestrated by the Kremlin. And that lack of transparency and accountability is puzzling. No one is blaming Schnucks for the hack attack. Cyber attacks are a way of life nowadays. Schnucks is being blamed for what seems indifference to or disregard for their customers."[20]

Weeks later, the company issued a flood of apologies including television commercials featuring CEO Scott Schnuck, but the damage appeared to be done as some customers said it was too little, too late. Several class action lawsuits were filed against Schnucks, and the overall cost of the attack is estimated to possibly reach US $80 million in Illinois alone, with Missouri litigation pending.[21]

Both Sony and Schnucks are great companies that produce terrific products, but they failed in the consumer's eyes to correctly respond to hacking incidents; their PR efforts failed. You should already have a PR plan ready to go in the event that your organization is the victim of a hacking incident. Include the following lessons learned from Sony, Schnucks, and others who have been hacked:

- Inform those affected immediately: Consumers consider it inexcusable when you delay notification that their personal and financial information is potentially at risk. Even if there is a suspicion that information is compromised, those affected want to know that you are attempting to look after their best interests by notifying them promptly so they can take appropriate action.

- Get yourself a world-class public relations consultant: As we pointed out in Section 3.4.3, even though you may have very capable in-house PR staff, their normal duties are akin to a medical general practitioner. What you need for a significant cybersecurity intrusion is a specialist. There are several highly competent firms and individuals that are experienced in dealing with such events. Do yourself a favor by including them in your disaster recovery and business continuity planning and provisioning them to launch into action when you need them the most.

- Get senior leadership out in front: Neither Sony nor Schnucks senior executives appeared to be fully engaged in leading the response to the hacks. While they may have been up to their eyeballs handling the events, the perception was that they

---

[20] Charles Jaco, "*Jacology: Schnucks Handling of Security Breach,*" April 10, 2013, http://kplr11.com/2013/04/10/jacology-schnucks-handling-of-security-breach/. Accessed on November 11, 2013.

[21] Georgina Gustin, "*Schnucks Credit-Card Breach Could Cost $80 Million Just in Illinois,*" May 21, 2013, http://www.stltoday.com/business/local/schnucks-credit-card-breach-could-cost-million-just-in-illinois/article_25804620-9ca5-5d0b-8b1c-1968a56012eb.html. Accessed on November 11, 2013.

were disengaged and the matter was not important enough to warrant their attention. Had the CEOs of Sony or Schnucks immediately taken ownership for the issues, made themselves available for interviews to update the public, and been transparent through the incident response, public perceptions may have been different.

- Be prepared to offset impacts on consumers: Many consumers are concerned about the impacts of hacking incidents on their personal and financial information. Offering to underwrite the costs of credit checks and security measures is often seen as a goodwill gesture that may avert lawsuits seeking damages. As a result, this measure may save you money in the long run.

- Keep those affected regularly updated and informed: While your CEO should be visible during hacking incidents with apologies and pledges to protect the consumer's interests, the organization's authorized spokespeople ought to be giving regular updates to the public and press during the recovery and remediation process. Transparency is highly valued and builds trust.

Your PR in the aftermath of a hacking incident is crucial to your bottom line. Most people have come to the realization that hacks are becoming a way of life. When they conduct business with your organization, they provide their personal and financial information to you as part of a transaction, and they expect you to be responsible custodians of that information. When you are the victim of a hack, that responsibility does not evaporate; it magnifies. Your PR effort must demonstrate that you are acting in a manner that relentlessly attempts to protect your consumers. You must never be portrayed as being blamed for *indifference to or disregard* for your customers. Get yourself that world-class public relations consultant.

9.4.2 Working with Law Enforcement

Hacking is criminal activity that should not be tolerated. It threatens the economic well-being of organizations and nations and will continue if not arrested. We believe that whenever you are the victim of a hacking incident, you should notify law enforcement officials and press charges so that the perpetrators can be apprehended and held accountable.

Making timely notification of hacking incidents to law enforcement officials protects your legal remedies in the aftermath of an attack and can help minimize the damage. Law enforcement officials now have access to increasingly sophisticated technologies and procedures that may reveal who attacked you, how they did it, and how to protect you from reattack. Importantly, calling in law enforcement officials early helps preserve evidence and makes sure that all investigative avenues are fully explored.

In the United States, law enforcement officials at the national, state, and local levels have varying degrees of capabilities to deal with computer crime. Typically, local police have limited capabilities, while state and national agencies have more robust investigative and forensic capabilities. Jurisdiction for the crime is dependent upon a variety of factors including who was involved (i.e., the victim and the perpetrator) and what was the crime (e.g., damages from hacking.). A state or local computer crime task force may be a more appropriate investigating organization for those cases that do not meet federal computer crime thresholds. As such, we recommend your first call should go to your local police department with the recognition that other law enforcement agencies later may become involved as details emerge.

Some people and organizations admit they do not have confidence in their local police department's ability to handle computer crime, especially those who are located in smaller communities. If you are one of these organizations, you may want to consider making notification of your attack to the Internet Crime Complaint Center (IC3), which is a partnership between the FBI, the National White Collar Crime Center, and the Department of Justice's Bureau of Justice Assistance. The IC3 will take your complaint and refer it to the authorities best aligned to handle your case. They encourage victims to make their reports through their web site at www.ic3.gov.

Outside of the United States, most countries also have established national computer crime task forces and procedures. For example, Canada has established the Canadian Cyber Incident Response Centre (CCIRC) to coordinate responses to cyber incidents. In cases where affected organizations believe a computer crime has been committed, the Canadian government recommends that the alleged victims contact their local law enforcement authorities. In cases where the alleged victim believes national security is at risk, they are instructed to contact the Canadian Security Intelligence Service. The CCIRC is often called in to review incidents and provide advice on whether to contact law enforcement or national security authorities.[22]

In the United Kingdom, the National Cyber Crime Unit (NCCU) was established to leverage national technical and law enforcement specialists in partnership with local law enforcement officials to respond quickly to address cyber threats and crimes. The NCCU addresses the most serious incidents of cybercrime and assists local law enforcement officials. As with the United States and Canada, British victims of computer crime are encouraged to notify local law enforcement officials to initiate the law enforcement investigative process.[23]

Other countries, such as Australia, France, Germany, and Japan have established similar constructs governing the involvement of law enforcement in the investigation of cybercrimes. For those organizations that operate in multiple countries, it is advisable to remember that every country has its own laws governing cybercrime. Normally, the law enforcement organizations of the countries where the perpetrator and victim are located will coordinate for any investigation and prosecution actions.

When to call law enforcement officials is a conscious decision. Many people and organizations elect not to inform law enforcement officials when they have been the victims of hacking and other computer crime. Most do so as they believe that the negative publicity resulting from law enforcement investigations and public disclosure will cost more than writing off the damages from the incident. While this is an understandable response, in many ways it is not socially responsible because it permits the perpetrator to continue to pursue their nefarious hacking activity. As such, we always encourage our

[22] Cyber Incident Management Framework for Canada, http://www.publicsafety.gc.ca/cnt/rsrcs/pblctns/cbr-ncdnt-frmwrk/index-eng.aspx#_Toc360619103. Accessed on November 11, 2013.

[23] United Kingdom National Crime Agency, http://www.nationalcrimeagency.gov.uk/about-us/what-we-do/national-cyber-crime-unit. Accessed on November 11, 2013.

clients to maintain their integrity, report criminal activity, and press charges when the perpetrators are apprehended. We believe in the long run that doing the right thing pays off positively as such actions are viewed by clients, partners, shareholders, and potential investors as appropriate, while failing to report criminal activity is viewed as complicity or condoning it. Furthermore, it conveys an image of weakness.

9.4.3 Addressing Liability

Earlier in this book, we addressed cybersecurity insurance and recommended you invest in a policy that best protects you and your organization based on your risk appetite. As Kevin Mitnick reminds us, hacking causes financial losses and hassles. In the aftermath of a hacking attack or similar cybersecurity incident, you will not have the opportunity to buy insurance retroactively, so you have to plan for the worst up-front and provide for sufficient insurance to offset your losses and minimize your risk.

Unfortunately, even insurance will not make the hassles associated with liability completely go away. You can reasonably expect numerous sources will call for accountability after a hacking incident; someone will have to pay.

Insurance is one of the most popular methods of addressing liability. Many companies offer specific cybersecurity insurance that will protect you and your organization in the event that you fall victim to hacking or other cybersecurity incidents. These tailored policies are a wise investment when the potential loss associated with such incidents is high and introduces unacceptable levels of risk; insurance transfers the insured risk to the insurer. However, be very careful with your insurance. Many people believe their general liability insurance will protect them against losses associated with hacking and cybersecurity incidents and are shocked to find that may not be the case.

As an example, recall our discussion of the Schnucks grocery store hacking incident detailed in Section 9.4.1. Facing damages in excess of US $80 million, Schnucks turned to its insurer, Liberty Mutual Insurance Company. Liberty Mutual, in court filings, said that Schnucks had a general liability policy that covers bodily injury and property damage but was not designed to insure against suits and claims associated with a data breach.[24] Hence, unless Schnucks is able to make some sort of agreement with its insurer, they will be liable for all damages associated with the breach.

Schnucks is not alone in finding that general liability insurance policies often do not cover the losses associated with hacking and other cybersecurity incidents. In fact, Sony currently is engaged in legal proceedings with its insurer, Zurich American, who says that it is not liable for customer lawsuits arising from the 2011 PlayStation Network hacking incidents cited in Section 9.4.1.[25]

You should expect that you will be attacked and your defenses breached. Therefore, you should make provisions to be properly insured. As with any type of insurance

[24] Danielle Walker, *Insurer to Schnucks: We Won't Pay for Lawsuits Related to Your Breach*," August 20, 2013, http://www.scmagazine.com/insurer-to-schnucks-we-wont-pay-for-lawsuits-related-to-your-breach/article/307960/. Accessed on November 11, 2013.

[25] Kari Timm, *Zurich Denies Coverage for Sony PSN Hacking Claims & Files Suit*," August 12, 2011, http://www.cyberprivacynews.com/2011/08/zurich-denies-coverage-for-sony-psn-hacking-claims-files-suit/. Accessed on November 11, 2013.

policy, read the fine print and know what your policy covers, as well as what it doesn't, *before* disaster strikes. Jason Weinstein, a partner at New York-based Steptoe & Johnson who specializes in data privacy and security litigation, says, "A lot of these policies were written before anyone knew what a data breach was. It's better to take a proactive review of your insurance coverage and confirm it's adequate before you need it."[26] That's good advice that everyone should act upon.

9.4.4 Legal Issues to Keep an Eye On

You should expect to get hacked. You also should expect to be sued.

Lawsuits in the aftermath of hacking and other cybersecurity incidents have become a fact of life. You and your organization should be well prepared to address lawsuits that will be filed shortly after the incidents.

Many lawsuits are filed alleging that due care and due diligence were not performed by your organization, thus permitting hackers to penetrate your defenses, put valuable client and organizational information at risk, and cause you to expend unplanned resources to remediate or repair the damages. These lawsuits often are filed by shareholders who believe their investments have been compromised by the hack and associated damages. Plaintiffs seek a measure of compensation for their perceived and actual losses. Such allegations will attempt to portray you and other executives as incompetent, ignorant, or worse. You will need thick skin and a good attorney to help you through the ensuing legal process.

You may expect other lawsuits directed at the corporate board itself alleging that the board did not institute sufficient oversight to ensure that adequate controls were implemented to protect the organization and its resources. If you are a member of the organization's board of directors, you already should have an insurance policy that protects you against liability incurred in your duties as a board member. Make sure you protect yourself by confirming that your policy covers the costs of lawsuits associated with hacking and cybersecurity incidents.

Don't be surprised if your business partners suddenly sue you for damages in the event of a hacking or cybersecurity incident. If they can prove that hackers who breached your defenses were able to leverage the trusted relationship between your two organizations to gain access to their networks and information, they may be able to prove that you and your organization are culpable for any damages wrought on them, their clients, and their associates. That is why it is important not only to pick your partners wisely but also to be very careful with whom and how you share information. Maintaining connections between networks using techniques referred to by technical personnel as *trusted relationships* permits information to flow freely between two network entities. Such an arrangement is considered very convenient by users and partners who share resources. To a hacker, this type of relationship is an information superhighway into your partner's network. Many great partnerships have been poisoned by one partner not adequately protecting against hackers and exposing both organizations to attack.

---

[26] Ibid.

Lawsuits also mean that there will be legal discovery and associated requests for information from the legal team suing you. We recommend that you carefully preserve all records associated with your information systems including your baseline hardware and software configuration records, all audit and system logs, and training records. Expect that technical experts will comb through all records to ensure that they are complete and that you and your staff indeed have practiced due care and diligence in the management of your information systems. Incomplete or missing records will put you at an extreme disadvantage and in some people's eyes is the equivalent of a confession of guilt. Even the very best-managed IT organizations can be hacked. When you can prove through your records that you and your organization did everything you could and should have done to protect your information from hacking and other cybersecurity incidents, you are best postured to protect yourself against lawsuits.

When you are the subject of a lawsuit in the aftermath of a hacking or cybersecurity incident, pick your counsel carefully. Your general counsel may be terrific in advising you on your particular business functions but may be a detriment in a court arguing the merits of cybersecurity law and culpability. Just as in selecting specialized PR experts, you will find that hiring a lawyer or firm that specializes in defending corporations in the aftermath of hacking or cybersecurity incidents is a wise investment.

We recommend that you don't wait until you are the victim of an incident to scramble around in the heat of battle to find the lawyer or firm you want to represent you when you have been victimized by a hacker. Ask your insurance company if they have any recommendations. Have your general counsel research who are the most successful defenders in known cybersecurity cases and consider interviewing candidates to see if they are a suitable match to represent you and your organization. If you find a good match, consider exploring options to put them on a limited retainer. You will find that legal specialists are invaluable in helping draft your disaster recovery and business continuity plan to safeguard your best interests, employ best practices to minimize risk exposure, and ensure that you emerge from a hacking incident with minimal damage.

9.5 FOOL ME ONCE...

Again, it is inevitable that you will be hacked, but that does not mean that your organization will be destroyed and you will face personal ruin. But it does mean that your organization *could* be destroyed and you *could* face personal ruin, if you do not prepare properly. Hacking and other cybersecurity incidents will cause financial losses and hassles for your organization, so you need to be prepared.

Executives are responsible to plan for the future, and your future includes the likelihood that a hacker, a hacktivist, an agent of a nation state, a disgruntled employee, or even a careless employee will cause you or your organization to be the victim of a cybersecurity incident. It is essential that you have a disaster recovery and business continuity plan that addresses what you and your organization will do in the event you are hacked or suffer some other form of cybersecurity incident.

We have seen a few organizations who utterly failed in their response to hacking incidents. In some of the most egregious cases, the victim organization was hacked and failed to fix the deficiency that allowed the hacker to enter in the first place. As a result, the hacker and some of their colleagues continued to exploit the deficiency and expose the organization to continued and intensified risk. We believe such instances indeed indicate a failure to maintain due care and due diligence. If you are hacked, you should fix the deficiency and prevent the hacker from ever exploiting you using that method ever again and do it promptly and expeditiously.

How you respond to a hacking or other cybersecurity incident says a lot about you and your organization. We recommend that you always be honest and forthright; that you have a plan; that you protect your shareholders, customers, and key stakeholders; and that you act quickly and deliberately in a transparent manner. As Charles Jaco said when referring to the Schnucks hacking incident detailed in Section 9.4.1, "No one is blaming Schnucks for the hack attack. Cyber attacks are a way of life nowadays. Schnucks is being blamed for what seems indifference to or disregard for their customers."[27] Cyber attacks indeed have become a way of life nowadays, and while you must do your best to prevent them, you must have a plan on what to do when they do occur. Some people will tell you that while the sin of exposure is bad and can be forgiven, the sin of indifference is much worse and will never be forgotten.

9.6 SUMMARY

- You will get hacked. Plan on what you will do when it happens.
- Hackers are criminals who will use the path of least resistance to deface and/or steal your information.
- You and your organization are under surveillance by hackers right now. They are looking at:
  - The type and strength of your physical security measures
  - The type and strength of your network security measures
  - The composition and configuration of your computers, network devices, and software
  - You and your people to see whether you practice proper security
- Your best defense against hacking is not only to defend against it with a cyber-hardened workforce but also to prepare for what to do when you do get hacked.

---

[27] Jaco, Ibid.

- You can minimize the impact of hacking incidents by doing the following before you are attacked:
  ○ Back up your information.
  ○ Maintain a current baseline.
  ○ Protect yourself with insurance.
  ○ Create a "Disaster Recovery and Business Continuity Plan" for hacking and cybersecurity incidents.
- When you are the victim of hacking or cybersecurity incidents, implementing the following ten *things to do* checklist will best protect you and your organization's interests.
  1. Don't panic.
  2. Make sure you've been hacked.
  3. Gain control.
  4. Reset all passwords.
  5. Verify and lock down all your external links.
  6. Update and scan.
  7. Assess the damage.
  8. Make appropriate notifications.
  9. Find out why it happened and who did it.
  10. Adjust your defenses.
- There are some key *foot stompers*[28] you must pay attention to:
  ○ The importance of PR.
    – Inform those affected immediately.
    – Get world-class PR help.
    – Get senior leadership out in front.
    – Be prepared to offset impacts on consumers.
    – Keep those affected regularly updated and informed.
  ○ Working with law enforcement:
    – Hacking is criminal activity and should not be tolerated.
    – Whenever you are the victim of a hacking incident, you should notify law enforcement officials and press charges so that the perpetrators can be apprehended and held accountable.
  ○ Addressing liability:
    – Many insurance companies offer cybersecurity policies that can protect you in the event you are hacked or suffer other cybersecurity incidents.

---

[28] *Foot stompers* are things that you need to pay attention to because you'll see them again on the professor's next test.

- Consider investing in cybersecurity insurance.
- Not all general liability insurance policies cover hacking and other cybersecurity breaches.
- Make sure that you conduct a thorough review of your insurance policies to make sure you are adequately covered against hacking and other cybersecurity incidents.

◦ Legal issues to keep an eye on:
  - You will get hacked and you will get sued.
  - Carefully preserve all records, especially logs, for the legal discovery phase of lawsuits.
  - Not all lawyers are adept at cybersecurity issues. Identify which lawyer is best prepared to defend you in the event that you are sued and consider putting them on retainer.
  - Never allow yourself to be victimized by the same vulnerability more than once after it has been detected.

10.0

BOARDROOM INTERACTIONS

Scene: Dr. Charles Clark, chairman of the board of Kilcawley Chemical Corporation, has convened a special meeting of the corporate board of directors in order to address the topic of cybersecurity. Kilcawley Chemical, a publicly owned company traded on the NASDAQ exchange, manufactures specialty chemicals, principally polymers and surfactants. It has plants in Eltopia, WA, and Middleburgh, NY. Corporate headquarters is in Cobleskill, NY. Its annual revenues have averaged about US $3.5 billion over the past three years. The board members are:

Dr. Charles Clark—emeritus professor of chemical engineering at Rensselaer Polytechnic Institute and an early investor in the company

Mark Austin—CEO of Kilcawley Chemical

Mary Francis—senior vice president of the Makamuny Investment Fund

Hortense Jefferson—vice president and director of public relations, the Basil Jaskowski Foundation

Harry Artur McGarry—CEO of the Fermoy Health System

Robert Shay—retired chairman of WTCX Corporation (an international manufacturer of water treatment chemicals)

Cybersecurity for Executives: A Practical Guide, First Edition. Gregory J. Touhill and C. Joseph Touhill.
© 2014 The American Institute of Chemical Engineers, Inc. Published 2014 by John Wiley & Sons, Inc.

Harold Sockem—managing partner, Roccim & Sockem, LLP (a law firm specializing in innovative technology matters)

Also attending the meeting at the chairman's request is **Keith Willis**, Kilcawley Chemical's CFO.

Charles: The reason I convened this special meeting is to address what Mark and I believe to be a crucial issue that could have tremendous impact upon our company. That issue is cybersecurity. About a month ago, Mark called to tell me that he just finished reading a great book entitled *Cybersecurity for Executives: A Practical Guide*. He asked me to read it because frankly he said it scared the heck out of him. I read the book and believe that we, as a board, have to take very seriously what the book says, so I have purchased copies for each of you and urge that you read it thoroughly and soon. Several of you have said that you already have finished the book. Good, that will help in making this meeting move faster.

The issue is so important, however, that rather than wait until you finish the book, I want to focus on the subject here today at this special meeting and have Mark share his impressions on what the book says and how we as a board have to react.

Mark: Thank you, Charles. It's always a pleasure to interact with the board because I find your input to be extremely helpful and clearly in the best interests of our shareholders. The topic of cybersecurity can be complex if viewed merely from a technical perspective, but from a business perspective, the conclusions are simple. First, everybody, no matter how sophisticated their IT systems and networks are, is under attack from greedy and opportunistic people who want to steal from us or harm our business. Second, these bad people at some time will penetrate our defenses. There is no way that we can be 100% risk-free. Third, we need to spend money to adopt a risk management strategy and plan to defend ourselves and to manage intrusions when they occur. Fourth, we have to let our shareholders know what we are up against. Fifth, we need you to consider and approve the actions that we expect to take to protect the assets of Kilcawley Chemical.

Robert: Could you give us an idea of where we stand now with our cybersecurity and IT protection program?

Mark: Robert, we have a good program that some industry people would regard as "acceptable." It is managed by Dan Kozusko, our CIO, who is an able executive as well as a skilled computer scientist. But based upon what you will read in the book that Charles has given you, we need to upgrade our program to minimize our risk profile. I intend to give you an overview of where I want to head with a rejuvenated cybersecurity approach. Nevertheless, while we currently have an approach that many believe is adequate, I believe that we have to do a great deal more to manage our risk.

Mary: Mark, do you have an idea of the magnitude of risk and threat that we face?

Mark: Mary, the potential risks to our business are immense. Hackers at this very moment are trying to penetrate our information systems. We know that based on the number of scans our network sees every day and the increasing number of sophisticated attempted accesses to our systems. They conceivably could take control of our processing and operating systems and effectively shut us down. They could interfere with our chemical formulations, thus tainting our products and exposing us to liabilities from damaged customers. If they penetrate our defenses, it is possible that they could ruin our reputation, expose us to immense liabilities and lawsuits, and drive us into bankruptcy.

Harold: You sure have my undivided attention!

Harry: All of you should be aware that the Fermoy Health System recently was attacked and the records of 100,000 patients were compromised, so I have personal and painful experience related to the malevolent work of some very bad people.

Charles: I don't want to be a pain by interrupting questions, but I am eager for Mark to take us stepwise through his approach to minimizing risks to our company. He shared with me a program that makes a great deal of sense, but it's going to cost money and the board will have to make some important decisions on the strategy Mark recommends. So try to hold your questions until Mark finishes laying out the essence of his program. Mark, have at it.

Mary: Charles, I understand your admonishment about questions, but I have got to ask before we move on (I'm sorry, Charles). Mark, in the last board meeting, you reviewed with us your preliminary conversations with Hawk Polymers about a potential merger or acquisition. What impact will our "new" cybersecurity program have on discussions with Hawk Polymers?

Mark: Excellent question, Mary. Recently, many M&A negotiations have been acutely affected by IT considerations. Nobody, neither us nor Hawk Polymers, wants to be sucked into a morass of liabilities due to breaches or intrusions into our sensitive data or operating systems. Just imagine if we decide to acquire Hawk and after the deal was completed we find out that the formula for their proprietary major product was stolen by a foreign company who was busily cranking out flawed products that are initially indistinguishable from the Hawk "formerly" secret formula. Furthermore, suppose these flawed products were plagued by questions of toxicity. No doubt that there would be lawsuits galore and the reputations of both companies would be destroyed. Do we have to be supercareful? You bet!

Mary: Thanks Mark. By the way, how is the exploration with Hawk Polymers going?

Mark: Not bad, Mary, although everything remains in the preliminary stage. Anyhow, let's get back to the layout of our cybersecurity program. I will make my presentation supported by some PowerPoint slides. I'd like to emphasize that even though we have done a lot of work on the various steps involved in the program, there's a lot left to do. For example, many of the steps require formal board approval; others, while not strictly needing board approval, are very sensitive, and thus, we would like the board's concurrence prior to moving ahead with implementation. Finally, we believe that it is in the company's best interests that the board be fully aware of our plans before implementation so that there are no misunderstandings or (forgive me for saying so) second guessing after the fact. Some decisions possibly can be made today, but we recognize that the board may correctly reserve judgment until they have had adequate time to study the cybersecurity book by Touhill and Touhill and to digest the documents we will hand out today. So let's begin with **Slide 1**.

KILCAWLEY CHEMICAL CORPORATION

1: Risk Management

- How much risk do we want to take?
- The "nuclear model" – the maximum credible accident
- The cost in terms of assets and reputation
- The incremental costs over where we are now
- Requires board approval

Pre-decisional Information; Not for Public Release

Cybersecurity experts point out that the likelihood of having our information systems invaded is essentially 100%. In his book on cybersecurity, General Touhill states, "If it is connected to the Internet, your system is exposed to countless risks. Cybersecurity is primarily about risk management. Minimize your risk through smart practices. Ensure your systems are properly configured, patched, and audited. Ensure your workforce is trained and regularly tested. Make cybersecurity part of your daily business practices."

I agree with General Touhill. Because we cannot eliminate risk, we have to focus on how we mitigate and minimize the risk of cyber intrusions. Moreover, because there are a range of remedies that we can take, we have to decide "how much risk do we accept?" Here's how my staff and I approached how to determine our risk appetite. We used a model that has been used for years in the nuclear industry. From the beginning, nuclear reactors were designed to anticipate the "maximum credible accident," for example, a large airplane crashing into the reactor containment vessel, an earthquake of significant magnitude close to the reactor, explosives planted by terrorists, or a similar catastrophic event.

So for us at Kilcawley Chemical, after considering a range of "maximum credible accidents" involving our data and information management systems, my staff and I decided that the worst things that could happen to us are threats and intrusions into our plant equipment and operating systems. Bad actors who compromise our equipment (pumps, mixers, valves, chemical feed machines, formulation tanks, hydraulic and piping systems, controllers) could cause us untold trouble. For example, you all are aware that we have an inventory of chlorine tanks at the plant. If they somehow fail, the release of significant amounts of chlorine gas could be devastating not only to our plant personnel but also to the surrounding community.

Clearly, there are many other vulnerable areas, and we cataloged them, but compromised processing equipment topped the list. Presently, we are investigating mitigation plans for this scary threat. One of the alternatives we are looking at involves isolating process control systems via an "intranet" that is physically and logically separated from the Internet.

We have and continue to examine what the cost to our assets and reputation would be if our processing control systems failed. Preliminary estimates run into the tens of millions (at least) and the potential exists for a permanent besmirching of our reputation as an industry leader. At our next meeting, we will have firm estimates of the magnitude of costs. In addition to these estimates, we will compare how much we are spending now for cybersecurity protection with these new estimates.

Charles: Mark, based on the last item on your PowerPoint **Slide 1**, do you really expect the board to formally approve some sort of action today?

Mark: No, Charles. I simply wanted to point out that when we bring to the board all of the facts and estimates, plans, and programs, there are major decisions that will have to be ratified by the board before my staff and I can move to the implementation of plans and programs.

Hortense: How far along are you in putting the required information together?

Mark: A lot already has been done; we expect to finish major elements of our strategy within two weeks. On the other hand, formulating detailed policies and procedures probably will take two or three months. But frankly, that will not be an impediment to getting the essence of the cybersecurity upgrade going. I'd like Keith Willis to comment on the finance and investing aspects of the upgraded cybersecurity program. Keith?

Keith: Thank you Mark. Currently, my staff and I are working on two levels of estimates of the financial costs of addressing the upgraded program and estimates for damage potential. The "satellite" level is an order of magnitude estimate and is intended to give the board a basis for making "first-cut" judgments. These will be completed by our next board meeting in a few weeks. The detailed estimates are contingent upon the specific plans developed by management and could take a couple of months. Some of these detailed estimates are crucial. For example, we will be approaching our liability insurance carrier to gain coverage for the costs of cybersecurity incidents. Rather than have the carrier simply guess a worst-case scenario, we want to be in a position where we can provide them with our assessment of the probability of a cyber event and our estimate of the dollar magnitude of damage. That way, we have a better degree of cost control.

Let me explain that from my perspective the really important cost figure is the one that defines how much we have to spend to minimize the odds of a catastrophe, and not the catastrophe itself, and then compare that number with what we are spending now.

Charles: Keith, do you have an idea of what we are spending on cybersecurity now?

Keith: Yes, I do Charles. Last fiscal year, the corporation spent US $6.4 million on measures to protect the information resources of the company.

Charles: Thank you Keith. Mark, we have a lot to cover, so let's move on.

Mark: Please focus your attention on **Slide 2**.

KILCAWLEY CHEMICAL CORPORATION

2: Strategy

- Strategy amendment and enhancement
- Revisit and revise the corporate strategy
- Revise mission and core value statements

Pre-decisional Information; Not for Public Release

In his book, Touhill says, "We contend there is only *one* strategy in your business, not an amalgamation of disparate strategies acting in concert or potentially in competition. Everything else, including marketing, investments, engagements, and even cybersecurity is done in support of the overall strategy." We agree with him. So our approach is not to build a separate strategy for cybersecurity, but rather to amend and enhance our existing strategy to accommodate the new demands of protecting our data and information assets and resources. Simply put, we are not going to change who we are just because of some bad actors; we are going to stay with our fundamental vision and do it better and more securely.

You as a board, together with your predecessors, spent much time and thought in developing and formulating a strategy for our fine company. Far be it from me to stray from that successful vision. It is my intent and that of my team to augment the corporate vision so that it will be protected from harm and diminishment. We are in the midst of completing that enhancement and will have it in your hands at the next board meeting. In conjunction with that effort, we also will have for your consideration updated mission and core value statements.

I want to acknowledge the help and advice that I received from two board members in thinking through the updates, revisions, and enhancements so far. Harry McGarry and Robert Shay both shared their considerable wisdom with me and my staff. They were especially helpful in the formulation of goals and objectives, the subject of **Slide 3**.

KILCAWLEY CHEMICAL CORPORATION

3: Formulate Goals and Objectives

- Strategic Level

- Operational Level

- Tactical Level

Pre-decisional Information; Not for Public Release

At the outset of our three-way conversation on this critical topic, Robert reminded me of the sage aphorism, "Directors direct, and management manages." He said that the board acts at the strategic level and the board expects me to successfully perform at the operational and tactical levels. Clearly, I understand that, but as I told them, their wisdom and accomplishments help to make my job easier. With their help, I am in a position to share finished strategic goals and objectives at the next meeting; a complete outline of operational goals and objectives will be presented at the next meeting (with a finished product within a month); and we will offer a first pass at when to expect completed tactical goals and objectives. Touhill states in his book that "Goals and objectives are at the heart of planning." I agree with that completely.

Any questions to this point? Seeing that there are none, I will continue with my presentation. Please direct your attention to **Slide 4**.

4: Policy and Procedures

- Make them lucid and unambiguous
- Clearly communicate to everybody
- Emphasize their importance
- Provide oversight and accountability

Pre-decisional Information: Not for Public Release

After amending and enhancing the corporate strategy; revising the vision, mission, and core value statements; and formulating goals and objectives (all related to addressing cybersecurity), the next step in our program is to promulgate relevant policies and procedures. We are eager for them to be clear and unambiguous. There should be no mistake on the intent. We will take our time in preparing them because of their importance. I said earlier that it will take two to three months to complete this task. Obviously, the policies drive the procedures, so these will be prepared first.

Once completed, we will communicate the policies and procedures to all employees. Additionally, where appropriate, we will distribute appropriate sections to our suppliers and contractors. In doing so, we will emphasize their importance and urge all who receive the policies and procedures to take them very seriously, because we surely will. To emphasize our seriousness, we will audit compliance, provide continuous oversight, demand accountability, and, where necessary, impose sanctions upon those who violate the rules.

Robert: Mark, we have talked recently about other essential programs that had a major impact on how we do business and how they share similar characteristics with cybersecurity risk management. Please relate to the rest of the board what we discussed.

Mark: Robert, your timing is impeccable. You have led into the material on the next slide, **Slide 5**.

5: Make Cybersecurity Risk Management Part of the Corporate Culture

- Safety
- HAZOPS
- Environmental considerations
  - Waste minimization
  - Recycling
  - Material acquisition

Robert and I kicked around the idea that it might be difficult to inculcate the concept of cybersecurity to keep the corporation "safe." Well, as it turns out, "safety" was a keyword. It sort of dawned on us both at the same time. "Hey! Kilcawley Chemical has a safety record that is the envy of the industry. Our employees are proud of that record, and are exceedingly happy to work in an environment that is essentially accident-free." Robert and I concluded that the success of our safety program should be a model of how the cybersecurity program should be implemented and embraced by our employees. As we congratulated ourselves on our brilliant insight, it occurred to us that there were other corporate programs that we implemented that have similar impact.

As you all know, we require all major processes to be subject to HAZOP evaluations. To refresh your memories, here's a good definition: "A **hazard and operability study** (HAZOP) is a structured and systematic examination of a planned or existing process or operation in order to identify and evaluate problems that may represent risks to personnel or equipment, or prevent efficient operation." Our safety auditor reports that several serious problems have been averted because of our strict attention to and implementation of HAZOP studies. The employees are aware of the benefits of such studies and appreciate our diligence in making their performance mandatory. We are not shy about communicating examples of the devastating effects of ignoring hazards at other chemical plant locations. We don't do it to scare workers; rather, we publicize the examples to emphasize the benefits and to keep everybody on their toes.

Ever since 3M initiated their "Pollution Prevention Pays (3P)" program in 1975 (conceived by one of Robert Shay's old buddies, Dr. Joseph Ling), our industry has increased its awareness of sound environmental management. From the outset, we have instituted waste minimization and recycling programs and have carefully purchased materials and supplies that reduce our environmental profile. As Dr. Ling showed more than 35 years ago, using such practices saves money both in the short and long term. You know, since I have become the CEO, nobody has ever questioned our commitment to sound environmental management.

I believe that the commitment our employees show for safety, HAZOP studies, and sound environmental management will easily be adopted for keeping Kilcawley Chemical "cyber safe."

Robert: Mark, you have done a good job of summarizing our long conversation on making cybersecurity an essential part of our corporate culture. Thank you.

Charles: Before the meeting, I took the liberty of having Harold keep track of key events and commitments, and the schedule for accomplishment, so at the conclusion of today's meeting we will know what's next. Knowing Harold's analytical and legal mind, I am certain that he will capture the central points admirably. Let's take a ten-minute break.

*****

Charles: Mark, please continue.

Mark: The next slide, **Slide 6**, moves us into the key elements of the plan we are developing.

KILCAWLEY CHEMICAL CORPORATION

6: Plan Execution

- What will be done
- Who is responsible for doing it
- How it will be done
- What resources are required
- Risk management
- Measuring progress and success

Pre-decisional Information; Not for Public Release

I believe that the first three items are fairly self-evident, assuming of course that we have identified the tasks that require accomplishment. The reason that I am not going into too much detail for those three items is that they fall under my management prerogatives, and at this point, I have a pretty good idea where I am headed. The next three items probably will need some discussion.

Harry: Earlier in this meeting you will recall that I said that Fermoy Health System had a very significant intrusion into our medical records system. Records of 100,000 patients were compromised. It took a great toll on us emotionally and financially, so if I probe a bit please be kind. The incident remains an open wound (every pun intended). So, if you could Mark, please expand upon the resources needed to execute the plan. Be aware that I am asking not only to find out what Kilcawley Chemical has in mind but also for my own benefit.

Mark: I understand Harry and sympathize with you greatly. If you don't mind, I will refer to the Touhill book to respond to your question because I believe that their approach is a good one. Here are the resources they cite as being important:

- "Information (critical information, access control requirements and procedures, etc.)
- Financial (cost benefit analyses, operating and capital investment options, etc.)
- Organizational Structures (policies, procedures, processes, management and decision structures, etc.)
- Personnel (staff, skills, management)
- Environment (facilities, power, physical security, etc.)
- Technology (types and capabilities of hardware and software, limitations and controls, etc.)
- Partnerships (alliances, supply chain, outsourcing, and other third party relationships, etc.)"

As you can see, there are a host of factors that have to be considered in building the plan. Some can be implemented quickly, while others will have to evolve over time. Does this answer your question, or do we need to drill down further?

Harry: The answer is adequate for the time being, but I am sure we will have to revisit it in the future. Now, how about "risk management"? What do you mean by that?

Mark: Basically, Harry I mean: How it is measured? How it is managed? Who makes risk decisions?

Harry: Ah ha! You crafty devil, you have led us into the next bullet. But I won't let you off easy. Tell us about measuring progress and success.

Mark: Succinctly, Harry, these are the questions I see as defining measurement: What to measure? When to measure? How to measure? How the measure will be used? Who is responsible?

Harry: Mark, I get the drift and I see how you are approaching the plan, but because of my bias, I would like to see the details of the plan as it develops. Nevertheless, I remember Robert's sage aphorism, "Directors direct, and management manages." Forgive me for being eager, and I will try hard to be aware of the aphorism.

Mary: What other key issues are you going to have to consider, Mark?

Mark: As it turns out Mary, the next and last slide, **Slide 7**, addresses your question.

KILCAWLEY CHEMICAL CORPORATION

7: Other Key Issues

- Cybersecurity audits
- Training
- Insurance
- Public relations and disaster recovery
- Personnel vetting

Pre-decisional Information; Not for Public Release

Currently, there is an emerging business in cyber auditing. It makes a great deal of sense to me because just as traditional CPA firms monitor the financial and physical assets of our company, the data and information assets of the company are equally important. Hence, the shareholders are eager to know that the financial and physical assets are accounted for and under control; isn't it reasonable that the data and information assets should be accounted for and under control? We plan to recommend undertaking a formal cyber audit annually and at such intervals that are appropriate and/or necessary.

Mary: Do you have anybody in mind to do such audits?

Mark: We have identified three nationally known firms that have excellent reputations in this field. Just as the selection of an accounting (CPA) firm is put to the vote of the shareholders, based upon a recommendation of the board of directors, we plan to select and elect a cyber auditing firm in the same way. By the way, Mary, based upon your knowledge and insights, could we prevail upon you to help us interview prospective candidates?

Mary: Certainly, I would be happy to.

Mark: In addition to cyber auditing firms, there are numerous emerging cyber training companies and associations that certify specialists in cybersecurity. We intend to interview some of these companies to determine if they can help us to train our employees. We have a few people on our computer and network staff who have comprehensive training and a couple who are certified. Hence, we need to figure out what we can do in-house and where we need outside help.

The next subject is insurance. It seems that nowadays we have insurance for darned near everything, and the costs are considerable. Because the field of insurance related to cyber events is very, very new, it's hard for the insurer and us to know what a fair premium is. That's why earlier we suggested that we make good estimates of our expected exposure to cyber attacks and the concomitant risk involved. That way, we can gauge the fairness of premiums and not rely upon some insurer's wild guess. It would be very helpful if we could get Harold to help us put together an approach to cyber insurance.

Harold: Mark, I would be pleased to assist.

Mark: Thank you, Harold. Hortense, the next two items are right up your alley, and we would like your help to deal with them: public relations and disaster recovery and personnel vetting. Let's start with public relations and disaster recovery. Harry tells me that the best decision he made in recovering from the cyber attack on Fermoy Health System was to get a world-class public relations firm with experience in cyber attack recovery. While Fermoy's reputation took a significant hit because of the attack, Harry tells me that effective communications internally and externally helped the sordid event to blow over quickly. Moreover, truthful and expeditious releases to Fermoy's board, patients, employees, the media, and, thereby, the public softened the impact of this incident.

Harry: Mark, sorry to interrupt, but your comments are right on target.

Mark: While we hope we never have need for such a public relations consultant, we need to have one standing by, just in case.

Hortense: Sure, I'd love to help. In fact, because Harry and I were introduced through our board membership at Kilcawley Chemical, Harry asked me for some help on that incident, so I'm familiar with the dirty world of cyber threats.

Tell me what you have in mind regarding personnel vetting?

Mark: Simply put, when we hire somebody, a clerk, a technician, an engineer, an accountant, a manager, a corporate officer, a contractor, or a subcontractor (vicariously through a contractor), how do we know we are getting a person who is trustworthy? Hortense, we would like you to work with our director of personnel to come up with a plan to reduce the risk of bad actors being hired into our corporation. We don't want you to develop the program, but we do need the benefit of your wisdom and experience in setting up such a program. Are you game?

Hortense: I am.

Charles: Is that the last slide, Mark?

Mark: It is Charles.

Charles: Are there any more questions? If not, then we will take a 30-minute break so that we can commit Harold's notes on actions, schedules, and responsibilities to paper.

*****

Charles: Welcome back. I will take a few minutes to tell you what Harold and I did with his notes. First, we established dates for meetings and schedule accomplishment that were consistent with what we perceived as indicative of Mark's intent in his presentation.

Today is October 1. I want to have our next special meeting on this topic (cybersecurity) two weeks from today, October 15.

Note that we eliminated ambiguities in dates. For example, the items promised for completion for the next meeting, or for the next board meeting, or in two weeks, or in a few weeks, or in a couple of weeks **are all due on October 15**.

Where task completion is indicated for the next couple of months, or in two to three months, I want them done in time for a subsequent special cybersecurity meeting that will be held on **December 17**. I am sorry to have to cut it so close to the Holidays, but I don't want this important effort to go beyond the end of our fiscal year. As everybody knows, year-end closings can be complicated, and we want to have the essence of the enhanced cybersecurity program articulated and communicated to our shareholders in conjunction with our annual report in management's discussion and analysis.

Second, we itemized the tasks that require completion at these next two special meetings.

Third, we identified the levels of board involvement in accomplishing these tasks. For example, these are the levels of board involvement we envision: **formal ratification, approval, concurrence, and review for purposes of information**. To make things simple, **formal ratification requires** the board to vote to accept the action; **approval** indicates that a majority of the board has reviewed management's recommendation and accepts it; **concurrence** suggests that the board has been made aware of management's plan and unless specifically indicating so in writing generally agrees with the approach; and **review for information** is a courtesy to the board that may or may not require a response.

Therefore, let's look at the list of what needs to be done.

Task	Due Date	Board Action
Present enhanced strategy including cybersecurity provisions	October 15	Formal ratification and approval
Update vision, mission, and core value statements	October 15	Formal ratification
Present strategic goals and objectives	October 15	Approval
Present an outline of operational goals and objectives	October 15	Concurrence
Present final operational goals and objectives	December 17	Concurrence
Present an outline of tactical goals and objectives	December 17	Review for information
Present current costs for existing cybersecurity program	October 15	Review for information
Prepare firm estimate of the magnitude of financial impact of worst-case scenario	October 15	Approval
"Satellite"-level estimate of the cost to implement the enhanced cybersecurity program	October 15	Concurrence
Detailed estimate of the cost to implement the enhanced cybersecurity program	December 17	Formal ratification
Formulate detailed policies and procedures	December 17	Approval
Evaluate management plans to hire cyber auditor, disaster relief public relations firm, training, insurance, and personnel consultants	December 17	Approval or concurrence, as appropriate
Board oversight of management plans	Continuing	Various

Harold and I have amended a few of the schedules for task completion and have been a bit aggressive in requiring plan elements to be completed sooner than indicated in Mark's presentation. Our reasoning is that we cannot afford delay in implementing the enhanced cybersecurity program especially because so much is at stake. In some cases, we may be placing a burden on Mark and his staff, but he should be assured that the board will accommodate any reasonable requests for relief, and we will provide the resources to make things happen.

Ladies and gentlemen, I believe that we have made a good start, so please read the book and come prepared to have an intense and productive next meeting.

*****

Epilogue

Scene: It is August 27, almost 11 months after the board meeting that we sat in on and reported earlier. This meeting is a press conference being held in the large lecture hall on the campus of the State University of New York at Cobleskill. In attendance are the board of directors of Kilcawley Chemical Corporation; many shareholders and employees of Kilcawley Chemical and Hawk Polymers, including Michael Greenwood, the Hawk CEO; local and state dignitaries; other interested parties; and an extensive contingent from the news media. Media representatives include Jack Reinhard, technology and cybersecurity editor of the *Wall Street Journal*; Christine Warren, *Forbes* magazine; Justin Tyme, *Fortune* magazine; Trudy Dolt, Albany *Times Union*; and staff reporters from the following television stations: WTEN (ABC), WRGB (CBS), WNYT (NBC), and Fox23.

Dr. Charles Clark: Good afternoon, ladies and gentlemen. I am Charles Clark, chairman of the board of Kilcawley Chemical Corporation, headquartered here in Cobleskill. The primary reason for this meeting is to formally announce our acquisition of Hawk Polymers International, the well-known manufacturer of specialty polymers used in various innovative materials applications. Joining me on the dais are Michael Greenwood, CEO of Hawk Polymers, who will retire from active management of Hawk and join our board as vice-chairman, and Mark Austin, CEO of Kilcawley Chemical. Each of us is available to the audience to answer any questions regarding why we are so excited about the amalgamation of our two fine organizations.

We believe that the acquisition benefits the interests of the shareholders and employees of both companies, and the welding of our proprietary technologies positions us for remarkable growth in the near and long term. In a few minutes, we will give you the good news about where we are headed; and believe me, the news is very good. Earlier this morning at breakfast, Jack Reinhard of the *Wall Street Journal* said to me, "Charles, this truly is a marriage made in heaven from a business viewpoint. If it wasn't for the fact that criminals conducted a cyber attack on Hawk Polymers in the midst of your due diligence, we wouldn't be here today. It still would be a great story, but not the 'detective novel' it turned out to be."

So our meeting today is divided into two parts: a short one that focuses on the financial aspects of the acquisition and a longer one that focuses on the cyber attack that Jack alluded to. Between you and me, I am absolutely thrilled we were able to get this deal done. It will benefit many, many people. However, from the excitement standpoint, the story of how we were able to parry the cyber attack and still get the deal done may make a wonderful movie thriller.

So without further ado, I will turn the meeting over to Mark Austin to give you the good news number-wise.

Mark: Thank you, Charles. The other day, our CFO, Keith Willis; Michael Greenwood; and I sat down and vetted all of the final numbers and projections for the acquisition. Operating together with the combined resources and talents that Kilcawley and Hawk bring to this new company, we anticipate significant growth, increased profits, more jobs, and even greater products for our customers. The table that follows presents the good news.

Fiscal Year	Revenue (US $ billion) for Kilcawley	Revenue (US $ billion) for Hawk	Combined Revenue (US $ bil)	Profit Margin as a % of Rev.	Number of Employees
2013	3.57	2.32	5.89	4.0	3000
2014	—	—	7.07	4.8	3300
2015	—	—	9.19	7.0	3700
2016	—	—	11.95	9.3	3900
2017	—	—	16.13	12.7	4100
2018	—	—	21.78	16.5	4300

We will provide you with more details on capital requirements in handouts at the end of the meeting.

Please note that we plan to consolidate the Kilcawley Chemical plant in Eltopia, WA, with the existing Hawk plant in Pasco, WA, without the loss of any employees. Moreover, because Eltopia is not that far from Pasco, we do not believe that the short commute to Pasco will present a problem to those currently working at Eltopia because most of the Eltopia employees live near Pasco already. The Pasco plant has sufficient capacity and area to accommodate the move, although there will be some moving expense and about a two-week hiatus in production during the move. We have taken these costs into account in our projections.

The Kilcawley Chemical facility in Middleburgh, NY, will remain as is except for some anticipated expansion over the next five years. Similarly, the Okmulgee, OK, plant of Hawk Polymers will remain and also will be expanding over the next five years. Neither Middleburgh nor Okmulgee will suffer any personnel reductions as a result of the acquisition.

The basis for these excellent projections comes from the great synergy of our respective proprietary technologies. Simply put, melding Hawk's innovative processes and formulations with Kilcawley's advanced equipment and facilities permits us to make a lot more product faster, cheaper, and higher in quality.

I would like to turn the microphone over to Mike Greenwood who wants to share with you his insights.

Mike: Ladies and gentlemen, it truly is a pleasure to be here today to tell you, our employees, and the investing public how fortunate we are to become a part of the new Kilcawley Chemical Corporation. Hawk Polymers will continue to operate with its name but as a division of Kilcawley Chemical. Our employees are proud of our heritage and are pleased that the Hawk brand will continue.

This merger came at the right time for Hawk Polymers, not only for the great business advantages this deal brings but for what we learned as we put the deal together.

Some people may not know, but this acquisition rescued Hawk Polymers from the brink of disaster, not because of any deficiencies in our technology or manufacturing prowess, but because some cyber criminals tried to sabotage and destroy our fine company. During Kilcawley's due diligence evaluation of our company, they noticed signs indicative of cyber attacks. This was brought to my attention, and we launched an investigation that uncovered what we believe to have been a deliberate cyber attack directed at our firm in an attempt to scuttle this deal and destroy Hawk Polymers. We were able to stop the criminals and fix the damage, and law enforcement officials are currently working to bring those responsible to justice. As a result, we have good news to report today as the fine folks at Kilcawley Chemical not only were able to put together an outstanding technological giant in the specialty chemicals business, but also they saved a wonderful company and the jobs of thousands of good people. I look forward to serving the new entity as vice-chairman and help to protect our employees and shareholders the way our friends at Kilcawley Chemical protected and saved us.

Mark: Mike, we appreciate your heartfelt thanks. That was a terrific introduction to the second part of our meeting, but before we move ahead, let me mention that we have put together an information package that gives you more data on our amalgamation and provides some important details of what lies ahead. As you know, the shareholders of both companies have cast overwhelming votes in favor of consolidation of our two companies. The SEC is in the final stages of approving the merger, and we have been in touch with those at the agency who are considering all aspects of the merger. They tell us they are almost done and have been very complimentary about the manner in which all parties approached the merger. Before we get to the part of the meeting that seems to be most eagerly awaited, let me entertain questions from the floor.

Christine Warren, *Forbes*: We are aware that merger discussions had been going on for some time prior to the actual final agreement to merge. Why did it take so long?

Mark: Thanks for the question, Christine. Last October, we began preliminary consolidation discussions with Hawk with the intention of moving smartly into serious due diligence efforts. Remember that Kilcawley Chemical and Hawk Polymers had become partners on some significant projects about two years prior to last October. It became clear to us that there were some tremendous advantages to the two companies working as a formal business entity rather than simply as partners on a project-by-project basis. We have not been shy about telling the investment community about our highly innovative process equipment and the way we have structured our facilities to make our products. Other people like to tell you about how they are "state of the art." Christine, you and most others in this room know that we invented the "state of the art." So when we partnered with Hawk Polymers, we discovered that they had unique proprietary processes and formulations that were ideally suited for our process and manufacturing facilities. It simply was a no-brainer that we would be much better off together. During the course of the two years we worked as partners, we wondered if there was a good fit or, as we facetiously call it, "chemistry." We asked ourselves, could we work together in a collegial and friendly way? What would happen to employees if we merged? How would we blend our management teams? We asked ourselves whether it would work smoothly. The answer was a resounding yes! So that takes us to last October, which coincidentally was the first meeting focused on our enhanced cybersecurity program.

Does that answer your question, Christine?

Christine: Very well. Thanks Mark.

Mark: Are there any other questions?

Jack Reinhard, WSJ: Mark, we all are very eager to hear about the detective story. Can you elaborate on what happened and where it stands?

Mark: By all means, Jack. For those of you who may not be aware, the cyber attack directed at Hawk Polymers and indirectly at Kilcawley Chemical is an ongoing criminal investigation being conducted by the New York State Department of Homeland Security and Emergency Services and several agencies of the federal government. Hence, we have to be extremely careful about what we say and how we say it. We have two objectives that in some cases may cause conflict: (1) we are eager to be transparent in all of our communications with our shareholders, the investing community, and the public at large; (2) we do not want to compromise forensic evidence being collected by the authorities so that the perpetrators of the cyber attack can be brought to justice.

Last December, as part of our enhanced cybersecurity program, we retained the eminent consultant, Gaynor Mercer, who has a world-class reputation in cybersecurity. He has been very successful for many organizations in dealing with the outcome of cyber attacks and the restoration of confidence in these victimized organizations. He understands our situation. He understands what can be shared during ongoing criminal investigations, and he is eminently qualified to achieve the objectives of being transparent and preserving the sanctity of forensic evidence. We asked him to join us today to help inform you on the cyber attack and ongoing investigation. So without further ado, I give the podium to Gaynor Mercer.

Gaynor: Thank you, Mark. Ladies and gentlemen, I am pleased to be here today because the story I have to tell has a happy ending, although the events leading up to the happy ending were very frightening. This story has three parts. First, had not Kilcawley Chemical recognized that increasing threats from cyber criminals could cause untold damage to their company, including bankruptcy and the firm's demise with the loss of all assets and jobs, and had taken aggressive action to protect against these criminals, the attacks may not have been detected in time to prevent disaster. Second, I will tell you how the initial cyber attack was designed to scuttle the consolidation of these fine companies. Third, I will tell you how subsequent attacks were thwarted. The end result is that there was damage, but it was mitigated in a sound and secure manner. The risk was contained, and your new company is in great shape.

On October 1 of last year, Dr. Clark, at the urging of Mark Austin, convened a special meeting of the board of directors of Kilcawley Chemical Corporation. Mark had just read the book *Cybersecurity for Executives: A Practical Guide* by Touhill and Touhill. The book alerted Mark to the fact that the cybersecurity program in place at that time needed to be improved. The board agreed that immediate steps should be taken to better protect the physical and human assets of the company. Over a period of five months, many important actions were taken to develop a strong cybersecurity protection program. First, the corporate strategy was enhanced to include cybersecurity concerns. Mark and his staff updated vision, mission, and core value statements in accordance with the enhanced strategy. The board discussed and adopted the strategy and amended vision, mission, and core value statements. Then they developed strategic and operational goals and objectives for purposes of formulating a plan to implement their program.

Next came the promulgation of policies and procedures that would guide execution of the plan. These policies and procedures were explicit and detailed. In order to make the plan fully operational, personnel had to be trained, new specialists had to be hired, and software and hardware had to be updated or purchased. Performance measures were instituted, and the program was aggressively implemented. As the program was implemented, there were many lessons learned, but Kilcawley Chemical soon developed a culture of where being "cyber smart" was expected of all employees and incorporated in every process and procedure. Every employee knew they were key stakeholders in maintaining a strong cybersecurity posture to not only protect their jobs and the company but also public safety due to the sensitive products made by Kilcawley Chemical.

It was clear in March that the program was working very well, and that management now could turn their attention to the due diligence process with the aim of merging with Hawk Polymers. This is the second part of the story. Before moving to that subject, are there any questions?

Trudy Dolt, *Times Union*: Gaynor, you are a noted public relations consultant who specializes in cybersecurity issues. You've been affiliated with many companies. In March, how well did the enhanced Kilcawley cybersecurity system stack up with industry standards?

Gaynor: Trudy, based upon my experience with dozens of other organizations across a broad spectrum of industries, I can safely say that the thoughtful approach taken by Kilcawley Chemical was the equal of the best. Moreover, in the specialty chemical business, Kilcawley clearly was the best. Fortunately, many other organizations are using Kilcawley Chemical as a model and are aggressively moving forward in the development of strong cybersecurity protection programs.

I have asked Mark Austin to make John Mortimer, Kilcawley Chemical's chief engineer, available to give us personal insight into what happened next. John is widely regarded as the man who detected the cyber attack. While not a cybersecurity expert, the training John received as part of Kilcawley's enhanced cybersecurity program gave him the knowledge needed to identify evidence of the carefully crafted and hidden cyber attack that currently is nearing the end of its criminal investigation. Because of the ongoing criminal investigation, John cannot share specific details on many items. Nonetheless, while I have briefed John on what he can and cannot say because of the ongoing criminal investigation, I believe it is instructive to hear the story firsthand. John?

John Mortimer, Kilcawley Chemical chief engineer: Thank you Gaynor. In March, Mark Austin called me into his office and told me that we were about to begin due diligence on a potential merger with Hawk Polymers. I was extremely pleased to hear that we were ready to move ahead smartly because I believed that it was a great idea to bring our two companies together. I had been in charge of the technical aspects of our company's partnership projects with Hawk for nearly two years. They are great people and they have wonderful proprietary formulations that dovetail amazingly well with our processing and manufacturing capabilities.

Mark, who had always been a great fan of Hawk, cautioned me about some issues that had been discovered. He told me that Hawk was in trouble. Their product quality had suffered recently, and it appeared that their formulations may be flawed. He told me he wanted me to go down to the Okmulgee plant to find out what's going on and leave immediately.

Wow! I thought to myself. That doesn't sound like the outfit that I have been working for nearly two years. So that evening, I flew into Tulsa and drove down to the plant. In the morning, I met with Mike Greenwood, the Hawk CEO, and Jerry Laskowski, his chief engineer. Jerry and I were good friends and colleagues and not only worked together on several joint projects, but we were classmates at RPI.

Mike was really down in the dumps. He told me, "John, I have no idea what's happening. We are extremely eager to join forces with Kilcawley Chemical, but the recent mishaps could cause your company to rethink the whole deal. I want you and Jerry to get to the bottom of our troubles, and I want you to do so as quickly and efficiently as possible." He instructed Jerry to tell me everything and in as complete detail as possible. He told us that nobody could afford to hide anything.

Gaynor is painting the picture of me being the hero in finding the cyber attack, but I share credit with Jerry and the cybersecurity experts at Kilcawley Chemical, who acted quickly to detect the attack, gain control, and preserve the evidence.

First thing Jerry and I did was to go to through the operating logs. Superficially, everything seemed in order. Then, we both saw something that caught our attention. One of the formulating chemicals (I can't tell you which because of the secret formula and because Gaynor says I can't) had suspicious inventory numbers. Jerry and I calculated what the dosage should be over time and then tried to match that with the expected inventory. It was low by a factor of three.

I said to Jerry, "Based upon my knowledge of your formulation regimen, if you were dosing at only one-third of the prescribed rate, then the solution viscosity would change measurably."

Jerry said I was correct and said if we took a hard look at the mixer speeds, we should be able to tell if that's the case. He said we have to make sure that the mixer logs, dosage logs, and other process parameters match.

We worked past midnight and discovered that there were significant discrepancies in matching up optimal processing data. Finally, Jerry said what I had been thinking. "John, somebody has been screwing with our system!"

While Jerry called Mike with the bad news, I got Mark out of bed in the middle of the night and told him, "Boss, they have been hacked!"

Mark reminded me about some guys he had on retainer to help us in the event of a major cyber attack. He called them the Red Adairs of the cyber world. He told me that he was going to call Mike and have them at Okmulgee by noon that day. He told me he would have them report to me and Jerry throughout their investigation and instructed me to keep him and Mike Greenwood fully informed as we proceeded.

I call the people who Mark referred to as the Red Adairs the "Cyber Commandos." These guys were good, very good. I can't tell you what they did, but I can tell you that they diagnosed all the intrusions, fixed the glitches, and restored proper process parameters. Moreover, they did it efficiently and quickly. I stayed at the plant at the direction of Mark and Mike, and Jerry and I put the remediated plant through three pilot runs and two prototype exercises to make absolutely certain everything was okay. In mid-April, the Hawk Polymer plant was humming along normally at high levels.

Justin Tyme, *Fortune*: John, that's a fascinating story. What happened next?

John: Mr. Tyme, I'd love to tell you more, but Mr. Mercer has advised me that I am unable to say more or answer questions because of the ongoing criminal investigations by law enforcement authorities. He said he would take over from here.

Gaynor: Justin, I can't go into a lot of details, but here's what I can tell you. We believe that the cyber attack was designed to scuttle the merger by making Kilcawley Chemical believe that the proprietary formulas and processes used by Hawk Polymers were flawed. It was done ingeniously by some very sophisticated people. Surely, it wasn't a bunch of kids fooling with big, bad industry. This was a focused and targeted attack designed by smart people with malevolent intent.

Trudy: Do you know who did it?

Gaynor: I can't say anything more because of the criminal investigation. Sorry.

Jack: Gaynor, what happened next?

Gaynor: The boards of both companies and their respective CEOs were convened by Dr. Charles Clark. In that meeting, it was concluded that the cyber attack was intended to scuttle the merger process and stimulate termination of due diligence by Kilcawley Chemical. To the boards and CEOs that meant that after the perpetrators found out the hack had been detected and remediated, there would be further attempts to harm both companies. This put everyone on high alert. The most important asset to be protected was Hawk's proprietary treasure. Suffice to say that our Cyber Commandos devised a way to place the treasure in an extremely secure and guarded place.

It was decided that regardless of the outcome of merger discussions, Kilcawley Chemical would make their enhanced cybersecurity program available to Hawk Polymers. No strings were attached.

Jack: When the enhanced cybersecurity program was implemented at Hawk, what did these so-called Cyber Commandos find?

Gaynor: Mike Greenwood and his management team at Hawk Polymers were great partners and fully embraced the Kilcawley Chemical enhanced cybersecurity program. Even though the merger was still being discussed, they recognized the need to thwart the ongoing attack and allowed our Cyber Commandos to come into their facilities to stop it dead in its tracks. When the Cyber Commandos were called in, they found evidence of numerous intrusions and took immediate action to prevent them from happening again. Their work restored control of the systems to Hawk Polymers and permitted the team to conduct other actions including restoration of proper processes, a full damage assessment, and cooperate with law enforcement authorities. Interestingly, a few days after the process and formula hack was fixed, the cybersecurity teams at both Kilcawley and Hawk detected a substantial increase in Internet-based probes against their networks. The Hawk Polymer and the Kilcawley Chemical information systems were being pinged at an alarming rate.

Christine: What does that mean?

Gaynor: We believe it means that the attackers were aggressively trying to penetrate the information systems of both companies.

Christine: Can you tell us how the companies reacted?

Gaynor: For obvious reasons, I cannot provide details, but I can give you an overview of how we approached the crisis. Basically, we followed the format recommended by Touhill and Touhill in their book, *Cybersecurity for Executives: A Practical Guide*, especially Chapter 9.0. I will list the actions, some of which were taken after the initial Hawk hack.

First, the damage of the attack was minimized because of some very prudent steps that both Kilcawley Chemical and Hawk Polymer had taken *before* the attack.

A crucial step was that all information had been backed up appropriately. After the forensic teams determined when the attacks began, having the backup information permitted the technical teams to restore the systems to a trusted configuration.

Next, both companies had well-documented baselines of their computer systems; they knew what their computers and networks were supposed to look like and how they were configured. Using this information, the forensic team was able to compare the baseline to what I call the "contaminated" systems in the aftermath of the attack to detect the attackers. Law enforcement personnel also are using this baseline information as part of their investigation and analysis.

Fortunately, both Kilcawley Chemical and Hawk Polymer invested in cybersecurity insurance. While Kilcawley did not suffer from an attack and did not make a claim against their policy, Hawk's policy proved effective in protecting the company from losses incurred because of these attacks. While I cannot get into details, the fact that Mike and Jerry detected the attack early helped minimize damage and resulting claims.

Most importantly, Kilcawley Chemical has a terrific Disaster Recovery and Business Continuity Plan that they were willing to share with Hawk Polymer. Using the Kilcawley plan, Hawk was able to move quickly to thwart the attack from causing any more damage, gain control, and protect people and resources. The fact that we are here today celebrating the merger of these two great companies may be attributed in large part to that Disaster Recovery and Business Continuity Plan.

Knowing what to do when being hacked helped Hawk Polymer, with assistance from its colleagues at Kilcawley Chemical, react extremely well to the attack. We decisively took the following actions that led to our success:

1. We didn't panic. Leadership at all levels were resolute in their determination to find and fix the problems resulting from this attack and hold those responsible accountable.

2. We confirmed that we indeed had been hacked and didn't overreact.
3. We regained control of our systems and ensured the attackers could no longer access them.
4. We reset all passwords to shut out attackers who may have compromised any accounts on our networks.
5. We verified and locked down all external links to make sure there wasn't any unauthorized digital traffic coming into or out of our networks.
6. We updated all security software and conducted scans to verify that our systems were free of infection.
7. We made a full assessment of damage to make sure that we allocated the right resources to ensure we restored our systems to a state that was effective, efficient, and secure.
8. We notified everyone who needed to know, especially law enforcement authorities.
9. We adjusted our defenses to prevent further attacks.
10. This is an essential element of the criminal investigation: we tried to find out why this happened and who did it.

The last item that I want to pass on to all of you is that the Kilcawley Chemical and Hawk Polymer cybersecurity posture is better than ever and is the envy of others in the industry. As the two companies are merging, so are their information systems to provide greater effectiveness, efficiency, and security. The strong corporate culture that values cybersecurity permeates all that we do and improves how we manage our information. That doesn't mean that we won't be hacked again. It does mean that we have made every effort to minimize and mitigate the risks we face. We are hopeful that other companies will profit from our experience and harden their defenses as well.

We appreciate your attendance and will share the complete story with you when the law enforcement investigation is concluded and the perpetrators have gone to jail.

We at the new Kilcawley Chemical are excited about the merger of these two great companies. We look forward to the future and meeting with you again to share our progress. Thank you for coming.

APPENDIX A: POLICIES

In Chapter 5.0, we enumerated many policies that we believe should be integral parts of an organization's cybersecurity program. This appendix is an annotated summary of these policies. In order to facilitate the usability of this appendix, enumerated policies and their subheadings are listed alphabetically.

Audit Security

- Audit logging: Some cyber attackers use techniques that hide their activities from normal security audits and scans. These frequently are referred to as "low and slow" attacks. Many cybersecurity professionals review audit logs over a long period of time to look for signs of these types of attacks. In the author's previous organization, log files were retained for over seven years. This proved extremely helpful in both vulnerability analysis as well as legal discovery. Establish your policy for the retention of logs and files. Recording when and what types of audits were performed provides better accountability, analysis, and useful "after-action" reports in case of an intrusion or other event.
- Vulnerability scanning: Your IT staff should continually conduct internal and external scans of your network to assess its vulnerabilities. Additionally, you should conduct similar physical security scans to identify any vulnerabilities. Establish a policy that mandates regular security-based audits to ensure that any vulnerabilities in your network or physical security are identified and addressed.

Computer Security

- Acceptable encryption: Some organizations encrypt every bit of information they have. Others don't encrypt anything. Encrypting information is advised to protect your vital information, both while it is in transit (e.g., in an email) and at rest (e.g., sitting on a hard drive in storage). Establish your policy for encryption. Address what information needs to be encrypted (i.e., all or some). Address how the keys are managed and controlled, including key escrow. Establish what level of encryption is considered acceptable for sensitive information (e.g., 128 bit, SSL).

Cybersecurity for Executives: A Practical Guide, First Edition. Gregory J. Touhill and C. Joseph Touhill.
© 2014 The American Institute of Chemical Engineers, Inc. Published 2014 by John Wiley & Sons, Inc.

- Acceptable use: As spelled out in Section 5.2.1.1, an acceptable use policy is essential to identify the rules clearly and completely that a user must agree to follow in order to be provided with access to a network or to the Internet. These policies have become common practice for many businesses, educational institutions, and government entities and require that all users physically sign an acceptable use policy before being granted network access. Protect your organization and your information by having a strong and unambiguous acceptable use policy. Make sure your general counsel has a hand in drafting and reviewing it prior to publication and have every employee acknowledge it in writing before granting them access to your network, its resources, or the Internet using corporate resources.

- Application service provider: With the advent of cloud computing, many organizations now procure services from external providers. It is important that these providers do not introduce vulnerabilities into your organization and jeopardize your vital information. Establish a policy that governs who can enter into agreements with application service providers, what your security requirements are for these agreements, and how security will be monitored and controlled.

- Computer disaster recovery: This is one of your principal business continuity documents and establishes procedures to be followed in the event of a catastrophic hardware or software failure to minimize the loss of data. With increased reliance on computers to acquire, store, and manage your vital information, you can't afford to have a hardware or software failure destroy your business. Establish a clear and easy-to-follow policy that sets the rules to follow to offset negative impacts (e.g., data backup and recovery procedures, alternate work practices in the event that systems are off-line, and data restoration and validation procedures).

- End-user encryption key protection: Many organizations now encrypt their information, both at rest and while it is in transit. This is a very good security practice that we recommend. Protect your information and your business whenever you can by encrypting your data too. When you do so, make certain you have a clearly defined policy on how employees manage and protect their encryption keys. Consider a key escrow process to protect keys to address situations such as when employees lose their encryption keys so you can recover encrypted information.

- Password protection: Passwords are the first line of access control and, in many organizations, the *only* means of access control. You should establish a policy that governs use of passwords and their protection. Standardized, enforceable rules for password length, strength, and periodic password changes are must-have policies for every organization. Your policy should clearly define your rules on how to protect the password and username combinations that grant access to your network and vital information.

- Software installation: Software is a huge enabler for your business. It also can be a source of malicious code that can torpedo your business if you are not careful and tightly control what software can be installed on your network and its devices. Establish your policy that spells out who has permission to install software on corporate systems in addition to a software vetting procedure. Make sure that all software being installed on your network and its devices has been properly scanned, tested, and evaluated for risk before it is installed.

- <u>Workstation security</u>: One of the primary entry and egress points for your corporate information is through your workstations. A disorganized and messy workstation environment is an inviting target for someone who wants to steal your password from a sticky note on which you've jotted your password, or to swipe confidential corporate information from an unattended paper, or to quickly copy personally identifiable information such as a social security number off a monitor screen that has been left on (hopefully you recognized these as unacceptable security practices!). Safeguard your sensitive information by establishing a clear policy that defines your rules for workstation security. Your policy should include password protection, clean desk, and physical security procedures.

Desktop Security

- <u>Clean desk</u>: Establish a policy that gives employees guidelines to prevent proprietary information or information that could provide access to proprietary information (e.g., usernames, passwords, safe combinations) on desks or unsecured areas. Direct a clean desk policy that secures all proprietary information in secured drawers, file cabinets, or safes, as appropriate. Include procedures to log off unattended computers and to turn off unused computer monitors.
- <u>Social engineering awareness</u>: Social engineering is a growing threat. Establish a policy to standardize your procedures regarding training your employees to recognize the threats, how to protect sensitive information such as passwords, and what to do when detecting a social engineering operation. Establishing your policies for your employees to identify and thwart a phishing scam or other social engineering operation before harm can be done is a great investment.

Email

- <u>Autoforward</u>: Automatically forwarding emails without human intervention is a dangerous practice that can inadvertently expose sensitive information. Nonetheless, there are many organizations that have an occasional need to use this feature. If you are one of these organizations, establish rules for when and how autoforwarding email filters should be used in order to ensure that sensitive emails are not distributed to those without clearance without hampering effective communication.
- <u>Email usage</u>: Business email is provided for official use only, yet many people use their work accounts for their personal use. Create a policy and have your employees sign in writing their acknowledgement that clearly establishes employee regulations for when the use of company email is appropriate and when it is restricted.
- <u>Email retention</u>: Emails are official records of your organization. Establish your policy of how long emails should be retained for both record keeping and security reasons.

Information Security

- Acquisition: Establish your rules regarding the acquisition of your information technologies including software, hardware, and mobile devices.
- Bluetooth: Define your rules for use of Bluetooth-enabled wireless devices, preventing bad actors from infiltrating your systems through an unsecured Bluetooth connection.
- Ethics: Create sustainable and responsible policies that specify the standards of conduct you expect of all employees. Define standards for the ethical use of information systems as well as the ethical use of both corporate and client information.
- Information sensitivity: This policy identifies the levels of confidentiality of your information, allowing security measures to be assigned based on the level of sensitivity.
- Internal lab security: Many organizations maintain their own information system test environment or lab separate from their operational or production environments. From a cybersecurity perspective, this is a very good practice. Unfortunately, some organizations do not follow the same security procedures in their internal labs as they do in their production environments. This policy establishes the security protocols to be followed in the internal lab or test environment.
- Remote access: Spell out your rules regarding remote access for employees into the network, including what kind of connection is allowed and limits on where the connection should originate (e.g., no libraries, coffee shops, or other unsecured or untrusted public connection).
- Technology disposal: As mentioned in the Section 5.2.1.7, your old IT likely will contain valuable information that you do not want exposed. Safeguard your information by creating a policy that specifies how hardware is disposed of both in a way that ensures the security of sensitive information and is ecologically friendly.

Internet Security

- Antivirus: Malicious code such as viruses is a fact of life in the Internet, and you must provide resources to protect against them. Establish a policy that spells out what antivirus client (software) employees are responsible for using on their corporate devices, as well as standard procedures for updating virus definitions and scheduled scans. Make sure you identify responsibilities for installation and maintenance of your antivirus software to include user responsibilities. Do this because many users believe that the IT shop is responsible for end-to-end management of antivirus software, while many devices require direct user intervention. Don't forget to include an auditing directive to ensure that your systems are in compliance!
- Digital signature: Using digital signatures to verify the identity of information senders is becoming an increasingly popular and security-smart activity. Consider implementing this technology and incorporating it into your cybersecurity program. Recall the READ program introduced in Section 3.2.3. Digital signatures

are an important defense against spear phishing and other emerging attack vectors. If you elect to use this technology, establish your rules for the use of digital signatures that provides confirmation of identity. Detail when to use the signatures, procedures for their use, and rules governing your organization's READ policy (some organizations now have a policy not to open emails that do not have a digital signature!).

- Employee Internet use monitoring and filtering: This is a critical must-have policy. Create a clearly elucidated policy informing employees when and how their use of the corporate network will be monitored and what sort of traffic is considered unacceptable and will be blocked (e.g., online shopping, pornography, social media). Because you should follow a "Deny All, Permit by Exception" policy to protect your resources, don't forget to include procedures on how to open filters to legitimate traffic. Make sure your general counsel has a hand in its creation and maintenance. This policy enforces your acceptable use policy and protects you and your organization by officially notifying your employees that you will monitor corporate resources for inappropriate use and that appropriate disciplinary action, up to and including termination, will be taken when inappropriate use is discovered. Make it part of your policy that all employees acknowledge understanding of this policy in writing. We recommend that you do not grant network access until they sign their acknowledgment of the acceptable use and employee Internet use monitoring and filtering policy.

- Extranet: Intranets are a rather generic term that refers to a corporate network that is firewalled (separated) from Internet connection so that only authorized users can access internal corporate resources. Extranets refer to networks or network segments that are established outside of that firewalled intranet to share information with outsiders (they also often have firewalls in place to protect them). Some people refer to extranets as "DMZs." Use of extranets is prudent when you are sharing and exchanging information with outsiders and want to protect your internal corporate resources. When using extranets, establish a policy that defines the rules regarding what information will be shared and exchanged, security protocols, and acceptable use for how employees should use the wider Internet outside of the internal corporate intranet.

- Internet use: The Internet is a powerful source of information, and every organization should be using it to the maximum positive effect. Unfortunately, it also is the source of malicious code, is the lair of bad actors, and hosts information that is inappropriate and incongruent with your business objectives, ethics, and values. Protect your business and its information by establishing your policy on proper Internet use. Reinforce that use of the Internet using corporate resources is for official use only. Spell out what Internet resources are acceptable and what are not in the conduct of official business. Make it clear what your policies are regarding the acquisition, storage, and dissemination of data, which is illegal or pornographic or negatively depicts race, sex, or creed. We recommend you ban such activities. We also recommend your policy explicitly prohibits the conduct of a business enterprise, political activity, engaging in any form of intelligence collection from your facilities, engaging in fraudulent activities, or knowingly disseminating false or otherwise libelous materials. Make sure your general counsel has a hand in the drafting and review of this policy to make sure that it is legally sufficient and clearly articulates your position on what is official and what is not. Make sure you monitor and control your Internet access in accordance with this policy and hold your employees (and yourself) accountable!

- Remote access: Many employees, perhaps including you, access corporate network resources while traveling or while at home. They usually do so through web-based access from various Internet connection points. This remote access is a valuable capability that increases productivity. It also increases the attack surface you present to potential attackers. Establish your policy regarding the proper management and control of remote access points. Detail the proper procedures to access your corporate resources via remote means, including the security protocols all employees should follow to harden remote access points from exploitation. Consider use of onetime passwords or dual-factor authentication when using remote access procedures.

Mobile Security

- Mobile device encryption: Unless you take specific steps to protect your mobile devices, you should assume that any conversation or data transmission is being broadcast in the clear and is susceptible to interception and potential exploitation. Many companies have recognized this and have provisioned for one of many commercially available encryption products to protect their sensitive information. If you want to protect the information on your mobile devices (and we hope you do), establish a policy that defines what level of encryption data on your mobile devices must have.

- Mobile device user responsibilities: More and more often, your employees rely on mobile computing devices such as smart phones and tablet computers to conduct their duties. Establish through policy what your rules are governing the appropriate use and security of these mobile devices. Define what level of protection must be used with mobile devices and the level of liability an employee assumes by using a mobile device in conjunction with the corporate network.

- Remote access, mobile computing, and storage devices: Your "road warriors" generate and consume a tremendous amount of sensitive information, not only regarding your organization but also about your clients. They use their mobile devices to connect to a variety of potentially dangerous sources, including public Wi-Fi and Bluetooth connections to other devices, none of which are protected by your corporate cybersecurity measures. Additionally, many industrial control systems have remote access capabilities that are intended to monitor and control vital infrastructure remotely. Storage devices often are used to archive sensitive information that may be required for legal discovery, regulatory compliance, or other important functions. Compromise of any of these devices poses a cybersecurity risk to your information. Protect your sensitive information and reduce your risk by establishing a policy that addresses the processes and procedures for the safe and secure use of remote access, mobile computing, and storage devices.

Network Security

- <u>Analog/ISDN Lines</u>: Way back in history (you remember the 1980s and 1990s, don't you?), people used to use dial-up service to access the Internet. The sound of modems synchronizing was a characteristic signal that you were connecting to the broader Internet. For a decreasingly few people, the use of analog or Integrated Services Digital Network (ISDN) phone lines remains a means of digital connection. You may even have a neglected modem still connected to your corporate network that could be providing a backdoor for a potential attacker (you ought to check to see if you do!). If you have the need to retain the capability to dial in to your corporate resources (including your ICS), you need to spell out your policy regarding the effective management and controls of these devices that ensure that your information is appropriately protected. Because these technologies generally are older and were not necessarily designed with security in mind, they require special attention and oversight. Frankly, we believe that you need to get rid of them.

- <u>DMZ lab security</u>: In our extranet policy discussion, we referred to the "DMZ" to mean that segment of your corporate network that is deliberately exposed to outside users to share or exchange information. Often, especially in organizations that engage in e-commerce, important software applications (such as the United Airlines web site mentioned in Section 6.4) ought to be thoroughly tested before it goes online. Many organizations that have the resources create test lab environments within their DMZs to support special applications that operate in the DMZ environment. If you are one of these organizations, safeguard your information by establishing a policy that clearly defines how this test environment is operated and its security protocols.

- <u>Remote access</u>: As mentioned earlier, your remote access provides a potential attack vector that could threaten your network and the information stored on it. Protect yourself by implementing and enforcing a strict remote access policy that details the process for granting remote access, procedures for using the capability, and how to monitor and control your remote access. As previously mentioned, consider one-time passwords or dual-factor authentication to better control remote access.

- <u>Router security</u>: Your routers are among your most important network devices. They make sure the right information gets to the right location in a timely and effective manner. They also have a host of information on them that would be invaluable to a potential attacker. For example, access to a router could permit the potential attacker to develop a sophisticated denial-of-service attack, harvest information that could lead to the compromise of credentials, or reroute vital information to a location of their choosing. Establish a policy that tightly controls both physical and logical access to your routers, ensuring that only authorized and well-trained personnel have access to your routers and the vital information they handle.

- <u>Third-party network connections</u>: Many businesses, especially those that engage in electronic data interchange (EDI) establish connections with other networks to facilitate information exchange. These types of exchanges are potentially risky for both parties, and due care needs to be exercised to ensure that the appropriate controls are in place to protect each of the parties. If you have a connection to a third-party network, establish a policy that clearly spells out your cybersecurity standard; how the connection will be maintained, monitored, and controlled; and what and when information will be exchanged.

• VPN security: Many organizations use VPNs to protect information while it is in transmission between two points. This is a good security practice that is cost-effective and reasonably secure. Establish a policy that defines when use of a VPN should be used, the protocol for when a VPN should be used, the level of encryption that the VPN should employ, as well as which VPN client should be used.

Physical Security

• Visitor and contractor access control: Standardize methods for identifying visitors and contractors, as well as protocols for escorts in areas with sensitive information. Establish a policy that clearly provides positive control of visitors and contractors who have access to your facilities and your information.

Server Security

• Removable media: Your sensitive and critical information can walk out the door on removable media if you do not have proper controls. Similarly, malicious code can bypass your security controls (e.g., your firewalls) when people plug unauthorized removable media (e.g., thumb drives, cameras, compact disks, and even cell phones) into your computer's USB ports. Control your information by creating regulations regarding what types of removable media, if any, are permitted to be used on systems that may store confidential information.

• Server malware protection: Many people focus on putting antivirus software on their desktop computers (as they should!) but forget that the server needs to be protected as well (if not more so!). Establish standard operating procedures designed to shield the corporate server architecture from malicious code.

• Server security: Your servers are the heart and soul of your computing architecture and need to be protected. Failure to protect your servers will compromise your information. Establish a server security policy that controls who has *both* physical and logical access to your servers. Be careful to limit access to your servers to only those who have a need to access the servers and the information they control.

Wireless Devices

• Wireless communications: Many organizations no longer string wire or fiber through their facilities and rely upon wireless capabilities to connect their employees to corporate network resources. They find that the wireless communications provide acceptable performance while offsetting the costs to wire offices and campuses. They also accept the risks associated with potential signal interference and mitigate interception and intrusion through the use of encryption and authentication (e.g., passwords). When usage of a wireless network is acceptable, safeguard your network and its information by establishing a policy that defines what type of network and frequencies should be used to prevent interference between different networks and the encryption protocol and password security that should be used. Make sure that your wireless communications policy clearly defines responsibilities for both the technical support staff as well as the user base.

APPENDIX B: GENERAL RULES FOR EMAIL ETIQUETTE
Sample Training Handout

Avoid the Caps Lock key

Typing using all capital letters in any sort of online correspondence usually is seen as the virtual equivalent of SHOUTING. For emphasis, bold type or italics always are preferable, because a caps-locked section of text is seen as both visually ugly and rude.

Keep your language clean

Even if your company culture is one in which the use of the occasional expletive for emphasis is acceptable, avoid any profanity or obscenity in emails coming from your business account. You never know who is going to forward your email outside of the intended circle of recipients. For example, Arkane Studios Creative Director Raphael Colantonio learned this lesson the hard way, disparagingly referring to journalists as in a profanity-laced email sent to the entire company after a leak regarding a new project.[1] Unfortunately for him, the email was immediately forwarded to those journalists, and the entire Internet was able to indulge its penchant for *schadenfreude*. As far as emails from your home account goes, keep your language clean there too. We believe that profanity is unprofessional and there are much better ways of expressing oneself than to use coarse and crude language.

Always double check your address fields

There are few virtual *faux pas* as embarrassing as forwarding an email with disparaging and snarky commentary, then realizing in horror that you accidentally hit "reply to all," allowing the entire company to laugh at your social ineptitude. Double checking to whom the email will be sent before hitting the launch button is a good practice, especially if you have the bad habit of disregarding the second tip in this appendix and

---

[1] Jason Schreier, "Leaked E-mails suggest Bethesda misled gamers about prey 2." *Kotaku*, August 15, 2013. http://kotaku.com/leaked-e-mails-suggest-bethesda-misled-gamers-about-pre-1149092622

Cybersecurity for Executives: A Practical Guide, First Edition. Gregory J. Touhill and C. Joseph Touhill.
© 2014 The American Institute of Chemical Engineers, Inc. Published 2014 by John Wiley & Sons, Inc.

indulging your potty mouth. Pay particular attention when replying to or sending to distribution lists.

Use blind carbon copies rather than carbon copies

You turn on your home computer and bring up your emails. Lo and behold, it's Aunt Gertie, and she's at it again! She's a good soul and well intentioned, but she really is a pain in the neck. You don't want to block her because she is a distant relative, and occasionally, she has some very important news to convey. Usually, however, it is an ancient joke or screed against Customanians or some goofy political party. The problem is that she sends it to her contact list of over 75 people, most of whom, like you, couldn't care less about her itch of the day. You have the pleasure of seeing the entire list of people with whom she is sharing this nonsense. Here's the bad part; if the email account of anybody on her list is compromised or if one of the honored recipients is a "bad guy," every email address on the list is captured in the database of somebody who will inundate you with ads for crap you don't want or don't need. Or worse yet, they will know now that you are a potential target for their nefarious schemes. Gertie doesn't comprehend the magnitude of her impropriety. In fact, you are not certain that she knows anything at all about computers.

Could she have spared you and her coveted list of contacts from the threat of intrusion and endangerment? Actually she could, and fairly easily. All she had to do was to send the email to the addressees as blind carbon copies (bcc), and the email you get will be marked "undisclosed recipient." You wouldn't know the others who have the privilege of being on her list, nor would you know how many other saps she sent the stuff. Moreover, this protects you when one of those other saps hits a "reply to all" button to respond to Gertie as their response will not be sent to you.

Does this have relevance in the workplace? It sure does! For personal emails, we suggest that if you need to send an email to more than four people, then you also should send to the group via the "undisclosed recipients" route. In fact, if the email is impersonal and for information only or conveys technical or sales data and information, even in the business world, you may want to use this method for all recipients.

Use a professional font and color

This is embarrassing. (In mint green)

As is this. (In flaming red)

And infamously, so is this. (In very, very pink)

Sticking to standard fonts like Arial, Times New Roman, or similar alternatives not only will make your correspondence more legible and professional looking but also will help to preserve your formatting, as pretty much every email client or word processor comes with them preloaded. As a general rule, if the font looks like it may have been designed for sympathy cards or for use in a comic strip text bubble, you probably shouldn't use it for business purposes. Please stick to black text of a reasonable size (e.g., no bigger than 16 point font; 12 point is preferable), which will save paper and ink when people elect to print it.

Keep your business and personal accounts separate

Use your business accounts only for business correspondence and your personal accounts for personal matters. This should help keep a mental buffer between when to use business etiquette and when to use more informal language and practices in your personal correspondence. This is a good habit to get into from a security perspective as well, as suspicious emails may be much harder to identify if you're getting spammed by chain letters and political diatribes from relatives and acquaintances on your business account.

Emoticons are unprofessional

While some may believe they are fine for personal email, texts, or instant messages, emoticons should be avoided in business correspondence. Emotes can be useful for indicating sarcasm and other nuances that don't read well in text, but their utility is negated by the fact that frankly they look silly. Skip the emoticons and stay professional.

Keep your attachments small

Many corporate and even educational email systems have strict inbox data limits, effectively locking out new emails when the upper threshold is reached. Sending somebody a 10 megabyte attachment could potentially fill their entire data allotment, keeping new emails out until they delete your bloated email. Uncompressed pictures can be an especially blatant offender. Even basic image editing software, such as MS Paint, will provide a variety of image file types and quality settings that can be used to balance image quality and file size. Saving an uncompressed image at a lower JPEG quality setting can reduce its file size drastically. For large groups of files or nonimage files, an encrypted archive format such as RAR can be used for moderate compression. However, it's generally polite to ask before sending a particularly large archive, and in some cases, a third-party data storage solution may be required, such as Google Drive service or Dropbox. As an alternative, you can burn a CD or fill a thumb drive and send them through the mail—overnight express if it's all that important.

Don't forward junk or unwanted email

How many silly emails do you receive from your Aunt Dottie, who recently learned how to use her Kindle Fire and is now clogging your inbox with dopey cat videos, horrible recipes, and incredibly stupid jokes? Or how about Uncle Gene, whom your cousin Ruth bought an iPad and is now bombarding you with racy or obscene jokes he received from his old Army buddies in the spirit of "sharing?" Chances are if you don't want these types of inane emails sent to you, others don't want them from you either. Never, ever, forward junk mail or non-business-related mail from your business email address. Don't do it from home either.

Use professional signature blocks

When sending an email, remember that it is official correspondence, just like a paper memo. We recommend you append at the end of your email a signature block that gives your title and contact information. Most people find signature blocks very helpful so that they can contact the sender via telephone and know who the person is and what their position is in the organization. Attaching your signature as an image file (often with a company logo) is an annoyingly common practice that should be put to pasture, as it only leads to bloated inbox data use. Here's an example of a good, simple signature:

> //signed//
> ABRAHAM LINCOLN
> Attorney-at-Law
> 413 S 8th St, Springfield, IL 62701
> (217) 492-4241

Avoid short unnecessary replies

Consider the following email exchange:

Sender at 2:58 p.m. on Tuesday: "Attached is the proposal for Contract X. Please review it and let me know if it is OK to send to corporate headquarters for processing."
Recipient at 2:59 p.m. on Tuesday: "Ok"

Did the recipient just give their approval to send the proposal to corporate headquarters, or did the recipient acknowledge receipt of the message? If it was the former, we doubt they actually read the proposal. If it was the latter and something is wrong with the proposal, both the sender and the recipient share in the blame. Brevity is appreciated, but clarity is more valuable.

Some people feel compelled to acknowledge receipt of every message they receive with short messages such as "Ok" and "Got it." This is a waste of resources and clogs up inboxes. Would you send paper memos like this? Email systems have become very efficient and reliable, and you do not need to send unnecessary and potentially confusing short replies. If the sender needs confirmation that you received or read the email, they can use a feature to automatically send a confirmation signal that it was delivered or read (and you should configure your system to hold off on returning these messages pending your approval).

GLOSSARY

4chan: The network often used by Anonymous to coordinate their activity. A network of "image boards," each covering its own topic ranging from video games to cars to pornography. The most popular 4chan board, and the most infamous, is the random board, or /b/. If you suspect you may be a target of Anonymous, your intelligence function should keep an eye on the 4chan /b/ board. *See also Anonymous.*

Acceptable Use Policy: An acceptable use policy establishes rules that a user must agree to follow in order to be provided with access to a network or to the Internet. These policies have become common practice for many businesses, educational institutions, and government entities and require that all users physically sign an acceptable use policy before being granted network access. Most acceptable use policies have some common attributes, requiring that the user agrees to adhere to specific guidance on what is acceptable use as well as clearly defined unacceptable use guidance. Often, the policy outlines what does and does not constitute acceptable uses of the corporate computer system as a way to shield the company from liability in the event of a termination or censure due to misuse. Also used in many public Internet hotspots such as airports, libraries, and cafes as a liability shield against illegal acts committed using the public connection.

Adware: Malicious software that is intended to hijack a system into displaying advertisements, sometimes for dubious services that themselves are malicious.

Anonymous: A blanket label for the loosely organized collective of hackers tracing their roots back to 4chan's /b/ board. Anarchic by nature, there are many factions under the Anonymous umbrella, each with their own agenda, and often a compartmentalized operational cell. Some hackers who self-affiliate with Anonymous have a political or ideological motivation to their attacks. *See also 4chan and Hacktivist.*

Backdoor: A concealed means of egress into a network or system, either inserted from outside for later use or from the manufacturer or other inside party for use as a fail-safe. *See also Trojan and Remote Access Trojan.*

Bot: A computer that has been remotely hijacked by a malevolent intruder, usually without the knowledge of the computer's owner. The system then can be used as a platform to stage an attack at the hacker's direction.

Cybersecurity for Executives: A Practical Guide, First Edition. Gregory J. Touhill and C. Joseph Touhill.
© 2014 The American Institute of Chemical Engineers, Inc. Published 2014 by John Wiley & Sons, Inc.

Botnet: A network of computers infected with malicious code, allowing a third party to take control of them, usually without the knowledge of their owners. Frequently used as attack points in a distributed denial-of-service attack. *See also Distributed Denial-of-Service Attack.*

C2: *See Command-and-Control Server.*

Certified Ethical Hacker: A professionally certified individual who, with permission of the system's owner, attempts to hack into a network for the purposes of exposing existing vulnerabilities. This is a professional credential overseen by the International Council of E-Commerce Consultants (EC-Council).

CISM: Certified Information Systems Manager.

CISO: Chief Information Security Officer.

CISSP: Certified Information Systems Security Professional, an accreditation provided and overseen by the International Information Systems Security Certification Consortium (ISC)^2.

Clean Desk Policy: Guidelines established by an organization regarding the securing of sensitive information (e.g., documents, computer systems, and physical media containing sensitive digital information such as disks or flash media) at the end of the workday, during an emergency, or any other times in which the person responsible for safeguarding that information is not at their workstation.

Cloud: A network of computers used for mass storage or processing power, existing outside of the user's local network.

Command-and-Control Server: A server configured to centrally manage and control activities of many other servers. Often referred to as a "C2" server. Hackers often use a C2 server to remotely control activities of zombie computers as they manage their botnets.

Cookie: A digital "breadcrumb" in the form of a small file left behind after visiting a web site. These files are sometimes beneficial to the user, storing usernames or passwords for faster access later, but sometimes can "phone home," tracking the user's Internet activity for advertising, data mining, or other purposes.

Cracker: Slang for a hacker who specializes in breaking encryption codes or digital rights management restrictions.

Cross-Site Scripting: An attack method used by hackers in which a third party's web site is infected, then injecting the malicious payload into any visitor's computers.

CSO: Chief Security Officer.

DDoS Attack: *See Distributed Denial-of-Service Attack.*

Defense-in-Depth: An information security method that uses multiple layers of security to prevent an intrusion or malicious action.

Degauss: The act of using a powerful magnet to eliminate another magnetic field, frequently used in the destruction of digital information in hard drives.

Denial-of-Service Attack: An attack that seeks to overwhelm a system with so many junk queries from a single point that legitimate traffic is not able to get through because of the overload. *See also Distributed Denial-of-Service Attack and Low Orbit Ion Cannon.*

Distributed Denial-of-Service Attack: Similar to a denial-of-service attack where the target is overwhelmed with junk traffic so it is overloaded but coming from a variety of different sources. These are often bystander computers that have been hijacked, forming a "botnet." Due to the fact that the attack is distributed and not attributable to a single source, it sometimes is much harder to counter by separating malicious traffic from legitimate traffic. *See also Denial-of-Service Attack, Botnet, and Low Orbit Ion Cannon.*

DNS: *See Domain Name System.*

Domain Name System: This is the "universal translator" of the Internet and translates Internet Protocol (IP) addresses (e.g., 108.166.31.131) into a plain language address that is easily understandable to the common user (e.g., www.post-gazette.com) and vice versa. Without the Domain Name System, all users of the Internet would have to type the IP addresses into their search browsers, which would be difficult and inconvenient. Cyber criminals repeatedly have attacked this capability, most notably in the DNS Changer case where a tiny (1.5 K) Trojan file would infect a computer, change the computer's Name Server registry file, and redirect all web traffic from that computer to a server of the criminal's choosing.

DoS Attack: *See Denial-of-Service Attack.*

EDI: *See Electronic Data Interchange.*

Electronic Data Interchange: This is a standard format for exchanging business data. Businesses use highly structured EDI messages to exchange critical information using preagreed formats. American National Standards Institute (ANSI) has defined the X12 standard for electronic data interchange commonly used across North America. The United Nations-sponsored Electronic Data Interchange for Administration, Commerce, and Transport (EDIFACT) is the only approved international EDI format.

Firewall: A virtual moat, a firewall is a program that filters traffic incoming and outgoing in a computer network using a set of permissions. Traffic that is permitted is free to travel through the firewall, while those that are restricted are unable to enter or exit the system. *See also White Listing.*

Freaker: An individual who experiments with accessing and manipulating telephone systems, much as a hacker accesses and manipulates computer networks. *See also Phreaker and Hacker.*

Hacker: A person who secretly gets access to a computer or data control system in order to unscrupulously obtain information and data with the intent to steal valuable information and data or to cause damage to the property and assets of the owner. Also, a person who hacks into a computer system.

Hacktivist: An individual who engages in malicious cyber activity under the pretense of political or ideological beliefs. Portmanteau (combination) of hacker and activist.

Hashed: This refers to a method where a discrete algorithm is applied against a string of information. The resulting unique "hashed" value is appended to the string of information and sent with the information to the distant end. The receiver applies the algorithm to the string of information and compares the hashed values. If they are the same, the receiver has reasonable assurance that the message has not been tampered within transit. This technique often is used to maintain the integrity of information.

Hotfix: A small patch usually pushed out to address an immediate threat or problem, often fast-tracked without full testing due to the severity of the issue. Hence, the fix is "coming in hot." *See also Patch*.

HTML: Short for **hypertext markup language**; this language and those descended from it form the backbone of the majority of web pages today.

Hub: Hubs are devices that provide a physical communication device that permits several computers and devices to communicate with each other. Hubs don't have the intelligence of routers, which read addressing and forwarding data to desired recipients. When a signal is received by a hub, it is broadcast to all the systems connected to the hub.

Internet Protocol Address: This is the numeric address for every computer or device attached to the Internet, each with a unique address. When the creators of the Internet originally defined IP addresses, they used 32 bits to define an address. This is called IPv4. Now, with the explosive growth in Internet use, that address space is rapidly running out, so a new standard called IPv6 is being implemented. IPv6 uses 128 bits to define addresses. Your technical staff should have already provisioned for IPv6 as part of your network management. Ask whether your network and all devices are IPv6 compliant. If they are not, soon you will have to recapitalize to maintain compliance.

IP: *See Internet Protocol Address*.

Keylogger: This is a malicious code that is designed to record every key stroke you make on a keyboard. Criminals like to use these programs to learn usernames and passwords in order to gain access to systems and information. Sophisticated hackers now are experimenting with keylogging software that they surreptitiously install on smart phones. Many people like to place their phones near their keyboards on their desk. Hackers know this and have developed software that uses sensors in your phone (the same ones that detect whether the phone is sideways or upright) to detect vibrations created when you type. Using a predicting algorithm based on the QWERTY keyboard, this keylogging software supposedly can detect what you are typing on your own smart phone.

LAN: *See Local Area Network*.

Local Area Network: This is computer network that connects computers in a specified area such as a home, office, laboratory, or school. LANs usually are defined by their geographic area, are managed by a single entity, and don't use leased telecommunications lines to connect network segments. Networks that extend across a wider area are referred to as Wide Area Networks (WANs), and those that extend across a community or campus with multiple nodes are referred to as Metropolitan Area Networks (MANs).

Low Orbit Ion Cannon: A tool freely available on the Internet that is configurable in order to allow the user to attempt to flood a system with connections in a denial-of-service attack. Due to its ease of use, it permits even those with a relatively low level of technical ability to engage in attacks or to contribute to a distributed denial-of-service attack. *See also DoS, DDoS, Denial-of-Service Attack, and Distributed Denial-of-Service Attack.*

Malware: Blanket term for any program that is intended to do harm to a system or the user. *See also Adware, Virus, and Trojan.*

MAC Address: The identifying code given to a specific device to identify that device on a local network.

Metasploit: A software program designed to identify vulnerabilities in networked systems, frequently used in penetration tests. *See also Nessus, Pen-Test, and Penetration Test.*

Memory Leak: An error in the code of a program that causes it to use up increasing system memory over time, potentially slowing or crashing the system. Because it takes time for the program to use up this memory, it is almost as though it is leaking away—hence memory leak.

Nessus: A freely available hacking tool that allows the user to scan a network for vulnerabilities such as open ports, default passwords on network configuration, and gateways. *See also Metasploit, Pen-Test, and Penetration Test.*

Netsploit: Tool that allows injection of a malicious backdoor in the Microsoft .NET framework. It permits bypassing of surface level antimalware scans.[1] *See also Backdoor and Rootkit.*

Nmap: Nmap is a tool used for security scanning of networks and devices. It originally was designed to send out a signal across a network to discover all devices and services residing on the network. The resulting "map" of the network would be used by administrators to configure and manage the network and devices appropriately. Nmap is used by administrators to conduct network inventories, asset management, maintenance, and network mapping. It also helps audit security by identifying what devices are connected and what ports and protocols are open and exposed to potential foes. Make sure your administrators are the only ones using Nmap on your network and nobody from the outside can use it against you.

Own: Hacker slang for infiltrating a system and exerting control, as though they "own" it. Sometimes written as "pwn" due to frequent typos resulting from the proximity of the o and p keys together on a standard QWERTY keyboard layout. *See also Pwn.*

---

[1] Metula, Erez, *.NET Framework Rootkits: Backdoors Inside Your Framework.* Black Hat Europe, https://appsec-labs.com/system/files/NET_Framework_rootkits.pdf. Accessed on March, 2009.

Patch: A revision to software or an operating system in an attempt to address a performance issue, to add features, or to eliminate a security exploit. *See also Hotfix.*

Penetration Test: An intentionally sanctioned hacking attempt, conducted by a certified ethical hacker, designed to expose existing vulnerabilities in a network. *See also Pen-Test and Certified Ethical Hacker.*

Pen-Test: *See Penetration Test.*

Phishing: Acquiring usernames, passwords, or other confidential information by pretending to be a legitimate source, typically either through email or via a fake log in prompt. *See also Spear-phishing and Whaling.*

Phreaker: *See Freaker.*

Ping: A digital "chirp" sent out to test the response of a system. Alternatively, shorthand for the amount of latency time between when the signal is sent and a response received.

Port: The digital "doors" through which data flows in and out. *See also Port Scan.*

Port Scan: A process to test various ports on a networked system, determining which of the virtual "doors" are unlocked. While not considered malicious by many in the hacker community, this is still an attempt to scout vulnerabilities and as such should send red flags. *See also Port.*

Proxy Server: A server that works as an intermediary between two different users or networks. While there are many legitimate uses of proxy servers, they often can be used to attempt to obscure the identity of systems during a malicious cyber action, for example, renting an anonymous PO box (or a system of PO boxes that automatically forward the mail in some cases).

Pwn: *See Own.*

RAT: *See Remote Access Trojan.*

Remote Access Trojan: Malware that infiltrates a system then lays dormant, waiting for the hacker who planted it to access it. This permits them to dial in (access) remotely, as though they were a user or system administrator. *See also RAT and Trojan.*

Rollback: The process of reverting a system back to an earlier configuration, sometimes in hardware but more frequently in software, in order to eliminate unexpected and unacceptable issues with a new configuration. Frequently used in the context of "rolling back a patch," wherein a software patch is removed from the system due to compatibility problems or other deleterious factors. *See also Patch and Hotfix.*

Rootkit: A form of malware that attempts to hide its existence deep inside a system's operating system, subverting every attempt to identify it or its progeny processes and programs. Due to the extremely stealthy nature of a well-designed rootkit, many system administrators prefer to wipe the system completely to ensure that it has been eliminated.

Router: A router is an intelligent appliance or dedicated computer that takes packets of information, reads addressing information, compares it against its access control lists, and routes the information toward its desired destination, if appropriate. Routers may filter information and block data that is not authorized. Routers move data toward destinations by the most efficient means available. If you send an email to a colleague, your data likely will pass through numerous routers until it reaches its intended destination.

Salting: This is a technique used to defeat password breaking attempts. Some hackers use programs that compare your encoded (or hashed) password against tables of precomputed words and phrases in the hope that they will find a match that will enable them to crack your password. Salting is done by adding additional random data to your password that extends the length and complexity of your password. That doesn't make your password bulletproof. It just makes it harder to crack. Salting is used as part of a defense-in-depth strategy.

Server: A computer or network that **serves** out data when requested, pushing out the data that forms web pages, routes your VOIP calls, etc. *See also Server Farm.*

Server Farm: A large group of server machines, typically positioned in rows using racks, drawing a parallel to crops in a field. *See also Server.*

ShadowCrew: An infamous hacker community best known for their breathtakingly large credit card theft, dissolved due to the actions of law enforcement in 2004.

Sniffer: A program used by a third party to intercept and analyze data between two systems or networks. While sometimes used for malicious purposes to attempt to spy on traffic or expose vulnerabilities, sniffers also can be used to identify suspicious behavior on a network.

Social Engineering: The human-to-human component of hacking, often taking the form of conning usernames, passwords, or other privileged information under a false identity or pretense. *See also Phishing and Whaling.*

Software Bloat: The slow increase in system resources used over time as new features are added to software through updates, eventually impacting system performance deleteriously.

Software License: A formal agreement authorizing an individual or an organization to use computer software by the creator. Physically possessing the disk is not enough to authorize use of a program; the correct license also must be acquired.

Spear-Phishing: Essentially the same definition as in "phishing." *See also Phishing.*

Spoofing: Attempting to conceal one's digital fingerprints or to gain access to privileged information by digitally masquerading as someone else. Spoofing can be as basic as the cyber version of a fake ID to buy beer or as complicated as pretending to be a bank manager in order to access the vault.

SQL: Often pronounced "sequel," SQL is a program language used in the management of databases. *See also SQL Injection.*

SQL Injection: An attack vector used by hackers that exploits a vulnerability in the SQL database running behind a web site or to attack a SQL database housing valuable data on its own. *See also SQL.*

Stuxnet: A coordinated cyber attack on the Iranian nuclear program using the eponymous malicious program. According to NSA leaker Edward Snowden, the creation of the Stuxnet code was carried out jointly by the American NSA and Israeli intelligence services.[2]

Switch: Networking hardware that filters traffic on a local area network.

Trojan: A type of malware specifically designed to create a backdoor into a system, often by presenting itself as legitimate or useful software much like the Greek Trojan Horse. *See also Remote Access Trojan and Backdoor.*

Twitterverse: All activity on the social network Twitter. "Gauging the Twitterverse" can be useful in determining current public opinion, as well as drawing attention to news as users share it through social media.

Two-Factor Authentication: A security protocol involving two different safeguards before access to a system is permitted. Used in everything from classified government and military intelligence information (the Department of Defense's combination of controlled access cards and passwords) to securing video game accounts.

Uninterruptible Power Supply: Essentially a large battery, an uninterruptible power supply, commonly referred to as a UPS, provides emergency power to a computer system in the event of an electrical outage. Many provide protection from other electrical events that may damage a system's hardware, such as electrical power surges.

UPS: *See Uninterruptible Power Supply.*

Virtual Machine: A secondary computing environment running its own operating system while on the same physical hardware. Essentially a cordoned-off environment running on your computer completely independent of your primary operating system. *See also Virtualization.*

Virtual Private Network: An encrypted network that exists within the open Internet, allowing remote access to networks or between networks to authorized individuals.

Virtualization: The use of virtual machines in a network environment, often as a way to create a secure testing environment for software updates, or in off-site, cloud storage. *See also Virtual Machine.*

Virus: Often colloquially used as a term for any malicious code, including rootkits, Trojans, and worms, viruses are a subset of malware that are specifically designed to be self-replicating by invading other files or programs, similar to viruses in nature.

VM: *See Virtual Machine.*

VPN: *See Virtual Private Network.*

VOIP: *See Voice over Internet Protocol*

---

[2] Iain Thomson, "*Snowden: US and Israel did create Stuxnet attack code,*" http://www.theregister.co. uk/2013/07/08/snowden_us_israel_stuxnet/. Accessed on July 8, 2013.

Voice over Internet Protocol: A form of digital telephony where live audio is encoded and passed over the Internet much like traditional phone signals. If you've used Skype or FaceTime, you have used a form of VOIP.

War Driving: A tactic where an individual drives around in a car with his computer, looking for vulnerable systems or unsecured wireless signals.

Whaling: A phishing tactic that specifically targets highly placed individuals in an organization or those who have large amounts of access, such as executives or system administrators. *See also Phishing.*

White Listing: A process through which all network traffic is issued a blanket denial into or out of the system, with the exception of programs that have been granted specific permission. This list of exceptions is known as the white list. *See also Firewall.*

Worm: A type of malware that attempts to spread itself between computers over a network, so named because of the huge amounts of bandwidth a successful worm can use as it attempts to reproduce. Much like biting into a wormy apple, the performance of a system riddled with worms will not be as expected.

XSS: *See Cross-Site Scripting.*

Zero-Day Exploit: An attack that occurs using a vulnerability or exploit that was previously unknown. As such, there are "zero days" to address the vulnerability as it has already been used.

Zombie: *See Bot.*

SELECT BIBLIOGRAPHY

Adams, Douglas, *The Hitchhiker's Guide to the Galaxy*, New York, NY: Harmony Books, 1979.

Axelrod, C. Warren, Jennifer L. Bayuk, and Daniel Schutzer, *Enterprise Information Security and Privacy*, Norwood, MA: Artech House, 2009.

Breyfoyle, Forrest W., *Implementing Six Sigma, Smarter Solutions Using Statistical Methods*, Hoboken, NJ: John Wiley & Sons, 2003.

Cohen, Gary B. *Just Ask Leadership: Why Great Managers Always Ask the Right Questions*, New York, NY: McGraw-Hill, 2009.

Collins, James, *Built to Last, Successful Habits of Visionary Companies*, New York, NY: Harper Business, 1994.

Collins, Jim, *Good to Great*, New York, NY: Harper Collins, 2001.

Drucker, Peter F., *Management Challenges for the 21st Century*, New York, NY: Harper Business, 1999.

Kaplan, Robert S. and David P. Norton, *The Balanced Scorecard: Translating Strategy into Action*, Cambridge, MA: Harvard Business Review Press, 1996.

Kaplan, Robert S. and David P. Norton, *The Strategy-Focused Organization*, Cambridge, MA: Harvard Business Review Press, 2000.

Keane, Michael, *Dictionary of Modern Strategy and Tactics*, Annapolis, MD: Naval Institute Press, 2005.

Kotter, John P., *Leading Change*, Cambridge, MA: Harvard Business Review Press, 2012.

Lewis, Michael, *Moneyball: The Art of Winning an Unfair Game*, New York, NY: W.W. Norton & Company, 2003.

Liang, Qiao and Wang Xiangsui, *Unrestricted Warfare*, Beijing: PLA Literature and Arts Publishing House, 1999.

Liang, Qiao and Wang Xiangsui, *On the Chinese Revolution in Military Affairs*, Beijing: New China Press, 2004.

Liker, Jeffrey K., *The Toyota Way, 14 Management Principles from the World's Greatest Manufacturer*, New York, NY: McGraw-Hill, 2004.

Ludlow, Peter, *High Noon on the Electronic Frontier*, Cambridge, MA: MIT Press, 1996.

Mitnick, Kevin P. and William L. Simon, *The Art of Deception, Controlling the Human Element of Security*, Indianapolis, IN: Wiley, 2003.

Mitnick, Kevin P. and William L. Simon, *The Art of Intrusion: The Real Stories of Hackers, Intruders and Deceivers*, Indianapolis, IN: Wiley, 2005.

Cybersecurity for Executives: A Practical Guide, First Edition. Gregory J. Touhill and C. Joseph Touhill.
© 2014 The American Institute of Chemical Engineers, Inc. Published 2014 by John Wiley & Sons, Inc.

Mitnick, Kevin P. and William L. Simon, *Ghost in the Wires: My Adventures as the World's Most Wanted Hacker*, New York, NY: Little, Brown, and Company, 2011.

Niven, Paul R., *Balanced Scorecard Step by Step: Maximizing Performance and Maintaining Results*, Hoboken, NJ: John Wiley & Sons, 2006.

Parmenter, David. *Key Performance Indicators: Developing, Implementing and Using Winning KPIs*. Hoboken, NJ: John Wiley & Sons, 2010.

Rose, Sarah, *For All the Tea in China, How England Stole the World's Favorite Drink and Changed History*, New York, NY: The Penguin Group, 2010.

Symantec Corporation, "*Internet Security Trends Report: 2013*," 2012 Trends, Volume 18, April 2013. Available at http://www.symantec.com/content/en/us/enterprise/other_resources/b-istr_main_report_v18_2012_21291018.en-us.pdf. Accessed on August 29, 2013.

Touhill, C. Joseph, Gregory J. Touhill, and Thomas A. O'Riordan. *Commercialization of Innovative Technologies: Bringing Good Ideas to the Marketplace*, Hoboken, NJ: John Wiley & Sons, 2008.

Von Clausewitz, Carl, *On War*. Edited and Translated by Michael Eliot Howard and Peter Paret, Princeton, NJ: Princeton University Press, 1984.

Walsh, Bill and Steve Jamison, *The Score Takes Care of Itself: My Philosophy of Leadership*, New York, NY: Portfolio Hardcover, 2009.

Welch, Jack and Suzy Welch, *Winning*. New York, NY: Harper Collins, 2005.

INDEX

Note: Page numbers in *italics* refer to Figures; those in **bold** to Tables.

acceptable use policy, 141–3, 169, 180, **197,** 246, 348, 351
  conscious agreement, 141–2
  improper nonofficial computer usage, 142
  Michael A. Smyth v. The Pillsbury Company (case), 142–3
  sample attributes, 141
  team effort, 143
  violations, 142
access control, 40, 42, 67, 72, 87, 121, 144, 166–70, 185, 244, 333, 348, 354
access control policy
  discretionary, 185
  mandatory, 185
  role-base, 185
  visitor and contractor
    agreement, 167
    badging and identification, 167–8
    check-outs, 168–9
    designated parking, 166
    electronics, 168
    emergencies and evacuations, 168
    enforcement, 169–70
    network or system access, 169
    reception, 166–7
    tours, 169
American Civil Liberties Union (ACLU), 21
annualized loss expectancy (ALE), 58
annualized rate of occurrence (ARO), 58, 62–3
anonymous, 25, 26, 62, 68, 230, 309, 311, 312
anti-malware, 178
antispyware, 245

antivirus, 43, 104, 178, 180, 184, 186–7, 207, 245, 263, 283, 304, 305, 350, 354
AOL Time Warner, 126
auditing, 29, 47, 80, 89, 187, 198, 241, 267, 297, 307, 334, 350
authentication, 48, 162, 178, 194, 244, 352, 353

backdoor, 35, 43, 152, 254, 305, 353
Balanced Scorecard, 265–6
"Big Data," 204–5
BigMIMS (medical information management system), 64, 67
BigRX
  BigMIMS, 64, 67
  business intelligence capability, 76
  CEO, recommendations, 75–6
  Corporation, 129–30
  event likelihood, 67, 69–70
  functional specifications, 132
  incident impact, 70–71
  new policy, 140
  planning division, 130–132
  Plieno Corporation, 64
  risk assessments, 65
  SQL injection vulnerability, 68, 129, 133, 135
  testing procedures, 134
  three-tier measurement techniques, 71
  web pages, specifications, 132–3
BigRX plan, reduced SQL injection risk
  additional resources, 137–8
  applicability, 135

Cybersecurity for Executives: A Practical Guide, First Edition. Gregory J. Touhill and C. Joseph Touhill.
© 2014 The American Institute of Chemical Engineers, Inc. Published 2014 by John Wiley & Sons, Inc.

BigRX plan, reduced SQL injection risk
    (*cont'd*)
  available resources, 137
  execution, concept
    contract modifications, 136
    promotion to live system, 136–7
    recurring performance management, 137
    software procurement, contracts, 135–6
    testing and acceptance, 136
  governance
    management reviews, 138–9
    penetration testing, 138
    vulnerability scanning, 138
  purpose, 135
board involvement, levels, 336
boardroom interactions
  cybersecurity audits, 334
  cybersecurity risk management, corporate
    culture, 331
  goals and objectives, 329
  hazard and operability study (HAZOP),
    331
  insurance, 334
  levels of board involvement, 336
  personnel vetting, 334
  plan execution, 332
  policy and procedures, 330
  "Pollution Prevention Pays (3P)"
    program, 332
  process control systems via intranet, 327
  public relations and disaster recovery, 334
  risk management, 326
  strategy, 328
  training, 334
botnet, 18, 19, 49, 302, 303

C2 *see* Command-and-control server (C2)
Canadian Cyber Incident Response Centre
    (CCIRC), 316
CEO or Chief Operating Officer, 189
Certified Associate in Software Testing
    (CAST) certification, 134
certified ethical hacker, 98, 195, 256, 306
Certified Information System Auditors
    (CISAs), 195
Certified Information Systems Manager
    (CISM), 234–6
Certified Information Systems Security
    Professional (CISSP)**,** 234–6

Certified Manager of Software Testing
    (CMST), 134
Certified Software Tester certification
    (CSTE), 134
"CF Disclosure Guidance: Topic 2,
    Cybersecurity" (CF DG 2), 79
4chan, 25, 26
change management
  backup plan, 220
  best practices
    asking for help, 221–2
    backup and back-out plan, 223
    communication, 220–221
    crummy products, 221
    flexibility, 224
    maintenance, 222–3
    monitor implementation, 223
    software patches, 223–4
    timing and needs, 221
  best value, 203–4
  catastrophe, 215
  COA (courses of action) selection
    and management approval,
    217–19
  competitive advantage, 204–5
  cyber attacks, 200
  description, 199
  downtime, 200
  evaluation, 217
  executives, 214, 215
  fundamental responsibility, 214
  identification, 217
  impacts
    capacity, 207–8
    contracts, 206–7
    people, 205–6
    security, 208–9
    SLAs, 206
  and internal controls, 209–14
  laws and regulations, 201
  manufacturers, 216–17
  monitoring and evaluation, 219
  obsolescence, 201–3
  organizational processes, 200
  patch installation, 215
  personnel, process and products, 200
  risk environment, 200
  steps, 215–16
  technical measures, 214

chief financial officer (CFO), 58–9, 64, 129,
133, 135, 159, 209, 233, 241, 263, 307
chief information officer (CIO), 42
Chief Information Security Officer (CISO),
46, 47, 90, 132–9, 189, 218, 219,
233–7, 240, 241, 263
"career-broadening" jobs, 236
CISSP and CISM certifications, 234–5
credentials and experiences, 234
effectiveness, efficiency and security,
233–4
programs and operational activities, 233–4
software management and IT systems, 233
chief operating officer (COO), 58–9, 64, 133,
135, 137, 233
chief risk officer (CRO)
CISSP/CISM certifications, 236–7
compliance programs, 90–91
corporate risk program, 90
risk appetite level, 91
Chief Security Officer (CSO), 196, 198, 233,
236, 237, 240
CIO *see* chief information officer (CIO)
CISM *see* Certified Information Systems
Manager (CISM)
CISO *see* Chief Information Security Officer
(CISO)
CISSP *see* Certified Information Systems
Security Professional (CISSP)
clean desk policy, 150–152, **197,** 285, 349
best policy practices, 151–2
HIPAA, 150–151
Privacy Act, 151
cloud, 20, 42, 44–5, 45, 134, 138, 152, 181,
207, 264, 296, 348
command-and-control server (C2), 19, 26, 30,
48, 50
communication
exceptional, 113
internal *see* risk management
lack of, 113–14
shareholders *see* shareholders
communications
Computer Ethics Policy
corporate ethics, 146–7
motivation, 146
Ten Commandments of Computer Ethics,
143–6
zero tolerance, 146

computer security
acceptable encryption, 347
acceptable use, 348
application service provider, 348
computer disaster recovery, 348
end-user encryption key protection, 348
password protection, 348
software installation, 348–9
workstation security, 349
confidentiality, 2, 214, 268, 294, 350
cookie, 154, 211
corporate espionage, 11
corporate ethics, 143, 146–7
cracker, 42
credentialing *see* employee credentialing
critical infrastructures
accident investigation, 252–3
ADS-B system, 251–2
audit, 258
automated control systems, 249–50
"bake in," 256
business and industrial safety systems,
255
Chinese hackers, 254
computers and networks, 249
cyber attacks, 250
cybersecurity risks, 256–7
disruption, services, 253–4
enemies, 255
executives, 254–5
frightened workers, 252
hackers and terrorists, 249
ICS controls, 253
insurance policies, 250
monitor and control, 257
organizational priority, 256
personnel training, 257
power generation, turbine, 252
practice, 257–8
"pr0f," 250–251
protection, 255
Sayano–Shushenskaya dam, 252
single points of failure, 257
"zero-defect" operations, 255
CRO *see* chief risk officer (CRO)
Cross-Site Scripting (XSS), 211–12
cryptography, **235**
CSO *see* Chief Security Officer (CSO)
Cyber Commandos, 343

cyber espionage, theft, and exploitation
    (checklist) *see* intellectual property
    and trade secrets
cybersecurity
    as business imperative, 2–3
    concerns
        "bad actors," 4
        CEOs, 7
        executive-level, 4
    definition, 2
    full-time activity, 7–8
    generic questions, 4–6
    incidents
        by category, 281, *281*
        cost, 281–2
        detection, 280–281
        responsible for, 282, *282*
    incorporating into strategy
        impacts in decisions, 119–20
        information, identification, 119
        as part of culture, 119, 331
        progress, measuring, 120–121
        risk management, 331
    in management processes, 4
    policies *see* policy(ies) complement plans
    risk management, 3
    strategy buiding, 95–7
*Cybersecurity for Executives: A Practical
    Guide,* 324
cyber threat sources, 12, 36, 55, 64, 68, 74,
    78, 85

dashboards
    acquisition, 285–6
    change management, 284
    competitions, 288–9
    costs, 287
    cybersecurity incidents, 281–2
    employees, current cybersecurity training,
        286
    Executive Cybersecurity Dashboard,
        289–90
    industry experts, 289
    industry groups and associations, 289
    information loss/tampering/destruction,
        consequences, 287–8
    network performance measures
        current security software, devices, 283–4
        previous measurements, 283, *283*

physical security, 285
reporting period, security violations, 286
risk exposure, 288
security incidents/violations, 286
software configuration management, 284–5
system administrator privileges, 286
technical staff, current certifications, 286
threats detection
    phishing messages, 278
    quantitative measure, 278, *279*
    "repeat offenders," charts, 277–8, **279**
    reverse lookup, 279
    scans and probing, 277–8
    spam filters, 278
value of information, 287
vulnerability *see* vulnerability(ies)
DDoS *see* distributed denial-of-service attack
    (DDoS)
defense-in-depth, 207, 257
degauss, 159
Delphi group technique, 70
demilitarized zones (DMZs), 244
denial-of-service attack, 19, 200, 278, 353
distributed denial-of-service attack (DDoS),
    18, 26, 68, 200, 302, 312, 362
DMZ lab security, 353
Domain Name System (DNS), 159, 210
dynamic test environment, 134

electronic data interchange (EDI), 41, 213,
    214, 353
electronic exchanges
    businesses, 213
    encryption, 213
    system configurations, software and web
        pages, 212
    third-party logistics providers (3PL), 213
    United Airlines web page, 214
Electronic Mail Policy
    account creation and removal, 175
    directories and personas, 175
    e-discovery, 173
    etiquette rule, 173–4
    internet-based resources, 172–3
    mail forwarding, 174
    mail retention, 175
email(s)
    address, double checking, 357–8
    attachments, 359

autoforward, 349
business and personal accounts, 359
caps lock key, 357
carbon copies, 358
emoticons, 359
junk/unwanted, 359
language, 357
professional font and color, 358–9
queries
    apathy, 51–2
    authentication, 48
    curiosity, 52–3
    digital signatures, 48
    ignorance, 50
    inadvertent disclosure, 49–50
    Koobface versions, 49
    lack of accountability, 54
    lack of leadership, 53–4
    negligence, 50–51
    relevant and expected message, 48
    social media, 48–9
    stupidity, 52
retention, 349
signature blocks, 360
unnecessary replies, 360
usage, 349
employee credentialing
    issuance, 170
    tail-gating (drafting), 170
    termination procedures, 171
    usage, 170
Employee Internet Use Monitoring and
        Filtering Policy
    applicability, 156
    corporate policies, 155–6
    enforcement, 158
    filter rule, 157–8
    monitoring, 156
    purpose, 156
    records, 156–7
    reports, 156
encryption
    acceptable, 347
    electronic exchanges, 213
    end-user, key protection, 348
equipment removal, 171–2
Estonia, 2007 cyber attack, 18–19
Executive Cybersecurity Dashboard,
        289–90

Facebook software, misconfiguration, 210
facility controls
    clutter, 165
    double check, 165
    locking information, 165
    in and out checking, 165
    rules, 165–6
    surveillance and control, 165
    temptation, 164–5
    vigilant and trained, 165
Financial Services Modernization Act, 1999
        see Gramm–Leach–Bliley Act (GLB)
firewall, 2, 21, 22, 40, 62, 71, 104, 186, 202,
        207, 241, 245, 287, 351, 354
Fortune, Robert, 9
freeware, 184

goals and objectives
    affordable and suitable, 108
    boardroom interactions, 329
    business, aspects of, 110
    definition, 108
    organizational, 240, 265–6, 282
    planning, 111, 121–2, 129, 190, 192,
            197, 298
    samples, 110
    SMART goals, 108, 124
Gramm–Leach–Bliley Act (GLB), 194,
        197, 263

hacking, managing
    checklist, plan implementation
        appropriate notifications, 307–9
        damage, assessment, 305–7
        defenses, adjustment, 310
        external links, verification, 304–5
        gaining control, 302–3
        isolation, 302
        panic, avoiding, 300–302
        passwords, reset, 303–4
        reason for hacking, 309–10
        stealth drive, 305
        update and scan, 304–5
    computers types, 294
    foot stompers
        insurance, 317
        law enforcement, working with, 315–17
        legal discovery and associated requests,
                319

hacking, managing (*cont'd*)
  legal issues, 318–19
  liability, addressing, 317–18
  public relations (PR), importance,
    310–315
  trusted relationships, 318
  2011 hack, of Sony Corporation's
    PlayStation Network, example,
    311–15
  insurance, 317
  Internet Crime Complaint Center (IC3), 316
  personnels, watch on, 295
  physical/network security measures, 293–5
  plan implementation, 299–310
  preparation *see* hacking, minimizing the
    impact
hacking, minimizing the impact
  baseline documents, maintenance, 296–7
  disaster recovery and business continuity
    plan, 298–9
  information, back up, 296
  insurance, protection, 297–8
  natural disasters, 299
  planning in advance, 299
  regular audits, 296
hacktivists
  Assange, 25
  cyber threat sources, 12, 36
  definition, 25
  Low Orbit Ion Cannon, 26
  Operation Payback, 26
  WikiLeaks, 25–6
hashed, 148
hazard and operability study (HAZOP), 331
Health Insurance Portability and
    Accountability Act (HIPAA), 64
  clean desk policy, 150–151
  legal compliance concerns, 1996, 193–4
human risk management
  email queries *see* email(s)
  spear phishing and whaling, 48
hypertext markup language (HTML), 145

industrial control systems (ICS), 47
  hacker, 250–251
  manufacturing processes, 238–9
  password, 251
  smelting and casting process, 238
  Stuxnet-like attack, 253

"industrial strength" disk wiping program,
    160
information security, 350
Information Systems Auditing Standards,
    195
insider threats
  cyber threat sources, 12, 36
  definition, 26–7
  example, 28
  spear phishing, 28
insurance, 297–8, 317
Integrated Services Digital Network (ISDN)
  phone lines, 353
integrity, 2, 30, 41, 83, 118, 186, 194, 209,
    212–14, 255, 271, 294
intellectual property and trade secrets
  cyber espionage, theft, and exploitation
    (checklist)
    competitors, 40
    computer storage, 40
    contracted system administration and
      software support, 42
    data backups, 41
    data feeds, 41
    DVD/CD read-write drives, 41
    intellectual property and trade secrets, 39
    internet access, 40
    off-site storage, 41
    USB connections, 40
  quantitative risk *see* quantitative risk
    assessment
internal controls
  electronic exchanges, 212–14
  organizations, 209
  policies and procedures, 209
  software, 210–211
  system configurations, 209–10
  web pages, 211–12
International Information Systems Security
    Consortium (ISC2), 234
International Standards Organization (ISO)
    27001 (Information Security
    Management) certification, 264
Internet Crime Complaint Center (IC3), 316
Internet Protocol (IP) address, 157, 210, 279,
    363, 364
Internet security
  antivirus, 350
  digital signature, 350–351

extranet, 351
   monitoring and filtering, 351
   remote access, 352
   use, 351–2
Internet Use Policy
   applicability, 153
   employee's actions, problems, 152
   internet resources, misuse, 152–3
   official business use, 153
   prohibited Internet use, 153–4
   purpose, 153
   threats, 153
intranet, 202, 217, 238, 250, 270, 327, 351
IP address *see* Internet Protocol (IP) address
ISC2 *see* International Information Systems
   Security Consortium (ISC2)
ISC2's ten domains, **235**

Joint Universal Lessons Learned System
   (JULLS), 245

Keep It Simple, Stupid (K-I-S-S), 105

law enforcement, 315–17
legal compliance issues
   Gramm–Leach–Bliley Act (GLB), 194
   HIPAA of 1996, 193–4
   Privacy Act of 1974, 194
   Sarbanes–Oxley (SOX) Act of 2002, 193
legal discovery and associated requests, 319
legal issues, 318–19
liability, addressing, 317–18
Low Orbit Ion Cannon, 26

malware, 16, 48, 152, 187, 270, 301, 313,
   354
management's discussion and analysis
   (MD&A), 81–2
mechanics of strategy
   core values, 107–8
   deliberations
      competencies, 106
      effectiveness, 106
      risk management, 106
      value, 106
   goals and objectives *see* goals and
      objectives
   Keep It Simple, Stupid (K-I-S-S), 105
   leadership, 98

mission statements, 107
   resources, cyber-related, 103–4
   simplicity, 105
   SWOT *see* SWOT analysis
   team effort, 98
   trustworthy computing, 105
   vision, 105–6
   vision statement, 107
memory leak, 208
mercenary hackers
   Hannibal Lecter of computer crime
      (Poulsen), 21
   independent contractors, 20
   Operation Firewall, 22
   ShadowCrew, 22
   SQL injection, 23
   threats, 21
   war driving, 23
messaging, 157, 186, 310
Metasploit, 200
metric categories, 277–88
metrics *see also* performance measures
   and C-suite
      compliance, 274
      cybersecurity questions, 275–7
      cyber threats, 274
      decision-making process, 275
      executives, metrics decisions, 274–5
      productivity and cybersecurity, 274
   cybersecurity controls, 264
   operational, 271–2
   organizational, 271
   performance measures, 262
   selection, 265, 267
   technical, 272
mission statements, 107
mobile device policy
   acceptable use, 180
   applications, 181
   back-up and recovery, 181–2
   business processes, 181
   data plans, 180
   devices, 179
   information and risk, 179
   legal considerations, 180
   maintenance, 181
   regulatory compliance, 179–80
   security, 180–181
   services, 180

mobile security, 352
modern IT, 128–9
multi-disciplinary "red teams," 195–6

National Cyber Crime Unit (NCCU), UK,
    316
National Cyber Security Alliance (NCSA), 51
National Industrial Security Program
    Operating Manual (NISPOM), 160
nation-state threats
    China
        Advance Persistent Threat 1, Mandiant's
            report, 15
        bad actors, 14
        botnet 'sleepers,' 19
        controversial activities, 16
        cyber attacks, factors, 14
        Cyber Corps, 13
        *Estimating the Cost of Cybercrime and
            Cyber Espionage,* 17
        Internet Security Report for 2013,
            Symantec Corporation, 16
    cyber threat sources, 12, 36
    DDoS attacks, 18
    Estonia, 2007 cyber attack, 18–19
    Russians, 19
natural disasters, 71, 299
Nessus, 200, 279
Netflix, 111–12
Netsploit, 14
network management policy
    application security, 187
    centralized logging, 188
    deny all, permit by exception, 186
    least privilege, 186
    malware filtering, 187
    "must-have" policy, 185–6
    secure operating systems, 186
    sensor architecture, 188
    threat/incident analysis, 188
    vulnerability management, 187
network security, 53, 72, 81, 294, 320, 353–4
Nmap, 200, 279

obsolescence
    dumpster diver, 202
    5.25 floppy disks and 3.5 diskettes, 201–2
    hackers, 202–3
    hardware and software, 202

    typewriter and carbon paper, 201
    web pages, 202
    Windows XP computer, 202
Open Web Application Security Project,
    129–30
Operation Edna, 116
Operation Firewall, 22
organized crime, 12, 20, 36

password(s) *see also* Password Protection
        Policy
    best practices, 45–7, 148–9
    creation and protection, 190, 206, 207, 244,
        245, 348
    cybersecurity risks, 45
    default, 245, 251, 295
    encryption and authentication, 355
    forgetting, 35
    one-time, 353
    password-protected network, 169
    practices, 45–6
    regularly changed and complex, 21
    remote access, 46–7
    resetting, 35, 303–4, 321, 345
    vulnerability scans, 46
Password Protection Policy
    best practices, 148–9
    "must-haves"
        account lock-outs, 149
        avoid transmitting passwords via email,
            150
        don't recycle passwords, 150
        force password expiration, 150
        separate administrative and user
            passwords, 149
    passphrases, 149
    risk, 147–8
patching, 67, 71, 187, 202, 208–9, 212,
    217–18
patch installation, 215
penetration test (Pen-test), 52, 66, 74, 101,
    162, 170, 195, 219, 251, 258
performance measures
    Balanced Scorecard concept, 265–6
    business drivers
        compliance, 270
        cost, 267–8
        quality, 269
        return on investment, 269–70

risk, 268–9
safety, 270–271
cybersecurity metrics *see* metrics
decision-making, 265
due diligence and due care, 262–3
employees, reward and discipline, 266
executive cybersecurity dashboard *see*
dashboards
information systems, 263
internal controls, 263
ISO 27001 certification, 264
nonpublic personal information, 263
partners and clients, agreements, 264
performance standards
accountability, 237
ICS, 238–9
malicious code, 238
measures, 238
monetary costs, 239
Plieno Steel Company, 238
product, 239–40
wisdom and judgment, 237
zero tolerance program, 237
Personal Health Information (PHI), 194
personally identifiable information (PII), 130,
158, 160, 194
personnel management
business function and activity, 228
critical infrastructure protection, 249–58
leaders, 228–9
organizational considerations
corporate sponsor, 240
Financial Review Boards, 241
leadership, 240, 242
plans, policies and procedures, 241
risk committees, 241
performance standards, 237–40
plans, policies and procedures, 227
responsibilities, 227–8
system administration, 228
team creation, 229–37
threats and vulnerabilities, 228
training, 242–9
personnel vetting, 334
phishing, 244, 278, 281, 349
physical security policy
emergencies and evacuations, 172
employee credentialing, 170–171
equipment removal, 171–2

facility controls, 164–6
port security, 163
security checks, 163
visitor and contractor access controls, 166–70
PII *see* personally identifiable information
(PII)
ping, 278, **279**
plan execution, 332
plan implementation, 299–310
planning *see also* goals and objectives
acceptance and payment process, 133
in advance, 299
auditing, 195–6
BigRX Corporation, 129–30
characteristics, 127
coding issue, 133
dynamic test environment, 134
flexibility, 128
great teams, 131
Hurricane Katrina, aftermath, 192
internal controls, 133
legal compliance concerns, 193–4
modern IT, 128–9
operational level, 122, 130
organizations, level of perfection, 191–2
plan of action
business, protection, 129
cost-effective, 129
management of information, 128
policies and procedures, 129
vigilance and updates, 129
policies *see* policy(ies) complement plans
procedures *see* procedures implement plans
risk management systems, 129
SMART model, 132
SQL injection attacks, risk, 129–30
strategic level, 122, 130
tactical level, 122, 130
testing software, 133, 134
title, 131
Plieno corporation
business intelligence capability, 76
revitalizing steel industry, 57, 213
risk management strategy, 73–6
policy(ies)
audit security, 347
boardroom interactions, 330
computer security, 347–9
desktop security, 349

policy(ies) (*cont'd*)
  email, 349
  information security, 350
  Internet security, 350–352
  mobile security, 352
  network security, 353–4
  physical security, 354
  server security, 354
  wireless devices, 354–5
policy(ies) complement plans
  BigRX, new policy, 140
  business rules and guidelines, 189–90
  CEO or Chief Operating Officer, 189
  corporate risk management processes, 188–9
  cybersecurity
    acceptable use policy, 141–3
    access control policy, 185
    clean desk policy, 150–152
    computer ethics policy, 143–7
    electronic mail policy, 172–5
    employee internet use monitoring and
      filtering policy, 155–8
    mobile device policy, 178–82
    network management policy, 185–8
    password protection policy, 147–50
    physical security policy, 161–72
    remote access policy, 177–8
    removable media policy, 175–7
    software policy, 182–5
    technology disposal policy, 158–61
    use of the internet policy, 152–5
Pollution Prevention Pays (3P) program, 332
port, 40, 162, 163, 175, 176, 186, 223, 304,
    354
port scan, 14, 141
Privacy Act of 1974, 194
procedures implement plans
  basic cyber hygiene, 191
  cybersecurity plans and policies, 190
  daily practice, 190–191
proxy, 40, 167, 187
Public Key Infrastructure (PKI), 45
public relations (PR)
  disaster recovery, 334
  importance, 310–315

qualitative risk assessment
  BigMIMS, 64
  HIPAA, 64

incident impact, definitions, 70–71
incident likelihood
  Delphi group technique, 70
  Microsoft products, 69
  plurality rules! techniques, 70
measurement techniques, 71
risk and asset value, 63
risk decisions
  acceptance, 72–3, 75, 76
  avoidance, 73, 75, 76
  BigRX CEO, 75–6
  cybersecurity issues, 76–7
  mitigation techniques, 71–2, 74–6
  Plieno CEO, 74–5
  transference, 72, 75, 76
threat and vulnerabilities, 68
threat identification, 64–5, **65**
vulnerability identification
  BigMIMS, 66
  BigRX, 67
  Pen-testing, 66
  scanning results, 66
  system administrators, 66–7
quantitative risk assessment
  acquisition or development, 56
  annual loss potential, 63
  assets value
    cost if unavailable (CU), 59
    cost to acquire or develop (CAD), 59
    cost to maintain (CM), 59
    cost to replace (CR), 59
    liabilities (L) issue, 59–60
    profit value (PV), 59
  CEO, 57–8
  cyber incidents, 55
  intellectual property and trade secrets
    acquisition or development, 56
    investors and competitors, 57
    liabilities, 56–7
    maintenance costs, 56
    profit value, 56
    replacement, and costs, 56
    risk analysis, 57
    unavailable cost, 56
  investors and competitors, 57
  liabilities, 56–7
  loss, estimation
    asset value (AV), 61
    risk exposure (RE), 61

SLE, 60–62
maintenance costs, 56
profit value, 56
replacement and costs, 56
risk analysis, 57
threat likelihood, estimation, 62–3
unavailable cost, 56

RAT *see* remote access trojan (RAT)
regulatory communication risk
cybersecurity disclosure, objectives, 83
disclose risks and incidents, reasons, 83–4
reporting mechanisms, 84–5
SEC CF DG 2 guidance
description of business, 82
disclosure controls and procedures, 82–3
legal proceedings, 82
MD&A, 81–2
risk factors, 80–81
SEC investigations, violations, 85–6
SEC regulations
administrative action, 86
civil action, 86
disclosure obligations, 79–80
voluntary disclosure program, 85
Remote Access Policy
anti-virus and anti-malware software, 178
authentication, 178
official business, 177
provision services, 178
training, 178
remote access trojan (RAT), 43, 50, 176,
207–8, 238
Removable Media Policy
automatic scanning, USB ports, 176
disable auto-play, 176
exfiltration of information, 175
media screening capability, 176
nonapproved media, USB ports, 176
in non-organization sources, 177
in organization, 176
risk appetite, 175
Universal Serial Bus (USB) ports, 175
value of information, 175
work force, training, 177
resources, cyber-related
finances, 104
information, 103
personnel, 104

plans, 103, 104
technology, 104
risk appetite level, 91
risk management
boardroom interactions, 326
committees
material enterprise risk, 89
nonmanagement directors, 89
quality decisions, 89–90
technical awareness, 90
threat awareness, 90
CRO, 90–91
cybersecurity risk, 38
human risks, 47–54
insurance, 38
intellectual property and trade secrets,
threats, 38–42
internal communications
Critical Information Reporting process,
78–9
Five W's, 78
metrics, 79
senior management, 78
standardized risk management process,
78
success breeds success, 79
threats and vulnerabilities, 79
qualitative *see* qualitative risk assessment
quantitative *see* quantitative risk assessment
regulatory communications *see* regulatory
communication risk
risk decisions, 71–7
shareholders communications, 38, 87–8
technical *see* technical risk management
threats, 37
router, 30, 62, 207, 278–9, 353

Sarbanes–Oxley (SOX) Act, 82, 193, 201,
246, 263
Securities and Exchange Commission (SEC)
Division of Corporate Finance, 79
security
measures, physical/network, 293–5
memory leakage, 208
patching, 208–9
server, 354
threats and vulnerabilities, 208
server, 40, 44, 45, 59, 73, 159, 161, 164, 187,
203, 207–8, 212, 266, 269, 304, 354

service-level agreements (SLAs), 206, 209
ShadowCrew, 22
shareholders communications
  crisis, 87–8
  information, 87
  preferences, 87
  SEC reports, 86
  stockholder meetings, 87
shareware, 184
SLAs *see* service-level agreements (SLAs)
SMART (specific, measurable, achievable,
      realistic, and timely) goals, 124, 132
sniffer, 23
social engineering, 35, 49, 52, 66, 148, 244,
      349
software bloat, 208
software license, 105, 182, 184
Software Policy
  acquisition, 183
  applicability, 183
  auditing, 183
  budget, 183
  copying and distribution, 184
  on home systems, 184–5
  inventories, 183
  licensing and registration, 183
  shareware and freeware, 184
  software installation, 183
  storage and documentation, 183
  upgrades, 183–4
spam, 53, 87, 141, 157, 174, 187, 244–5, 263,
      278, 283
spam filters, 278
spear-phishing, 28, 48, 49, 53, 99, 149, 173,
      187, 210, 244, 263, 269, 350–351
spoofing, 251
spyware, 157
SQL injection *see* structured query language
      (SQL) injection
strategic planning, 125
strategy
  boardroom interactions, 328
  cybersecurity, 95–7
  definition, 96
  failure *see* strategy failure, avoiding
  incorporating cybersecurity, 119–20
  leaders, examples, 96
  mechanics *see* mechanics of strategy
  operational level planning, 122

strategic level planning, 122
  and strategic planning, difference, 97
  tactical level planning, 122
strategy failure, avoiding
  communication, 113–14
  human element, understanding, 116
  leadership and oversight, lack of, 117–18
  Netflix, example, 111–12
  Operation Edna, example, 116
  poor plans, poor execution, 111–12
  resistance to change, 114–16
  strategies, implementation, 117
Structured Query Language (SQL) injection,
      129–33, 135, 140, 187, 211
  BigRX Plan, 135–9
  reducing BigRX's exposure, 129–30, 133–4
Stuxnet, 17, 40, 43, 47, 75, 175, 250, 253, 255
substandard products and services
  Boeing 787 Dreamliner, 31
  counterfeit products, use of, 30
  cyber threat sources, 12, 36
  help desks, 34–5
  loss, business, 34
  outsourcing, 35–6
  recognizing, 29–30
  smart systems and sensors, 32
  social engineering, 35
  technology, 33
Supervisory Control and Data Acquisition
      (SCADA) systems, 47
switch, 76, 220, 285, 302–3, 305
SWOT analysis
  opportunities
    data sharing, 102
    partnerships, 102
    payments, 102
    presence, 101
  strengths
    information, 99
    plans, 100
    policies, 100
    presence, 99–100
    processes, 100
    resources, 100
    talent, 100
    technology, 100
  threats
    benchmarking, 103
    standards, 103

technology changes, 103
weaknesses
  priorities, 101
  resources, 101
  risk analysis, 101
  third-party assessment, 101

tail-gating (drafting), 170
team creation
  business and employees, 229
  cybersecurity leaders
    business processes, 236
    CIOs and CISOs, 233–4
    CISSP and CISM certifications, 234–5
    CRO and CSO, 236–7
    C-suite, 236
    cyber smart, 237
    technical staffs, 236
  Kolb, Jon, 230
  long snappers, 229–30
  right leaders
    attitudes, 231
    bad reputation, 231
    cyber smart, 233
    e-commerce, 232
    education, 232
    information-enabled activities, 230–231
    professional career, 230
    safeguard, 231–2
    slick salesman's, 232
    technical experts, 231
team selection, 131–2
technical risk management
  common technical risks (checklist)
    cloud, storage, 44–5
    hacking, previous incidents, 43
    malicious code, 43
    mobile devices, 45
    passwords, 45
    probing, 43–4
    software currency, 44
  passwords
    practices, 45–6
    remote access, 46–7
    vulnerability scans, 46
Technology Disposal Policy
  construct (model), categories, 160
  enforcement, 161
  financial or logistics department, 161

information, categories, 159–60
PII disclosure, 158
printers and copiers, 160–161
risk management program, 159
THAAD system, 158
Ten Commandments of Computer Ethics,
    143–6
Terminal High Altitude Area Defense
    (THAAD) system, 158
threat
  awareness, 90
  detection
    phishing messages, 278
    quantitative measure, 278, *279*
    repeat offenders, charts, 277–8, **279**
    reverse lookup, 279
    scans and probing, 277–8
    spam filters, 278
  identification, 64–5, **65**
  incident analysis, 188
  insider
    cyber threat sources, 12, 36
    definition, 26–7
    example, 28
    spear phishing, 28
  mercenary hackers, 21
  nation-state *see* nation-state threats
  risk management, 38–42
  sources, 12, 36, 55, 64, 68, 74, 78, 85
  SWOT analysis *see* SWOT analysis
  and vulnerabilities, 68, 79, 208
Time Warner, 126
training
  employee
    accomplishment, 246
    CEO client, 243
    certification/decertification, 246
    common mistakes, 245
    computer safety, 242
    cybersecurity and risk management,
      243–4
    ethics, 246
    foot stompers, 244–5
    home and office, protection, 245–6
    mistake, organizations, 243
    principles and techniques, 242
    privacy, 244
    procedures, 244
    purpose, 243

training (*cont'd*)
  executives
    computer crime, 247
    decision-making processes, 246
    ethical behavior and business practices,
      247
    globalization and Internet, 248–9
    intellectual property and trade secrets,
      248
    legal and regulatory requirements,
      246–7
    privacy and anonymity, 247–8
    Sarbanes–Oxley Act, 246
    violation, 247
trojan, 42, 43, 254
trusted relationships, 318
trustworthy computing, 105
Twitterverse, 26
two-factor access procedures (dual factor)
      authentication, 45, 178, 352–3, 355

uninterruptible power supply (UPS), 163
Universal Serial Bus (USB) ports, 39, 175
Unrestricted Warfare, 13
UPS *see* uninterruptible power supply (UPS)
U.S. Department of Homeland Security
      (DHS), 7, 52, 62

values, 107
virtualization, 203
virtual machine (VM), 134
virtual private network (VPN), 23, 354
virus, 43, 46, 54, 103, 175–7, 238, 250, 253,
      269, 281, 303, 350
vision, 105–6
vision statement, 107
VM *see* virtual machine (VM)
voice over Internet protocol (VOIP), 21

VOIP *see* voice over Internet protocol (VOIP)
VPN *see* virtual private network (VPN)
vulnerability(ies)
  beyond projected life span, 280
  dashboards *see* vulnerability(ies)
  discovered on systems, 279–80
  identification
    BigMIMS, 66
    BigRX, 67
    Pen-testing, 66
    scanning results, 66
    system administrators, 66–7
  management, 187
  passwords, scan, 46
  residual, 280
  scanning, 138
  SQL injection, 68, 129, 133, 135
  system, 279
  threats and, 68, 79, 208, 228
  time frames, 280

war driving, 23
web pages
  Cross-Site Scripting (XSS), 211–12
  digital storefront, 212
  products and services, advertise, 202
  United Airlines, 212
  ZNews Network, 211
whaling, 28, 48, 244
white listing, 187
wireless devices, 354–5
worm, 42, 43, 49, 103, 176

zero-day, 216, 290
zero tolerance, 146
ZNews Network, 211
zombie, 60, 176, 200, 302
zone locations, 204

Printed and bound by CPI Group (UK) Ltd, Croydon, CR0 4YY

04/06/2023

03223921-0005